General Relativity and Gravitation

Fundamental Theories of Physics

A New International Book Series on The Fundamental Theories of Physics: Their Clarification, Development and Application

General Relativity and Gravitation

Invited Papers and Discussion Reports of the 10th International Conference on General Relativity and Gravitation, Padua, July 3-8, 1983

edited by

B. Bertotti
Department of Nuclear and Theoretical Physics, University of Pavia, Italy

and

F. de Felice and A. Pascolini
Department of Physics, University of Padua, Italy

D. Reidel Publishing Company

A MEMBER OF THE KLUWER ACADEMIC PUBLISHERS GROUP

Dordrecht / Boston / Lancaster

Library of Congress Cataloging in Publication Data

CIP

International Conference on General Relativity and Gravitation
(10th: 1983: Padua, Italy)
General relativity and gravitation.

(Fundamental theories of physics)
Conference was sponsored by the International Society on General Relativity and Gravitation.
Bibliography: p.
Includes index.
1. General relativity (Physics)–Congresses. 2. Astrophysics–Congresses. 3. Gravitation–Congresses. 4. Quantum gravity–Congresses. I. Bertotti, B., 1930- . II. De Felice, F., 1943- . III. Pascolini, A., 1944- . IV. Title. V. Series.
QC173.6.I57 1983 530.1'1 84–15137
ISBN 90–277–1819–9

Published by D. Reidel Publishing Company
P.O. Box 17, 3300 AA Dordrecht, Holland

Sold and distributed in the U.S.A. and Canada
by Kluwer Academic Publishers,
190 Old Derby Street, Hingham, MA 02043, U.S.A.

In all other countries, sold and distributed
by Kluwer Academic Publishers Group,
P.O. Box 322, 3300 AH Dordrecht, Holland

Printed in The Netherlands

TABLE OF CONTENTS

EXPERIMENTAL GRAVITATION

QUANTUM GRAVITY

HIGHLIGHTS OF THE CONFERENCE

PREFACE

The Tenth International Conference on General Relativity and Gravitation (GR10) was held from July 3 to July 8, 1983, in Padova, Italy. These Conferences take place every three years, under the auspices of the International Society on General Relativity and Gravitation, with the purpose of assessing the current research in the field, critically discussing the progress made and disclosing the points of paramount importance which deserve further investigations.

The Conference was attended by about 750 scientists active in the various subfields in which the current research on gravitation and general relativity is articulated, and more than 450 communications were submitted. In order to fully exploit this great occurrence of experience and creative capacity, and to promote individual contributions to the collective knowledge, the Conference was given a structure of workshops on the most active topics and of general sessions in which the Conference was addressed by invited speakers on general reviews or recent major advancements of the field. The individual communications were collected in a two-volume publication made available to the participants upon their arrival and widely distributed to Scientific Institutions and Research Centres.

The Editors aimed to construct the Proceedings as a comprehensive book on General Relativity and Gravitation. Each of the four Sessions of the Conference corresponds to a Part of the Proceedings; the invited lectures and the reports on the workshops written by their chairmen have been ordered in the most consequential and self consistent way possible. The Editors fully realize that their scope will probably be inadequately reached, however they hope that the Proceedings will be useful not only to specialists, but also to those who are entering this field of research and to scientists active in other fields who want to keep their knowledge up to date in this fundamental component of contemporary scientific knowledge.

Following the advice of an International Scientific Committee, composed by E.Amaldi, P.Bender, P.Bergmann, B.Bertotti (chairman), C.Clarke, Y.Choquet-Bruhat, F.de Felice, M.Demianski, J.Ehlers, L.D.Faddeev, S.Ferrara, A.E.Fisher, M.Francaviglia, Y.Fujii, J.B.Hartle, A.Held, M.MacCallum, I.D.Novikov, M.J.Rees, E.Schmutzer, J.Tiomno, C.V.Vishveshwara, the subject matter has been divided into four parts:

A. Classical Relativity,
B. Relativistic Astrophysics,
C. Experimental Gravitation,
D. Quantum Gravity,

and the following workshops have been selected:

A1. Equations of motion, gravitational radiation and asymptotic structure of spacetime,
A2. Exact solutions and singularities,
A3. Initial value problem,
A4. Numerical relativity,
A5. Classical and quantum alternative approaches to gravitational theories,
B1. Physics and astrophysics of black holes,
B2. The early universe,
B3. Cosmological parameters and cosmological models,
C1. High frequency gravitational waves: attempts to laboratory detection,
C2. Space experiments,
C3. Laboratory experiments,
D1. Quantum gravity and quantum field theory in a curved space,
D2. Supergravity.

In Session A the Conference witnessed an impressive advance in Computational Relativity, a field where individual mathematical skill is enhanced by the capability of sophisticated computer codes to perform the numerical integration of Einstein's equations and of new computer languages to allow for algebraic manipulation of all kinds of mathematical expressions.

The Conference also gave an important contribution to the clarification of a major and much debated issue of Classical Relativity, the problem of gravitational radiation in isolated systems and the 'Quadrupole Formula'. A special session within workshop A1 and two invited lectures, by T.Damour and M.Walker, were devoted to this topic. The report on the special session by its chairman J.Ehlers, gives a reasoned and critical account of the discussion and of the present view on the problem.

Part two is mostly devoted to Relativistic Cosmology a field traditionally open to new developments. An extensive and exhaustive account of the many aspects of the problem is contained in G.F.R.Ellis' lecture, while the most recent advances in specific areas of relativistic astrophysics are discussed in the subsequent talks and related workshops.

Part three acknowledges the growing body of contributions to gravity afforded by experiments and makes assessments of the present research programs. The results of the Viking gravitational experiments were presented at Padova by R.Hellings for the first time and constitute a major advance in the accuracy in the measurement of the relativistic parameters as well as a confirmation of Einstein's theory.

S.Hawking's paper in Part four presents an entirely new and original way of setting up the initial quantum conditions of the universe and possibly of explaining its peculiar present state. The lectures and papers on quantum gravity and supergravity stress the great and increasing importance of gravitation for the interpretation of microscopic phenomena concerning elementary particles and the constructions of the physical laws which govern them.

Finally J.A.Wheeler considering the achievements of the Conference and the status of the field, suggests new goals for physics in our age.

The Editors and the Publisher agreed on the importance of a rapid publication of these Proceedings, and for this reason a camera-ready composition was chosen. The Editors regret that some delay in the publication was caused by the desire to have all the lectures and workshop reports included in order to give a full account of the results reached by the Conference.

The Editors have made efforts to ensure uniformity of presentation and graphical homogeneity of the book: these tasks would have been inconceivable without the expert help of several, dedicated persons. The Editors acknowledge the essential role of Dr. Massimo Calvani in all the stages of editing the text, up to proof reading. Dr. Maria M. Sperotto was of valuable help in the preparation of the references and Prof. Mary Prosperi Evans assisted in suggesting correct use of the English language. The delicate task of quick and correct typing of the final camera-ready script was carried out by Mrs. Ariella De Rossi and Mrs. Annarosa Spalla, and Mr. Angelo Rampazzo carefully redrew a considerable part of the figures and completed the lettering of the formulae and of the text.

Finally a special thanks is due to Prof. Claudio Villi, whose experience and advice the Editors have often relied on, and to Dr. Laura Salvadori, whose contribution to the Conference has been invaluable.

Padova, March 1984

B. Bertotti
F. de Felice
A. Pascolini

THE PROGRAMME OF THE 10th INTERNATIONAL CONFERENCE ON GENERAL RELATIVITY AND GRAVITATION (Padova, 3-8 July, 1983)

SUNDAY, JULY 3

Palazzo della Ragione
Registration
Guided visit to the Exhibition "Galileo e Padova"
Refreshment

MONDAY, JULY 4

Palazzo della Ragione
Opening Ceremony addressed by
B. Bertotti, Conference Chairman
S. Gottardo, Mayor of the Town of Padova
L. Merigliano, Rector of Padua University

The mathematical theory of black holes
S. Chandrasekhar

On stochastic properties of relativistic cosmological models in the vicinity of a singularity
E.M. Lifshitz

Biology Department
Parallel workshops
A3 Initial value problem
B2 The early universe
C1 High frequency gravitational waves: attempts to laboratory detection

A2 Exact solutions and singularities
B1 Physics and astrophysics of black holes
A5 Classical and quantum alternative approaches to gravitational theories

TUESDAY, JULY 5

Palazzo della Ragione

Testing relativity with solar system dynamics
R.W. Hellings

The quadrupole approximation to gravitational radiation
M. Walker

Prospects for laser interferometer gravitational radiation detectors
R.W.P. Drever

Asymptotic properties of isolated systems: recent developments
A.A. Ashtekar

Biology Department

Parallel workshops
A3 Initial value problem
B2 The early universe
C1 High frequency gravitational waves: attempts to laboratory detection

A2 Exact solutions and singularities
B1 Physics and astrophysics of black holes
D1 Quantum gravity and quantum field theory in a curved space
C1 High frequency gravitational waves: attempts to laboratory detection

S. Sophia Church

Concerto of the Ensemble Vivaldi

WEDNESDAY, JULY 6

Palazzo della Ragione

The motion of compact bodies and gravitational radiation
T. Damour

Positive energy and Kaluza-Klein theory
E. Witten

Biology Department

Parallel workshops
A1 Equations of motion, gravitational radiation and asymptotic structure of spacetime

D2 Supergravity
C2 Space experiments
A2 Exact solutions and singularities
D1 Quantum gravity and quantum field theory in a curved space
C3 Laboratory experiments
B3 Cosmological parameters and cosmological models
C2 Space experiments
A5 Classical and quantum alternative approaches to gravitational theories

THURSDAY, JULY 7

Palazzo della Ragione
Algebraic computing in general relativity
I. Cohen

Quantum cosmology
S.W. Hawking

The cosmic censorship problem
D. Christodoulou

Nonlinear wave propagation in relativistic hydrodynamics and cosmology
A.M. Anile

Biology Department
Parallel workshops
A4 Numerical relativity
B3 Cosmological parameters and cosmological models
A1 Equations of motion, gravitational radiation and asymptotic structure of spacetime
C2 Space experiments

Physics Department
Assembly of the GRG Society

Conference Banquet at Villa Giovanelli

FRIDAY, JULY 8

Palazzo della Ragione
Unified theories of gravitation and free differentiable algebras
T. Regge

Quantum gravity
B.S. DeWitt

Voids in the expanding universe
H. Sato

Relativistic cosmology: its nature, aims and problems
G.F.R. Ellis

Biology Department
Parallel workshops
A4 Numerical relativity
D2 Supergravity
C3 Laboratory experiments

Physics Department
Highlights of the Conference
Round Table
J.A. Wheeler (chairman)
M. MacCallum, I.D. Novikcv, P. Bender, P. van Nieuwenhuizen (participants)

INTRODUCTORY TALK

Bruno Bertotti

Dipartimento di Fisica Nucleare e Teorica
Università di Pavia
Pavia (Italy)

In opening this tenth Conference on General Relativity and Gravitation I would like to put on record a feeling of scientific stress, which I felt vividly during its preparation. Like several other areas in contemporary physics, our subject is torn between the need of sophisticated, even exasperate, specialization in highly technical subdisciplines and the strife for a unified picture of the material world, at all scales of length and time. On the one hand our work progresses mostly in small communities numbering not more than one of two hundred people, who work in an intense, coordinated manner, communicate their results quickly by voice or by preprints and set the standard in their own restricted endeavour. On the other hand the work of these communities would be sterile and would soon wither unless pursued within wider research programs and a global perspective. Any reasonable decision about priorities and research strategy must be taken having in mind related possible advances, other ways of attacking the problem and a broad, final purpose. Thus for example, the community of experts in exact solutions will have to make a determined effort to identify the relevance of their work in black hole astrophysics; theoretical cosmologist will have to go into the intricacies of relativistic galaxy formation to ensure a firm basis to their work; and so forth.

Compliance with these two different levels of preoccupation and discussion at the same time is necessarily a source of difficulties and conflicts. The scientific organization of this Conference was based upon the assumption that these two levels - technical spe-

B. Bertotti et al. (eds.), General Relativity and Gravitation, 1–2.

cialization and broader outlook - complement each other and must be made the meat of our work here. This is an ambitious task and it relies heavily for its fulfilment upon the unselfish collaboration of all participants. On this basis we have reduced the number of simultaneous workshops to three, have asked each chairman to coordinate and possibly to report on the contributions and will include his assessment in the Proceedings. It is clear that the authors of the communications will not necessarily be asked to present their work personally.

One can feel this scientific tension also going through the subject matter of the Conference, where the asymptotic coalescence between cosmic physics and microphysics is an obvious underlying theme. Over the past twenty years general relativity has greatly extended its field of fertility: from a mostly mathematical-physics endeavour it has now become the crucial interpretative key in different areas of physics and the source of impressive research programs, for example in experimental gravitation. And the mathematical doctrines which have been invented and developed in this framework have outstanding power and value. Those who can claim to understand all subjects dealt with here this week are a vanishing race; and one could even question the opportunity of a conference with such a wide scope. We took the opposite view. We have taken care, in fact, to mix lectures and workshops of different nature, so as to facilitate and to suggest the exposure of the participants to ideas and perspectives in which they are not competent. We thereby make people feel ignorant, and run the risk of wasting time with silly questions; and we take up some time of the precious, technical in-depth investigation of single subjects. As a result, the program of the Conference is relentless, with praticably no time off, which certainly would be deserved in the heat of the Po Valley.

I think this is not an easy Conference for the organizers, but even less for the participants. The Scientific Committee has decided to run it like this because, I believe, of a stubborn feeling which Einstein called heilige Neugier, holy curiosity. I trust that the participants share this feeling and will thereby make the Conference an outstanding scientific event.

Classical Relativity

THE MATHEMATICAL THEORY OF BLACK HOLES

S. Chandrasekhar

The University of Chicago
Chicago, Illinois 60637, USA

In a course of lectures on the 'underlying mathematical structures of classical gravitation theory' given in 1978, Brandon Carter began with the statement 'If I had been asked five years ago to prepare a course of lectures on recent developments in classical gravitation theory, I would not have hesitated on the classical theory of black holes as a central topic of discussion. However, the most important developments in gravitational theory during the last three or four years have not been in the classical domain at all...' Carter is undoubtedly right in his assessment that the mathematical theory of black holes has not been in the mainstream of research in relativity since 1973. I therefore find it difficult to understand why the organizers of this meeting should have chosen precisely this topic for the opening talk of this meeting. But I am grateful to them for their courtesy in assigning to me this privilege.

The mathematical theory of black holes has held my attention during the past ten years because of its classical severity and because also of its relevance to our understanding of real physical objects in the universe around us. As I have stated elsewhere, 'the black holes of nature are the most perfect macroscopic objects there are in the universe'. They are the most perfect almost by definition since the only elements in their construction are our concepts of space and time. And since the general theory of relativity provides only a single unique family of solutions for their description, they are the simplest objects as

B. Bertotti et al. (eds.), General Relativity and Gravitation, 5–26.

well.

In my consideration of the mathematical theory of black holes I shall restrict myself to exhibiting some features of the theory which reveal its richness. And as in a hurried guided tour of a museum, I shall draw your attention only to a very few chosen exhibits; and I shall not have time to dwell at any length even on the chosen exhibits.

Let me begin by defining a black hole. For my purposes, it will suffice to say that a black hole is a time-independent asymptotically flat space-time satisfying Einstein's vacuum field-equations which partitions the three-dimensional space into two regions; an external region and an internal region separated by a smooth convex two-dimensional surface - the event horizon - in such a way that no point in the interior is communicable to the space outside. By a theorem due to David Robinson the only solutions of Einstein's equations, compatible with these requirements, are those which belong to the Kerr family with its two parameters, the mass of the black hole and the angular momentum of the black hole. The static solution with zero angular momentum is the Schwarzschild solution. Besides the Schwarzschild and the Kerr solutions, I shall also consider the Reissner-Nordström solution which is a solution of the Einstein-Maxwell equations and represents the charged counterpart of the Schwarzschild solution.

May I digress for a moment to point out that it is now exactly twenty years since Kerr discovered his solution. In my judgement, it is the most important discovery in general relativity since its founding; and I should like to congratulate Kerr on this occasion.

The mathematical properties to which I shall refer derive from a consideration of the perturbations of the black holes. It is convenient to consider the perturbations as resulting from the incidence of waves of different sorts such as gravitational waves, electromagnetic waves, and massless or massive spin-$\frac{1}{2}$ particles. Stated simply and somewhat narrowly, the goal is to determine the scattering and the absorption which the incident waves will experience. The problems of the incidence of gravitational waves on uncharged black-holes and of gravitational or electromagnetic waves on charged black-holes are in some sense the most relevant since they will result in the perturbation of the space-time itself and therefore of the metric coefficients as well.

I shall consider first the problem of the reflexion and transmission of incident gravitational-waves by the Schwarzschild black-hole.

The Schwarzschild metric in its standard form (in the units (c=G=1) is

$$ds^2=\frac{\Delta}{r^2}(dt)^2-r^2\left[\frac{(dr)^2}{\Delta}+(d\theta)^2+\sin^2\theta(d\varphi)^2\right], \quad (1)$$

where

$$\Delta=r^2-2Mr \quad (2)$$

is the 'horizon function'. In considering the perturbations of this space-time, one must remember that the metric (1) is written in a coordinate system so adapted to its spherical symmetry that it takes its simplest form. The perturbed space-time will not have any special symmetry and the coordinate conditions that have been imposed to achieve the simplicity of the static metric must be relaxed. However, since we are considering the perturbations of a spherically symmetric background, we may without any loss of generality suppose that while the system is non-stationary it has, nevertheless, symmetry about some chosen axis. And it can be shown that the metric appropriate to a non-stationary axisymmetric space-time can be reduced to the form

$$ds^2=e^{2\nu}(dt)^2-e^{2\psi}(d\varphi-\omega dt-q_2dx^2-q_3dx^3)^2$$

$$-e^{2\mu_2}(dx^2)^2-e^{2\mu_3}(dx^3)^2, \quad (3)$$

where ν, ψ, μ_2, μ_3, ω, q_2, and q_3 are all functions of t, $r(=x^2)$, and $\theta(=x^3)$.

The Schwarzschild line-element, considered as a special case of the line-element (3) has the metric coefficients

$$e^{2\nu}=e^{-2\mu_2}=\Delta/r^2,\ e^{\mu_3}=r,\ e^{\psi}=r\sin\theta, \quad (4)$$

and

$$\omega=q_2=q_3=0. \quad (5)$$

Consequently, a general perturbation of the Schwarzschild black-hole will result in ω, q_2, and q_3 becoming small quantities of the first order of smallness and the functions ν, μ_2, μ_3, and ψ experiencing small increments $\delta\nu$, $\delta\mu_2$, $\delta\mu_3$, and $\delta\psi$. It is clear that perturbations leading to non-vanishing values of ω, q_2, and q_3 and perturbations leading to increments $\delta\nu$, $\delta\mu_2$, $\delta\mu_3$, and $\delta\psi$ are of very different kinds: the former induce a dragging of the inertial frame and impart a rotation to the black hole while the latter impart no such rotation. For this reason we shall call them *axial* and *polar* perturbations, respectively. As one may expect the two classes of perturbations decouple in the sense that they can be considered independently of one another.

In our further considerations it will be assumed that the various quantities representing the perturbations have a time-dependence given by

$$e^{i\sigma t} \qquad (\sigma>0); \tag{6}$$

and that it is with the *amplitudes* of the perturbations (with this time-dependent factor) that we shall be concerned. With this assumed time-dependence, we find that, after the separation of the variables, the various equations governing the perturbations can be reduced to a pair of one-dimensional wave-equations of the forms,

$$\left(\frac{d^2}{dr_*^2}+\sigma^2\right) Z^{(\pm)} = V^{(\pm)} Z^{(\pm)}, \tag{7}$$

where the superscripts "+" and "-" distinguish the polar and the axial perturbations, $V^{(\pm)}$ represent short-range potentials[1] (positive everywhere) and r_* is a radial coordinate defined by

$$dr_* = (r^2/\Delta)\,dr, \tag{8}$$

or, explicitly,

$$r_* = r + 2M \ln(r/2M-1). \tag{9}$$

The solution to the perturbation problem is thus reduced to the elementary one of barrier penetration in one dimension with which we are familiar in quantum theory. Precisely, we must seek solutions of the wave equations which satisfy the boundary conditions

$$Z^{(\pm)} \to e^{+i\sigma r_*} + R^{(\pm)}(\sigma) e^{-i\sigma r_*} \qquad (r_* \to +\infty),$$
$$\to \qquad T^{(\pm)}(\sigma) e^{+i\sigma r_*} \qquad (r_* \to -\infty). \tag{10}$$

And the required reflection and transmission coefficients for incident gravitational waves are given by

$$\mathcal{R}^{(\pm)}(\sigma) = |R^{(\pm)}(\sigma)|^2 \quad \text{and} \quad \mathcal{T}^{(\pm)}(\sigma) = |T^{(\pm)}(\sigma)|^2. \tag{11}$$

Now the potentials $V^{(\pm)}$ belonging to the axial and the polar perturbations are given by

$$V^{(+)} = \frac{\Delta}{r^5} \frac{\mu^4(\mu^2+2)r^3 + 6M\mu^4 r^2 + 36M^2\mu^2 r + 72M^3}{(\mu^2 r + 6M)^2} \tag{12}$$

and

$$V^{(-)} = \frac{\Delta}{r^5}\left[(\mu^2+2)r - 6M\right], \tag{13}$$

where

$$\mu^2 (\ell-1)(\ell+2). \tag{14}$$

It is, in the first instance, a remarkable (and an unexpected) fact that in spite of their entirely dissimilar appearances, the potentials $V^{(+)}$ and $V^{(-)}$ are very simply related: they are included in the simple expression,

$$V^{(\pm)} = \pm\beta\frac{df}{dr_*} + \beta^2 f^2 + \kappa, \tag{15}$$

where $\beta = 6M, \quad \kappa = \mu^2(\mu^2+2)$,

and $f = \Delta/[r^3(\mu^2 r + 6M)]$. (16)

The fact that the potentials $V^{(+)}$ and $V^{(-)}$ are related in the manner (15) has an important consequence: given a solution $Z^{(-)}$ belonging to $V^{(-)}$, the expression,

$$Z^{(+)} = (\kappa + 2\beta^2 f) Z^{(-)} + 2\beta dZ^{(-)}/dr_*, \tag{17}$$

provides a solution for $Z^{(+)}$ belonging to $V^{(+)}$. From the relation (17) it follows that a solution $Z^{(+)}$, derived from a solution $Z^{(-)}$, satisfying the boundary conditions (10) (appropriate for $Z^{(-)}$), satisfies the boundary conditions (10) (appropriate for $Z^{(+)}$), with a reflexion and a transmission amplitude that are related to those for $Z^{(-)}$ by

$$R^{(+)}(\sigma) = \frac{\kappa - 2i\sigma\beta}{\kappa + 2i\sigma\beta} R^{(-)}(\sigma) \quad \text{and} \quad T^{(+)}(\sigma) = T^{(-)}(\sigma). \tag{18}$$

The reflexion amplitudes thus differ only by a phase while the transmission amplitudes are the same. We conclude that *the reflexion and the transmission coefficients for the axial and the polar perturbations are the same for all incident frequencies.*

The principal features of the perturbed Schwarzschild space-time we have described are manifested in the more general context of the perturbed Reissner-Nordström space-time.

The metric of the Reissner-Nordström space-time (in its standard form) differs from the Schwarzschild metric (1) only in the definition of the horizon function, Δ: in place of (2), we now have

$$\Delta = r^2 - 2Mr + Q_*^2, \tag{19}$$

where Q_* denotes the charge of the black hole. But one important difference between the two space-times is clear from the outset. The Reissner-Nordström space-time is a solution of the coupled Einstein-Maxwell equations and not of Einstein's vacuum-equations. On this account, one should expect that the incidence of a pure gravitational-wave on the black hole must result in the excitation of electromagnetic waves and conversely.

It is clear that thé perturbations will again fall into two non-combining groups having axial and polar characteristics. And we find that in spite of the additional complexities resulting from the requirement of treating the coupled system of the Einstein--Maxwell equations, the various equations governing the perturbations can be reduced to two pairs of one--dimensional wave-equations with potentials $V_i^{(\pm)}$ (i=1,2 and "+" and "-" distinguish the polar and the axial perturbations) belonging to the same class (15) as in the context of the Schwarzschild space-time. We find

$$\left(\frac{d^2}{dr_*^2}+\sigma^2\right) Z_i^{(\pm)}=V_i^{(\pm)} Z_i^{(\pm)} \qquad (i=1,2), \tag{20}$$

where r_* is now defined, in terms Δ given by equation (19), by the same equation (8); and the potentials, $V_i^{(\pm)}$, occurring in equation (20) are given by

$$V_i^{(\pm)}=\pm\beta_i \frac{df_i}{dr_*}+\beta_i^2 f_i^2+\mu^2(\mu^2+2)f_i, \tag{21}$$

where

$$\begin{aligned}\beta_1&=3M-(9M^2+4Q_*^2\mu^2)^{\frac{1}{2}},\\ \beta_2&=3M+(9M^2+4Q_*^2\mu^2)^{\frac{1}{2}},\end{aligned} \tag{22}$$

and

$$f_i=\frac{\Delta}{r^3(\mu^2 r+\beta_i)} \qquad (i=1,2). \tag{23}$$

It follows that the reflexion and the transmission amplitudes for the solutions $Z_i^{(\pm)}$, satisfying boundary conditions analogous to (10), must be related in the manner (cf. equation (18))

$$R_i^{(+)}(\sigma)=R_i^{(-)}(\sigma)e^{\delta_i} \quad \text{and} \quad T_i^{(+)}(\sigma)=T_i^{(-)}(\sigma), \tag{24}$$

where $$e^{i\delta_i}=\frac{\mu^2(\mu^2+2)-2i\sigma\beta_i}{\mu^2(\mu^2+2)+2i\sigma\beta_i} \quad (i=1,2)$$

However, it should be noted that the reflexion and the transmission coefficients derived from the amplitudes $R_i^{(\pm)}(\sigma)$ and $T_i^{(\pm)}(\sigma)$ do not refer to incident gravitational or electromagnetic waves: the amplitudes of these waves are not given by $Z_1^{(\pm)}$ and $Z_2^{(\pm)}$ but by linear combinations of them. Suppressing the distinguishing super-scripts and denoting by H_1 and H_2 the amplitudes of the prevailing electromagnetic and gravitational waves, we have

$$H_1=Z_1\cos\psi-Z_2\sin\psi$$

and

$$H_2=Z_2\cos\psi+Z_1\sin\psi, \tag{25}$$

where $$\sin 2\psi=\frac{2Q_*\mu}{(9M^2+4Q_*^2\mu^2)^{\frac{1}{2}}} . \tag{26}$$

It is clear that in terms of the relations (25) we can express the coefficients of reflexion and transmission of electromagnetic and gravitational waves for any incident arbitrary superposition of such waves. The most convenient way to express these coefficients is by means of a scattering matrix, $\mathbf{S}$, which relates the amplitudes of the fluxes $(\vec{H}_1^{(\ell)}, \vec{H}_2^{(\ell)})$ and $(\overleftarrow{H}_1^{(r)}, \overleftarrow{H}_2^{(r)})$ incident on the potential barriers from the left and the right, respectively, to the amplitudes of the fluxes $(\overleftarrow{H}_1^{(\ell)}, \overleftarrow{H}_2^{(\ell)})$ and $(\vec{H}_1^{(r)}, \vec{H}_2^{(r)})$ reflected to the left and to the right, respectively:

$$\mathbb{S}\cdot\begin{vmatrix} \overleftarrow{H}_1^{(r)} \\ \overrightarrow{H}_1^{(\ell)} \\ \overleftarrow{H}_2^{(r)} \\ \overrightarrow{H}_2^{(\ell)} \end{vmatrix} = \begin{vmatrix} \overleftarrow{H}_1^{(\ell)} \\ \overrightarrow{H}_1^{(r)} \\ \overleftarrow{H}_2^{(\ell)} \\ \overrightarrow{H}_2^{(r)} \end{vmatrix} \tag{27}$$

We find that the scattering matrix, expressed in terms of the amplitudes $R_i(\sigma)$ and $T_i(\sigma)$ (appropriate for the solution, Z_i) is given by[2]

$$\mathbb{S} = \begin{vmatrix} T_{11} & \tilde{R}_{11} & T_{21} & \tilde{R}_{21} \\ R_{11} & T_{11} & R_{21} & T_{21} \\ T_{12} & \tilde{R}_{12} & T_{22} & \tilde{R}_{22} \\ R_{12} & T_{12} & R_{22} & T_{22} \end{vmatrix} \tag{28}$$

where

$$\begin{aligned} R_{11} &= R_1\cos^2\psi + R_2\sin^2\psi, \\ R_{22} &= R_1\sin^2\psi + R_2\cos^2\psi, \\ T_{11} &= T_1\cos^2\psi + T_2\sin^2\psi, \\ T_{22} &= T_1\sin^2\psi + T_2\cos^2\psi, \\ R_{12} &= R_{21} = (R_1 - R_2)\sin\psi\cos\psi, \\ T_{12} &= T_{21} = (T_1 - T_2)\sin\psi\cos\psi. \end{aligned} \tag{29}$$

By virtue of these definitions, the scattering matrix is unitary and time-symmetric. The form of $\mathbb{S}$ implies, consistently with the expectation stated earlier, that any perturbation of the metric must unavoidably result in the excitation of both electromagnetic and gravitational waves. Further, from the unitarity and the time symmetry of the scattering matrix, it follows that the fraction of the energy, incident as pure gravitational waves, that is converted into electromagnetic energy (in the form of reflected electromagnetic waves) must be the same as the fraction of energy, incident as pure electromagnetic waves, that is converted into gravitational energy (in the form of reflected gravitational waves).

We now turn to the reasons for the equality of the reflexion and the transmission coefficients associated with $Z^{(+)}$ and $Z^{(-)}$ in the context of the Schwarzschild black-hole and with $Z^{(+)}$ and $Z^{(-)}$ (i=1,2) in the context of the Reissner-Nordström black-hole. I do not know if there is any clear physical reason for these equalities. But the mathematical base for them can be adequately explained by considering the alternative formulation of the perturbation problem in the framework of the Newman-Penrose formalism. Let me preface the explanation by a few remarks concerning the formalism itself.

In the Newman-Penrose formalism, one sets up at each point in the space-time a basis for the tangent space consisting of two real null-vectors, ℓ and n, and a complex-conjugate pair of null vectors m and $\bar{m}$ orthogonal to both ℓ and n. One then projects onto this basis all the quantities of interest. Thus, by projecting the Weyl tensor, for example, one replaces its ten components by five complex scalars Ψ_0, Ψ_1, Ψ_2, Ψ_3, and Ψ_4. When one is dealing with type-D space-times, as the space-times of black holes are, one chooses for the real null-vectors, ℓ and n, the two shear-free congruences of null geodesics which the space-time allows. With this choice of basis, the Weyl scalars Ψ_0, Ψ_1, Ψ_3, and Ψ_4 vanish in the stationary background and Ψ_2 is the only non-vanishing scalar. In the perturbed state, when gravitational waves will be present, Ψ_0, Ψ_1, Ψ_3, and Ψ_4 will cease to vanish. The aspect of the Newman-Penrose formalism, which makes it so adaptable to the treatment of the perturbations of type-D space-times, is that it provides, directly, equations for Ψ_0 and Ψ_4, without any reference to the metric perturbations; and the radiative aspects of the perturbations can be ascertained directly by solving for Ψ_0 and Ψ_4. In particular, the reflexion and the transmission coefficients for incident waves can be inferred without reference to any other features of the perturbation; and the coefficients one obtains in this manner must agree with what one derives from other perturbation schemes. It is this last fact which guarantees the equality of the reflexion and the transmission coefficients derived from the separate considerations of axial and polar perturbations. Let me consider how this arises in the specific mathematical contexts.

In the context of the Schwarzschild space-time we find that, with the substitution

$$\Psi_0 = \frac{r^3}{\Delta^2} Y(r) S_{+2}(\theta), \tag{30}$$

where $S_{+2}(\theta)$ is a suitably defined angular function, the radial function $Y(r)$ satisfies the equation

$$\Lambda^2 Y + P\Lambda_- Y - QY = 0, \tag{31}$$

where

$$\Lambda_\pm = \frac{d}{dr_*} \pm i\sigma, \quad \Lambda^2 = \frac{d^2}{dr_*^2} + \sigma^2, \tag{32}$$

$$P = \frac{d}{dr_*} \ln \frac{r^8}{\Delta^2} \quad \text{and} \quad Q = \frac{\Delta}{r^5}(\mu^2 r + 6M). \tag{33}$$

Since we should be able to deduce the reflexion and the transmission coefficients for incident gravitational waves in terms of the solutions of equation (31) and its complex conjugate, it follows that it must be possible to transform equation (31), simultaneously, to either of the two equations governing $Z^{(+)}$ and $Z^{(-)}$. The existence of such *dual transformations* is guaranteed by the coefficient Q in equation (31) satisfying a certain non-linear ordinary differential equation of the second order. Thus, by letting (cf. equation (16))

$$Q = \frac{\Delta^2}{r^8} F = \frac{\Delta^2}{r^8} \frac{1}{f}, \tag{34}$$

we can verify that

$$\frac{1}{F}\left(\frac{dF}{dr_*}\right)^2 - \frac{d^2F}{dr_*^2} + \frac{\Delta^2}{r^8} F^2 = \frac{36M^2 (=\beta^2)}{F} + \mu^2(\mu^2+2)(=\kappa). \tag{35}$$

By virtue precisely of this equation (and, in particular, by the occurrence of β^2 on the right-hand side), we are enabled to make the following dual transformations,

$$Y=V^{(\pm)}Z^{(\pm)}+(W^{(\pm)}+2i\sigma)\Lambda_+Z^{(\pm)} \tag{36}$$

and

$$\frac{\Delta^2}{r^8}K^{(\pm)}Z^{(\pm)}=QY-(W^{(\pm)}+2i\sigma)\Lambda_-Y, \tag{37}$$

where

$$W^{(-)}=\frac{2}{r^2}(r-3M),W^{(+)}=2\ \frac{\mu^2r^2-3\mu^2Mr-6M^2}{r^2(\mu^2r+6M)}, \tag{38}$$

$$K^{(\pm)}=\mu^2(\mu^2+2)\pm 12i\sigma M, \tag{39}$$

and $V^{(+)}$ and $V^{(-)}$ are given by equations (12) and (13). The content of equations (36) and (37) is that with $Z^{(+)}$ and $Z^{(-)}$ defined as solutions of the respective wave equations, Y given by equation (36) satisfies equation (31); and conversely with Y defined as a solution of equation (31), $Z^{(+)}$ and $Z^{(-)}$ given by equation (37) satisfy the respective wave-equations. It is the existence of these transformations that accounts for the remarkable relationships that we encountered in our earlier consideration of the axial and the polar perturbations of the Schwarzschild black-hole.

Turning to the Newman-Penrose equations governing the perturbations of the Reissner-Nordström space-time, we find that their *decoupling* depends crucially on two facts; the spherical symmetry of the background space--time which ensures their separability *and* the special circumstances that the angular functions, S_1 and S_2, belonging to the spin-1 Maxwell-field and spin-2 gravitational-field are related very simply: S_1 being a simple linear combination of S_2 and its derivative, and conversely. (It is by virtue of this last fact that the common characteristic-value parameter μ occurs in the definition of β_1 and β_2, and of the "mixing" angle ψ given by equation (37).) Because of these special circumstances the equations of the Newman-Penrose formalism governing the perturbed Reissner-Nordström space-time are reducible to the pair of independent equations,

$$\Lambda^2 Y_i + P_i \Lambda_- Y_i - Q_i Y_i = 0 \qquad (i=1,2) \tag{40}$$

where

$$P_i = \frac{d}{dr_*} \ln \frac{r^8}{D_i}, \quad D_i = \Delta^2 \left(1 + \frac{2\beta_j}{\mu^2 r}\right) \quad (i,j=1,2; i \neq j), \tag{41}$$

and

$$Q_i = \mu^2 \frac{\Delta}{r^4} \left(1 + \frac{2\beta_j}{\mu^2 r}\right)\left(1 + \frac{\beta_i}{\mu^2 r}\right) \quad (i=1,2; i \neq j). \tag{42}$$

And equation (40) allows dual transformations, as in the context of the Schwarzschild space-time, by virtue of Q_i satisfying a certain nonlinear ordinary differential equation. Thus, with the definition,

$$F_i = \frac{1}{f_i} = \frac{r^8}{D_i} Q_i = \frac{r^3}{\Delta} (\mu^2 r + \beta_i), \tag{43}$$

F_i satisfies the equation (cf. equation (33))

$$\frac{1}{F_i}\left(\frac{dF_i}{dr_*}\right)^2 - \frac{d^2 F_i}{dr_*^2} + \frac{D_i}{r^8} F_i^2 = \frac{\beta_i^2}{F_i} + \mu^2(\mu^2+2) \quad (i=1,2), \tag{44}$$

where the β_i's are defined in equations (21).

Thus, while the mathematical structure of the theory clarifies the various underlying relationships and in particular the origin of the potential barriers belonging to the class (15) (which guarantees the equality of the derived reflexion and the transmission coefficients), it cannot be said that the deeper (?) physical significance of the relationships has been understood.

I now turn to a consideration of the Kerr space-time.

The Kerr space-time is stationary and possesses an axis of symmetry. On these accounts one naturally expects that the equations of mathematical physics, written in this geometry, will separate in the variables t and φ. But one does not expect to separate the variables r and θ; and the fact that they do endows the Kerr geometry with an aura of the miraculous.

We list below the equations and the authors who separated them:

i) The Hamilton-Jacobi equation; B.Carter (1968)
ii) The scalar wave-equation; B.Carter (1968)
iii) Maxwell's equations; S.Teukolsky (1971)
iv) Gravitational equations; S.Teukolsky (1971)
v) Dirac's equation; S.Chandrasekhar (1976).

The mathematical theory underlying these various equations is inherently so complex that, within the allotted time, I can hardly give even a glimmering of the richness of the subject. In my account I shall restrict myself to the propagation of gravitational waves in the Kerr space-time; this is only one aspect of the larger problem of the metric perturbations.

In Kerr geometry as in Schwarzschild's geometry, the equations governing the Weyl scalars Ψ_0 and Ψ_4 decouple from the rest and can be considered independently of the others; and as Teukolsky showed in 1971, these equations can be separated. For perturbations with a t- and a φ-dependence given by

$$e^{i(\sigma t+m\varphi)} \qquad (m=0,\pm 1,\pm 2,\ldots), \tag{45}$$

one finds that with the substitutions,

$$\Psi_0=R_{+2}(r)S_{+2}(\theta), \quad \Psi_4=\frac{1}{(\bar{\rho}^*)^4}R_{-2}(r)S_{-2}(\theta), \tag{46}$$

where

$$\bar{\rho}=r+ia\cos\theta, \quad \bar{\rho}^*=r-ia\cos\theta, \tag{47}$$

the separated equations are

$$(\Delta\mathscr{D}_{-1}\mathscr{D}_0^{\dagger}-6i\sigma r)(\Delta^2 R_{+2})=\lambda(\Delta^2 R_{+2}), \tag{48}$$

$$(\Delta\mathscr{D}_{-1}^{\dagger}\mathscr{D}_0+6i\sigma r)R_{-2}=\lambda R_{-2}, \tag{49}$$

$$(\mathscr{L}_{-1}^{\dagger}\mathscr{L}_2+6a\sigma\cos\theta)S_{+2}=-\lambda S_{+2}, \tag{50}$$

$$(\mathscr{L}_{-1}\mathscr{L}_2^{\dagger}-6a\sigma\cos\theta)S_{-2}=-\lambda S_{-2}, \tag{51}$$

where λ is a characteristic-value parameter (determined by the requirements that S_{+2} and S_{-2} are smooth at the poles) and $\mathscr{D}_n$, $\mathscr{D}_n^\dagger$, $\mathscr{L}_n$ and $\mathscr{L}_n^\dagger$ are differential operators defined by the equations,

$$\mathscr{D}_n = \partial_r + [iK + 2n(r-M)]\,\Delta^{-1}, \quad \mathscr{D}_n^\dagger = (\mathscr{D}_n)^*$$
$$\mathscr{L}_n = \partial_\theta + Q + n\cot\theta, \quad \mathscr{L}_n^\dagger = -\mathscr{L}_n(\pi-\theta), \tag{52}$$

where

$$K = (r^2+\alpha^2)\sigma, \quad Q = a\sigma\sin\theta + m\,\mathrm{cosec}\,\theta,$$
$$\alpha^2 = a^2 + am/\sigma \quad \text{and} \quad \Delta = r^2 - 2Mr + a^2. \tag{53}$$

It is clear from the foregoing definitions that $\Delta^2 R_{+2}$ and R_{-2} satisfy complex-conjugate equations, while S_{+2} and S_{-2} satisfy equations, one of which can be obtained by the other by replacing θ by $\pi-\theta$.

The solutions of the pair of equations (48) and (49) and (50) and (51) satisfy certain remarkable identities which derive from the following directly verifiable commutation relations:

$$\Delta^2 \mathscr{D}_0\mathscr{D}_0\mathscr{D}_0\mathscr{D}_0(\Delta\mathscr{D}_{-1}^\dagger\mathscr{D}_0 + 6i\sigma r) = (\Delta\mathscr{D}_{-1}\mathscr{D}_0^\dagger - 6i\sigma r)\Delta^2\mathscr{D}_0\mathscr{D}_0\mathscr{D}_0\mathscr{D}_0, \tag{54}$$

and

$$\mathscr{L}_{-1}\mathscr{L}_0\mathscr{L}_1\mathscr{L}_2(\mathscr{L}_{-1}^\dagger\mathscr{L}_2 + 6a\sigma\cos\theta) = (\mathscr{L}_{-1}\mathscr{L}_2^\dagger - 6a\sigma\cos\theta)\mathscr{L}_{-1}\mathscr{L}_0\mathscr{L}_1\mathscr{L}_2. \tag{55}$$

These commutation relations have far-reaching consequences for the theory. We shall consider some of them.

The commutation relation (54) implies, for example, that if $\Delta^2 R_{+2}$ is a solution of equation (48), $\Delta^2\mathscr{D}_0^\dagger\mathscr{D}_0^\dagger\mathscr{D}_0^\dagger\mathscr{D}_0^\dagger(\Delta^2 R_{+2})$ is a solution of equation (49); and conversely, if R_{-2} is a solution of equation (49), $\Delta^2\mathscr{D}_0\mathscr{D}_0\mathscr{D}_0\mathscr{D}_0 R_{-2}$ is a solution of equation (48). In other words, we may write

$$\Delta^2\mathscr{D}_0\mathscr{D}_0\mathscr{D}_0\mathscr{D}_0 R_{-2} = \mathscr{C}_1\Delta^2 R_{+2} \tag{56}$$

and $$\Delta^2\mathscr{D}_0^\dagger\mathscr{D}_0^\dagger\mathscr{D}_0^\dagger\mathscr{D}_0^\dagger(\Delta^2 R_{+2}) = \mathscr{C}_2 R_{-2}, \tag{57}$$

where $\mathscr{C}_1$ and $\mathscr{C}_2$ are two complex constants. Similarly, the commutation relation (55) allows us to write

$$\mathscr{L}_{-1}\mathscr{L}_0\mathscr{L}_1\mathscr{L}_2 S_{+2} = D_1 S_{-2} \tag{58}$$

and $$\mathscr{L}^\dagger_{-1}\mathscr{L}^\dagger_0\mathscr{L}^\dagger_1\mathscr{L}^\dagger_2 S_{-2} = D_2 S_{+2}, \tag{59}$$

where D_1 and D_2 are two real constants. Equations (56)--(59) are the Teukolsky-Starobinsky identities. At this point, one may ask why the simple act of taking the complex conjugate of a solution of equation (48) (or (49)) is equivalent to applying a fourth-order differential operator to it; or, why the simple act of substituting θ by $\pi-\theta$ in a solution of equation (50) (or (51)) is equivalent to applying a fourth-order differential operator to it. The answers to these questions are sheathed in enigmas.

Further consequences of equations (56)-(59) are:

$$\begin{aligned}\Delta^2\mathscr{D}^\dagger_0\mathscr{D}^\dagger_0\mathscr{D}^\dagger_0\mathscr{D}^\dagger_0\Delta^2\mathscr{D}_0\mathscr{D}_0\mathscr{D}_0\mathscr{D}_0 &= \mathscr{C}_1\mathscr{C}_2 \\ &= |\mathscr{C}|^2 \text{ (say)} \quad \text{mod} \quad \Delta\mathscr{D}^\dagger_{-1}\mathscr{D}_0 + 6i\sigma r - \lambda = 0\end{aligned} \tag{60}$$

and $$\begin{aligned}\mathscr{L}^\dagger_{-1}\mathscr{L}^\dagger_0\mathscr{L}^\dagger_1\mathscr{L}^\dagger_2\mathscr{L}_{-1}\mathscr{L}_0\mathscr{L}_1\mathscr{L}_2 &= D_1 D_2 \\ &= D^2 \text{ (say)} \quad \text{mod } \mathscr{L}^\dagger_{-1}\mathscr{L}_2 + 6a\sigma\cos\theta + \lambda = 0,\end{aligned} \tag{61}$$

where $|\mathscr{C}|^2$ and D^2 are certain 'universal' constants. We shall call them the Starobinsky constants.

From the manner in which equations (60) and (61) have been derived, it would appear that there is no alternative to the evaluation of the Starobinsky constants except by a direct (but enormously tedious) reduction of the left-hand side of these equations and the successive application of the modulo conditions. By such reductions, one finds that

$$\begin{aligned}|\mathscr{C}|^2 = \lambda^2(\lambda+2)^2 - 8\sigma^2\lambda\,[\alpha^2(5\lambda+6) - 12a]^2 \\ + 144\sigma^4\alpha^4 + 144\sigma^2 M^2\end{aligned} \tag{62}$$

and

$$D^2=\lambda^2(\lambda+2)^2-8\sigma^2\lambda\left[\alpha^2(5\lambda+6)-12a^2\right]$$
$$+144\sigma^4\alpha^4=|\mathscr{C}|^2-144\sigma^2M^2, \tag{63}$$

where it may be recalled that

$$\alpha^2=a^2+am/\sigma. \tag{64}$$

At this point one may ask: why does the evaluation of $|\mathscr{C}|^2$ require so much detailed calculation; and what, indeed, is the origin of the simple relation (63) between the Starobinsky constants D^2 and $|\mathscr{C}|^2$? Let me digress briefly to answer, in part, these last questions.

An evaluation of $|\mathscr{C}|^2$,'without tears', emerges from a consideration of the so-called 'algebraically special' perturbations. These are perturbations in which only one of the scalars, Ψ_0 and Ψ_4, is non-vanishing, while the other vanishes. From equations (46), (56) and (57), it follows that the requirement, $\Psi_0\neq0$ (i.e., $R_{+2}\neq0$), and $\Psi_4\equiv0$ (i.e., $R_{-2}\equiv0$), for example, can be met if and only if

$$\mathscr{C}_1=0, \tag{65}$$

$$\mathscr{D}_0^\dagger\mathscr{D}_0^\dagger\mathscr{D}_0^\dagger\mathscr{D}_0^\dagger(\Delta^2R_{+2})=0. \tag{66}$$

and Δ^2R_{+2} satisfies equation (48).

The general solution of equation (66) can be rea-readily written by making use of the fact that

$$\mathscr{D}_0^\dagger=\frac{\tilde{\omega}^2}{\Delta}\left(\frac{d}{dr_*}-i\sigma\right), \tag{67}$$

where $\tilde{\omega}^2=r^2+\alpha^2$, (68)

and r_* is a new variable derived in accordance with the equation[3],

$$\frac{dr_*}{dr}=\frac{\tilde{\omega}^2}{\Delta}. \tag{69}$$

It follows from equation (67) that

$$\mathscr{D}_0^\dagger(r^ne^{+i\sigma r_*})=nr^{n-1}e^{+i\sigma r_*}. \tag{70}$$

Accordingly, the general solution of equation (66) is given by

$$\Delta^2 R_{+2} = (Ar^3+Br^2+Cr+D)e^{+i\sigma r_*}, \tag{71}$$

where A, B, C, and D are constants of integration. The requirement now that $\Delta^2 R_{+2}$ satisfies equation (48) leads to the conditions

$$\begin{aligned} &\lambda A+2i\sigma B=0, \\ &6(M-i\sigma\alpha^2)A+(\lambda+2)B+4i\sigma C=0, \\ &6a^2A+4i\sigma\alpha^2B-(\lambda+2)C-6i\sigma D=0, \\ &2a^2B+2(M+i\sigma\alpha^2)C-\lambda D=0. \end{aligned} \qquad (72) \qquad (72)$$

The determinant of this system of equations must therefore vanish; and a simple evaluation of the determinant leads to the condition,

$$|\mathscr{C}|^2=0. \tag{73}$$

We have thus an alternative derivation (without tears!) of the Starobinsky constant.

Returning to equations (56) and (57), we observe that in view of $\Delta^2 R_{+2}$ and R_{-2} satisfying complex-conjugate equations, we may so normalize them, relatively, that they satisfy the complex-conjugate equations,

$$\Delta^2 \mathscr{D}_0\mathscr{D}_0\mathscr{D}_0\mathscr{D}_0 P_{-2} = \mathscr{C} P_{+2}$$

and $$\Delta^2 \mathscr{D}_0^\dagger\mathscr{D}_0^\dagger\mathscr{D}_0^\dagger\mathscr{D}_0^\dagger P_{+2} = \mathscr{C}^* P_{-2}, \tag{74}$$

where we have written P_{+2} and P_{-2} in place of $\Delta^2 R_{+2}$ and R_{-2} to signify the chosen normalization of these functions. It is important to emphasize that even after this normalization, there is an ambiguity in the solutions for P_{+2} and P_{-2} which the physical problem demands. For, it is clear on general grounds that the solution of the entire perturbation-problem must be unique apart from a single real constant of proportionality. Therefore, there can be no ambiguity in P_{+2}, once P_{-2} has been specified in accordance with some selected boundary condition. In the context of equations (74), with the chosen relative normalization of P_{+2} and P_{-2} , the ambiguity resides in the relative phases of $P_{+2}$and P_{-2}. The resolution of *this* ambiguity will consist in determining the real and the imaginary parts of $\mathscr{C}$ (consistent with $|\mathscr{C}|^2$ given by equation (62)); but it will still leave open the question of the numerical scale-factor by which P_{-2} must be multiplied to obtain the proper solution of the physical problem. The solution of these two problems - that of determining the real and the imaginary parts of $\mathscr{C}$ and that of ascertaining the numerical scale-factor - are necessary prerequisites for determining the reflexion and transmission coefficients, for incident gravitational--waves, in terms of the solutions of equations (48) and (49); and they require the consideration of the entirety of the equations provided by the Newman-Penrose formalism.

The Newman-Penrose formalism provides a formidable array of equations: a set of 18 complex equations expressing the components of the Riemann tensor (derived from the Ricci identity) in terms of the directional derivatives of the spin coefficients (the Ricci rotation-coefficients in the context) and the Weyl and the Ricci scalars; a set of 10 complex equations (or 8 equations for a vacuum) expressing the Bianchi identities in terms of the directional derivatives of the Weyl and the Ricci scalars and the spin coefficients; and a set of 12 complex equations expressing the Lie brackets of the basis vectors with the structure constants written out in terms of the spin coefficient; altogether an array of 40 (or 38, for the vacuum) coupled complex-equations. The consideration of this array of equations, for a complete treatment of the gravitational perturbation of the Kerr space-time, is a task of prolixious complexity. I shall not attempt to describe how this task has been accomplished; but I will state how the solution to the two problems I have enunciated earlier emerges from an integrability condition. The mere statement of the integrability condition and of the identities which follow from it will give some indication of the nature of the realm of mathematical analysis that one has yet to explore.

First, I must introduce a *'bracket notation'* in which the equations of the theory are most compactly written. We define various 'brackets' by

$$[P]^{\pm}=P_{+2}\pm P_{-2} \quad ; \quad [S]^{\pm}=S_{+2}\pm S_{-2},$$
$$[\mathscr{D}P]^{\pm}=\mathscr{D}_0^{\dagger}P_{+2}\pm\mathscr{D}_0P_{-2} \quad ; \quad [\mathscr{L}S]^{\pm}=\mathscr{L}_2S_{+2}\pm\mathscr{L}_2^{\dagger}S_{-2}, \tag{75}$$
$$[\mathscr{D}\mathscr{D}P]^{\pm}=\mathscr{D}_0^{\dagger}\mathscr{D}_0^{\dagger}P_{+2}\pm\mathscr{D}_0\mathscr{D}_0P_{-2}; \quad [\mathscr{L}\mathscr{L}S]^{\pm}=\mathscr{L}_1\mathscr{L}_2S_{+2}\pm\mathscr{L}_1^{\dagger}\mathscr{L}_2^{\dagger}S_{-2}.$$

In many ways the central problem in the solution of the Newman-Penrose equations is the integrability condition of the equations,

$$\Delta^{3/2}\frac{\partial}{\partial r}\frac{\Psi}{\Delta^{\frac{1}{2}}}=-\Bigg[\Big\{Q[\mathscr{L}\mathscr{L}S]^{-}-2a\sigma[\mathscr{L}S]^{-}\cos\theta\Big\}r\,[P]^{+}$$
$$-ia\Big\{Q[\mathscr{L}\mathscr{L}S]^{+}\cos\theta+2(Q\sin\theta-a\sigma\cos^2\theta)\,[\mathscr{L}S]^{+}$$
$$-6a\sigma\,[S]^{+}\sin\theta\cos\theta\Big\}[P]^{-}\Bigg], \tag{76}$$

and

$$\frac{\partial\Psi}{\partial\theta}=i\Bigg[\Big\{rK[\mathscr{D}\mathscr{D}P]^{-}-2(r^2\sigma+K)\,[\mathscr{D}P]^{-}+6r\sigma\,[P]^{-}\Big\}[S]^{+}$$
$$-ia\Big\{K[\mathscr{D}\mathscr{D}P]^{-}-2r\sigma\,[\mathscr{D}P]^{+}\Big\}[S]^{-}\cos\theta\Bigg], \tag{77}$$

where Ψ is some function defined in the theory. I shall leave it to your imagination as to how one is to confront the problem presented by these equations. But the remarkable fact is that the integrability condition of equations (76) and (77) determines precisely the real and the imaginary parts of $\mathscr{C}$ and the scale factor. In fact, we find

$$\mathscr{C}=\mathscr{C}_1+i\mathscr{C}_2, \quad \text{where} \quad \mathscr{C}_1=D \quad \text{and} \quad \mathscr{C}_2=-12\sigma M; \tag{78}$$

and

$$\Delta^2\Psi_0=P_{+2}S_{+2} \quad \text{and} \quad \Psi_4=\frac{1}{4(\bar{\rho}^*)^4}P_{-2}S_{-2}, \tag{79}$$

where S_{+2} and S_{-2} are the normalized proper solutions of equations (50) and (51) (in which case $D_1=D_2=D$ in equations (58) and (59) with D given by equation (63)).

The solution of the integrability condition of equations (75) and (76) provides at the same time an explicit evaluation of the following two indefinite integrals:

$$\mathscr{R} = \Delta^{\frac{1}{2}} \int \left(-i\,[P]^- + \frac{4\lambda\sigma}{\mathscr{C}_1 + \Gamma_1}\, r\,[P]^+ \right) \frac{dr}{\Delta^{3/2}}\,, \tag{80}$$

and

$$\mathscr{S} = \int \left\{ [S]^- \cos\theta + \frac{4\sigma(\lambda\alpha^2 - 6a^2)}{a(\mathscr{C}_1 + \Gamma_1)} [S]^+ \right\} d\theta, \tag{81}$$

where

$$\Gamma_1 = \lambda(\lambda+2) - 12\sigma^2\alpha^2 \text{ and } \mathscr{C}_1^2 - \Gamma_1^2 = -16\sigma^2\lambda(\lambda\alpha^2 - 6a^2). \tag{82}$$

We find, in fact, two alternative expressions for $\mathscr{R}$ and $\mathscr{S}$ given by

$$\frac{1}{4}(\mathscr{C}_1 + \Gamma_1)\mathscr{R} = K[\mathscr{D}\mathscr{D}P]^+ - 2r\sigma[\mathscr{D}P]^+,$$

$$\sigma(\lambda\alpha^2 - 6a^2)\mathscr{R} = i\left\{ rK[\mathscr{D}\mathscr{D}P]^- - 2(r^2\sigma + K)[\mathscr{D}P]^- + 6r\sigma[P]^- \right\} \tag{83}$$

and

$$-\lambda a\sigma\mathscr{S} = Q[\mathscr{L}\mathscr{L}S]^- - 2a\sigma[\mathscr{L}S]^- \cos\theta,$$

$$-\frac{1}{4}(\mathscr{C}_1 + \Gamma_1)\mathscr{S} = Q[\mathscr{L}\mathscr{L}S]^+ \cos\theta + 2(Q\sin\theta - a\sigma\cos^2\theta)[\mathscr{L}S]^+$$
$$-6a\sigma[S]^+ \sin\theta\cos\theta. \tag{84}$$

I may also note that other equations of the Newman-Penrose formalism lead to the following differential equations for $\mathscr{R}$ and $\mathscr{S}$:

$$\frac{\Delta^2}{K}\left[\frac{d}{dr}\left(K\frac{d}{dr}\frac{\mathscr{R}}{K}\right) + \left(\frac{4\alpha^2\sigma^2}{K^2} - \frac{K^2}{\Delta^2}\right)\mathscr{R}\right]$$
$$= 2\left([P]^+ + i\,\frac{4\lambda\sigma}{\mathscr{C}_1 + \Gamma_1}\, r\,[P]^-\right), \tag{85}$$

and

$$\frac{1}{Q}\left[\frac{d}{d\theta}\left(Q\frac{d}{d\theta}\frac{\mathscr{S}}{Q}\right) - \left(\frac{4\alpha^2\sigma^2}{Q^2} - Q^2\right)\mathscr{S}\right]$$
$$= -2\left\{[S]^+\cos\theta + \frac{4\sigma(\lambda\alpha^2 - 6a^2)}{a(\mathscr{C}_1 + \Gamma_1)}[S]^-\right\}. \tag{86}$$

One does not need to be told that the identities expressed by equations (83) and (84) are astonishing; they provide a measure of the mathematical richness of the theory.

As I said at the outset, it has not been my intention in this lecture to give any coherent account of what one may properly consider as the content of the mathematical theory of black holes. I have tried, rather, to suggest that there are aspects of the theory which require a deeper understanding [4].

NOTES

(1) Short-range potentials are potentials whose integrals over the range of r_* are finite.

(2) The quantities distinguished by a tilde refer to the time-reversed situation. Thus $\tilde{R}(\sigma)=R(-\sigma)$.

(3) The introduction of this variable r_* is crucial to much of the subsequent developments.

(4) For additional details on the subject matter dealt with as well as for references to the literature see the author's recently published book on 'The Mathematical Theory of Black Holes' (Oxford, Clarendon Press, 1983).

GRAVITATIONAL COLLAPSE OF A DUST CLOUD AND THE COSMIC CENSORSHIP CONJECTURE

Demetrios Christodoulou

Courant Institute of Mathematical Sciences
New York University
New York, NY 10012, USA

O.Tolman found in 1934 a class of solutions of Einstein's equations which represent inhomogeneous spherically symmetric dust clouds (Tolman, 1934). In 1939 Oppenheimer and Snyder studied, as an idealized model of gravitational collapse (Oppenheimer and Snyder, 1939), the special case of Tolman's class of solutions which corresponds to a homogeneous spherical dust cloud. They analyzed the behaviour of the outgoing light rays and were thus led to the black hole concept. The Oppenheimer-Snyder study, although treating only a very special case, was highly important for providing the intuition which guided the approach to more general problems; the concept of a trapped surface plays a central role in the Penrose-Hawking singularity theorems (Hawking and Ellis, 1973). Then, a conjecture was introduced, derived again from the Oppenheimer--Snyder example, namely that no singularities which are visible from infinity can develop from regular initial data (Hawking and Ellis, 1973). This is the weaker form of what is now called "the cosmic censorship conjecture". There is a number of important results, among them the area theorem of black holes (Hawking and Ellis, 1973), which assume the truth of this conjecture. Finally, Penrose (1979) introduced a stronger form of the cosmic censorship conjecture which states that any singularities that arise from regular initial data are not even locally visible. This stronger conjecture is verified in the Oppenheimer-Snyder example.

Although much has been learned from the example of the homogeneous spherically symmetric dust cloud,

B. Bertotti et al. (eds.), General Relativity and Gravitation, 27–36.

the general inhomogeneous Tolman class of solutions has not been given the attention it deserves despite the fact that it is the only known infinite dimensional family of asymptotically flat exact solutions of Einstein's equations. The difficulties of such an investigation lie in solving the differential equation for the light rays.

Here I shall present an analysis of the behaviour of the outgoing light rays in the general inhomogeneous Tolman class. The results are rather surprising, leading to a different picture of gravitational collapse: for an open subset of initial density distributions, the first singular event, which occurs at the center of symmetry, is the vertex of an infinity of future null geodesic cones which intersect future null infinity. The frequency of the corresponding light rays is infinitely redshifted.

We consider the evolution of a spherically symmetric dust cloud which starts from rest. The description is given in comoving coordinates τ and R, where R, which takes values in the nonnegative real numbers, labels the spherical shells of dust, and τ is the proper time along the lines R=const., the world lines of the dust. Let $A(\tau,R)$ be the area of the 2-sphere τ=const., R=const. - One defines the radius $r(\tau,R)$ by setting $A=4\pi r^2$. As it is possible to make an arbitrary relabelling of the spherical dust shells by $R\to f(R)$, we fix the labelling by requiring that on the initial surface $\tau=0$, R coincides with the radius:

$$r(0,R)=R.$$

We denote by $\rho(R)$ the initial mass density and by $m(R)$ the initial mass distribution:

$$m(R)=4\pi\int_0^R \rho(S)S^2 dS.$$

Let $a(R)$ denote mean density within a sphere of radius R on the initial surface:

$$a(R)=3m(R)/4\pi R^3.$$

The radius $r(\tau,R)$ is given by the parametric equations:

$$\tau=(3/32\pi a)^{\frac{1}{2}}(\eta+\sin\eta)$$

$$r=\tfrac{1}{2}R(1+\cos\eta);\quad 0\leqq\eta<\pi.$$

The value of $\eta=0$ of the parameter corresponds to the initial surface $\tau=0$ while the value $\eta=\pi$ corresponds to the final singular surface on which we have $r=0$. We are considering the evolution in the future of the initial surface. Denoting partial differentiation with respect to R by a prime, the evolution of the mass density $\varepsilon(\tau,R)$ is given by:

$$\varepsilon=m'/4\pi r'r^2=(\rho/r')(R/r)^2.$$

The spacetime metric is given by:

$$ds^2=-d\tau^2+e^{2\omega}dR^2+r^2d\Sigma^2$$

where $d\Sigma^2$ is the canonical metric of the 2-sphere and

$$e^{\omega}=r'\cdot(1-2m/R)^{-\frac{1}{2}}.$$

We assume that $2m/R<1$ which is equivalent to assuming that the initial hypersurface contains no trapped surfaces. We also assume that the mass density on the initial surface is smooth, that is, ρ is such that if extended to the negative R-axis as an even function is a C^∞ function on the entire real line. Since we wish to consider only finite dust clouds, we assume in addition that ρ is of compact support. Let R_0 be the radius of the support of ρ. For $R>R_0$ the solution is the Schwarzschild solution with mass $m_0=m(R_0)$.

The surface $\eta=\pi$ or

$$\tau=\tau_s(R)=[3\pi/32a(R)]^{\frac{1}{2}}$$

in the Tolman solution is an essential singularity; not only do the mass density and curvature invariants blow up there, but the metric itself admits no continuous extension. The only other set where the solution is not regular is the set where $r'=0$ and $\eta<\pi$. This is the set of points where the spherical shells of dust cross each other. These shell crossing singularities are inessential (Papapetrou and Hamoui, 1967); although the mass density and curvature invariants blow up there, the metric is in fact continuous. This is seen if we replace the coordinates (τ,R) by the coordinates (τ,r). In these coordinates the Christoffel symbols belong to L^∞ locally in the region $0\leqq\eta<\pi$ and the Riemann curvature tensor and the energy momentum tensor are well defined distributions in this region. Since r' is given by

$$r'=\frac{1}{2}(1+\cos\eta)-\frac{1}{4}\,\frac{\sin\eta(\eta+\sin\eta)}{(1+\cos\eta)}\,\frac{Ra'}{a}\ ,$$

on a world line $R=\text{const.}>0$, we have a shell crossing point if and only if $a'(R)>0$; there is no shell crossing on the central world line $R=0$. We wish to avoid considering the shell crossing singularities so that we can concentrate on the structure of the essential singularity. We therefore assume that the initial mass density ρ is a monotonically non-increasing function of R. This implies that the same is true for the initial mean density a and we have no shell crossing.

An outgoing light ray satisfies the differential equation

$$d\tau/dR=e^{\omega(\tau,R)}\,.$$

Along an outgoing light ray we have:

$$\frac{dr}{d\tau}=\left(1-\frac{2m}{R}\right)^{\frac{1}{2}}-\left(\frac{2m}{r}-\frac{2m}{R}\right)^{\frac{1}{2}}.$$

Hence $dr/d\tau>0,=0,<0$ according to whether $r>2m,=2m,<2m$. Thus the surface defined by:

$$r(\tau,R)=2m(R)$$

or equivalently,

$$\tau=\tau_a(R)=\left(\frac{3}{32\pi a(R)}\right)^{\frac{1}{2}}\left\{\cos^{-1}(4m(R)/R-1)+\left[1-(4m(R)/R-1)^2\right]^{\frac{1}{2}}\right\}$$

is the locus of turning points of the outgoing light rays, namely, the set of events where the outgoing light rays stop diverging and start to converge. This is the apparent horizon. The apparent horizon lies to the past of the singular surface except at the center R=0 where there is a second order contact point:

$$\tau_a(0)=\tau_s(0)=(3\pi/32\alpha)^{\frac{1}{2}}=\tau_0,$$

$$\tau_a'(0)=\tau_s'(0)=0$$

$$\tau_a''(0)=\tau_s''(0)=\frac{3}{10}(3\pi/8\alpha^2)^{\frac{1}{2}}\beta,$$

where $\alpha=\rho(0)$ is the central density and $\beta=-\frac{1}{2}\rho''(0)$. The apparent horizon is null and future directed in the vacuum region $R\geqq R_0$, while it is either spacelike or past directed for $0<R<R_0$. In the vacuum region the apparent horizon coincides with the event horizon of the Schwarzschild solution $r=2m_0$. Any light ray which intersects the apparent horizon goes to the singularity within a finite affine parameter interval. The outgoing light ray which in the vacuum region coincides with the curve $r=2m_0$ is the generator of the boundary of the region of the spacetime which can communicate with infinity, that is, the generator of the event horizon. Any light ray which is later than the event horizon intersects the apparent horizon inside the cloud.

It can be easily seen that no light ray can emanate from the singular surface except from the singular point at the center. In view of this, strong cosmic censorship is equivalent to nonexistence of light rays which emanate from the singular point at the center and weak cosmic censorship is equivalent to nonexistence of such light rays which go to future null infinity. The following is readily shown:

Proposition 1: If $\rho''(0)=0$, then strong cosmic censorship holds.

In view of this we consider from now on initial densities with $\rho''(0)<0$.

Let us be given a nonnegative nonincreasing (even) C^∞ function $\rho_1(R)$ of compact support such that $\rho_1''(0)<0$. Let:

$$\alpha_1=\rho_1(0)$$

$$\beta_1=-\tfrac{1}{2}\rho_1''(0)$$

$$R_1^1=(\alpha_1/\beta_1)^{\frac{1}{2}}$$

and let R_0^1 be the radius of support of $\rho_1(R)$. We set

$$\rho_0(x)=\rho_1(R)/\alpha_1 : R=xR_1^1.$$

Then $\rho_0(x)$ is a nonnegative nonincreasing (even) C^∞ function of compact support and

$$\rho_0(0)=1$$

$$\rho_0''(0)=-2$$

and $\ell=R_0^1/R_1^1$ is the radius of support of $\rho_0(x)$. We then define a 2-parameter family of nonnegative nonincreasing (even) C^∞ functions $\rho(R)$ of compact support in the following way: Given $\alpha,\beta>0$ we set:

$$\rho(R)=\alpha\rho_0(x) : x=R/R_1,\ R_1=(\alpha/\beta)^{\frac{1}{2}} .$$

We have

$$\rho(0)=\alpha$$

$$\rho''(0)=-2\beta$$

and $R_0=\ell R_1$ is the radius of support of $\rho(R)$. Thus each $\rho_1(R)$ generates a 2-parameter family $\rho(R;\alpha,\beta)$ and this family has a standard member $\rho_0(x)$ which corresponds to $\alpha=\beta=1$. The ratios:

$$R_0/R_1=\ell$$

$$a(R_0)/\alpha=\xi$$

are common to all members of the family; ξ is the ratio of the initial mean density of the cloud as a whole to the initial central density. We introduce the dimensionless constant

$$\gamma = \frac{8\pi}{3}\frac{\alpha^2}{\beta}$$

Defining

$$\eta = \max_{0 \leqq x \leqq \ell} x^2 a_0(x) \quad \text{(common to the family)},$$

the requirement $2m(R)/R<1$ is equivalent to $\gamma<\gamma_0=1/\eta$.

Together with the dimensionless radial coordinate x we define a dimensionless time coordinate ζ by:

$$\zeta = (32\pi\alpha/3)^{\frac{1}{2}}(\tau-\tau_0).$$

The aim is to find a form of the differential equation for the outgoing light rays in which the singular point at the center $x=\zeta=0$ appears as a regular singular point. This is achieved by introducing in the role of ζ the variable θ defined by:

$$\zeta = \frac{3\pi}{10} x^{7/3}\ \theta$$

and in the role of x the variable u, where

$$x = u^3.$$

Using then in the role of γ the dimensionless constant γ defined by:

$$\lambda = (15/\pi)^{1/3}\ \gamma^{\frac{1}{2}}:\ \lambda<\lambda_0=(15/\pi)^{1/3}\ \gamma_0^{\frac{1}{2}}$$

I bring the differential equation for the outgoing light rays to the desired form

$$d\theta/du - 7(\lambda-\theta)/u = 2\lambda(u,\theta;\lambda)$$

$$f(u,\theta;\lambda) = \frac{1}{u}\left\{[1-(\pi/15)^{2/3}\ \lambda^2 u^6 a_0(u^3)]^{-\frac{1}{2}} f_1(u;\theta) - 7\right\}. \quad (1)$$

Here $f_1(u;\theta)$ is a function which is common to all members of the 2-parameter family and which is C^∞ and nonnegative in the strip $0\leqq u\leqq \ell^{1/3}$, $0\leqq\theta<\sigma(u)$, where

$\sigma(u)$ corresponds to the singular surface minus the singular point at the center. Furthermore we have

$$f_1(0;\theta)=7.$$

It follows that $f(u,\theta;\lambda)$ is C^∞ in the above strip and $f(u,\theta;\lambda)\geq -7/u$. I then prove:

Theorem 1: For every $0<\lambda<\lambda_0$ there exists an $a_2\in]0,\ell^{1/3}]$ and a solution $\theta\in C^\infty[0,a_2]$ to the differential equation (1).

This solution is nonnegative, satisfies $\theta(0)=\lambda$ and is the only solution of the equation which is continuous at $u=0$.

The solution uniquely extends as a C^∞ solution to an interval $[0,b[$ where either 1) $b=\infty$ or 2) b is finite and $\theta(u)\to\sigma(u)$ for $u\to b$.

The solution given by the above theorem represents an outgoing light ray n_0 which emanates from the singular point at the center and which either (case 2) intersects the apparent horizon inside the surface of the cloud $x=\ell$, in which case it reaches the singular surface, or (case 1) it arrives at the surface of the cloud earlier than the apparent horizon and therefore goes to future null infinity. This leads to:

Proposition 2: Through any event to the future of n_0 there passes an outgoing light ray which emanates from the singular point at the center.

There is therefore an infinity of outgoing light rays emanating from the singular point at the center. I then prove:

Theorem 2: n_0 is the earliest of all light rays which emanate from the singular point at the center.

By this theorem, the light ray n_0 is the generator of the boundary of the region of spacetime which is uniquely predictable by the given data on the critical surface, that is, the generator of the Cauchy horizon. The proof of the theorem is by showing that there exist x_0 and η_0 positive such that for $0<\eta\leq\eta_0$ the curve $n_0-\eta$ defined by $\zeta(x)=\zeta_{n_0}(x)-\eta$ is spacelike for $0\leq x<x_0$.

I then study the global behaviour of n_0. This leads to:

Theorem 3: In any given 2-parameter family of initial density distributions there exists a λ_c, $0<\lambda_c<\lambda_0$, $\lambda_c\to 0$ for $\xi\to 1$, such that 1) for all $0<\lambda<\lambda_c$, n_0 goes to future null infinity, 2) for $\lambda=\lambda_c$, n_0 coincides with the event horizon, 3) for all $\lambda_c<\lambda<\lambda_0$ we have a regular event horizon.

Note: The fact that $\lambda_c\to 0$ for $\xi\to 1$ means that the region of naked singularities tends to zero for a sequence of standard initial density distributions tending to homogeneity.

The proof of the theorem is based on the following sequence of lemmas:

Lemma 1: In any given 2-parameter family of initial density distributions there exists a $\lambda_1>0$ such that for all $0<\lambda<\lambda_1$: $\zeta_{n_0}(\ell;\lambda)<\zeta_a(\ell;\lambda)$, that is, n_0 arrives at the surface of the cloud earlier than the apparent horizon.

Lemma 2: In any given 2-parameter family of initial density distributions there exists a $\lambda_2<\lambda_0$, $\lambda_2\to 0$ for $\xi\to 1$, such that for all $\lambda_2\leqq\lambda<\lambda_0$ all outgoing light rays which emanate from the singular point at the center intersect the apparent horizon in the interior of the cloud.

Lemma 3: In any given 2-parameter family of initial density distributions, $\zeta_{n_0}(x;\lambda)$ is a monotonically increasing function of λ at each x.
Also, $\zeta_a(x;\lambda)$ is a monotonically decreasing function of λ at each x.

Finally, the proof of Theorem 2 leads to:

Theorem 4: Consider a dust cloud satisfying the condition of case 1 of Theorem 3 and consider an observer following a world line $r=r_0$ in the vacuum region who receives light rays emitted from the central world line R=0 of the dust cloud. If ω_e is the frequency of emission of a light ray at the center and ω_r the frequency of its reception by the observer, then the ratio ω_r/ω_e is a discontinuous function of the proper time T of the observer; this function is smooth and greater than a positive constant for $T<T(n_0)$ and is zero for $T>T(n_0)$ where $T(n_0)$ is the proper time at which the light ray n_0 is received by the observer.

NOTE ADDED

After this work was completed I was informed that singularities similar to those analyzed here have been found in computer studies by Douglas M.Eardley and Larry Smarr: 1979, "Time Functions in Numerical Relativity: Marginally Bound Dust Collapse", *Phys.Rev.*, D19, 2239.

REFERENCES

Hawking, S.W., and Ellis, G.F.R.: 1973, *The Large Scale Structure of Spacetime*, Cambridge University Press, Cambridge.

Oppenheimer, J.R., and Snyder, H.: 1939, "On Continued Gravitational Contraction", *Phys. Rev.*, 56, 455.

Papapetrou, A., and Hamoui, A.: 1967, "Surfaces caustiques dégénérées dans la solution de Tolman. La singularité physique en Relativité Générale", *Ann.Inst. Henri Poincaré*, 6, 343.

Penrose, R.: 1979, "Singularities and Time Asymmetry", in S.W.Hawking and W.Israel (eds), *General Relativity: An Einstein Centenary Survey*, Cambridge University Press, Cambridge, pp. 581-638.

Tolman, R.A.: 1934, "Effect of Inhomogeneity on Cosmological Models", *Proc. Natl. Acad. Sci.USA*, 20, 169.

ASYMPTOTIC PROPERTIES OF ISOLATED SYSTEMS: RECENT DEVELOPMENTS

Abhay Ashtekar*

Laboratoire de Physique Théorique
Institut H. Poincaré, 11, rue P. et M. Curie,
75 231 Paris Cedex 05, France
and
Physics Department, Syracuse University
Syracuse, N.Y. 13210, USA

1. INTRODUCTION

Since this report belongs to the general plenary session, I thought it would be best to present a review of the current status of the field in broad terms. Thus, I will omit all proofs and try to convey the essential ideas instead. Furthermore, in the main text, technical points will be occasionally skipped, and, in some cases, results presented will not be the most general ones available today. The hope is that the style adopted will facilitate non-experts to get a good feeling for the type of issues that are discussed in the field. I must say, however, that the discussion is *not* evenly distributed among the various topics of interest: the emphasis is on the material which was published during the last three or four years and which is not covered in other talks at this conference. The experts will find that some of the material is presented from a new angle - the relation between the conformally rescaled and physical space-time quantities is stressed - and that a few of the ideas are new (in the sense that they have not appeared in print before).

Let me begin with some motivaing remarks. Physical applications of classical general relativity fall into two broad categories: cosmology and the study of isolated gravitating systems. In the first, one is

B. Bertotti et al. (eds.), General Relativity and Gravitation, 37–68.

interested in the large scale structure of the universe as a whole, while in the second, one is interested in the behaviour of individual systems such as two colliding black holes, or a system of binary stars orbiting around each other. In the study of isolated systems, one needs a conceptual framework containing basic notions such as energy-momentum, angular momentum, multipoles, and presence or absence of gravitational radiation; without such notions, it is not even possible to pose - let alone answer - meaningful questions of physical interest. Now, it is *not* a priori clear that even these basic notions can be introduced unambiguously in exact general relativity. The problem is that, while in Minkowskian field theories one has access to stress-energy tensors, and hence to locally defined energy density, etc., the analogous notions cannot be invariantly defined in the case of the gravitational field. Another avenue is to analyse the asymptotic properties of gravitational fields produced by systems of interest: one expects that, as one goes far away from sources, the structure of the gravitational field would simplify and one would be able to pick-up physically interesting information in an unambiguous way. Now, one knows from field theories in Minkowski space that such a program can be successful only if suitable boundary conditions are imposed on the fall-off of physical fields. In the present case, the choice of realistic boundary conditions is a delicate matter since the space-time geometry, in terms of which these notions are normally formulated, is itself the dynamical field of interest. Therefore, if one adopts the 'asymptotics' avenue one's task is first to find suitable boundary conditions for gravitational fields of isolated systems, then to introduce physically interesting notions using these boundary conditions and to prove theorems relating them, and finally, to apply the framework so obtained to physical problems.

How has one fared in this task? As we shall see, a rather elaborate conceptual framework *is* now available to read physically interesting information from the asymptotic behavior of the gravitational field. Furthermore, there exist a number of theorems asserting that the framework is 'correct'. What is sorely missing, however, are approximation methods which can put this exact, geometrical framework at the service of physical problems involving dynamical evolution of systems(1). Thus, for example, while one knows how to compute the energy emitted by, say, a binary system in the form of gravitational waves once the asymptotic form of the metric is known up to the retarded instant of time under consideration, one does not know how to predict the back-reaction effect of this emission on the orbits of the system[(2)]. More generally, one knows very little about the relation between the behavior of sources and that of their asymptotic fields. The conceptual framework, however, has paid off handsomely in unexpected directions in quantum gravity as well as in other branches of Mathematical Physics which have nothing to do with gravitation[(3)]. In my opinion, these applications are substantial in their own right.

The plan of this report is as follows. Section 2 recalls basic definitions and explains the motivation and meaning of their requirements. Section 3 summarizes the first-order structure available at infinity and Section 4 reviews the higher order structure. Some of the important open problems are discussed in Section 5. Appendices contain more technical material which has not appeared in print before.

2. BASIC DEFINITIONS

One can move away from sources in different directions - space-like, null and time-like - to explore the structure of the gravitational field. In each case, the structure reveals a different aspect of the isolated system. Thus, by moving away in space-like directions, one can explore the 'coulombic' aspects and read-off quantities like the total energy-momentum and angular momentum from the asymptotic gravitational field, while by moving away in null directions one can investigate the 'radiative' aspects and compute fluxes of energy-momentum and angular momentum carried away by gravitational waves. The structure at time-like infinity is more complicated since there are time-like directions along which one always remains 'close' to sources, i.e., along which the curvature does not fall-off. Consequently, time-like infinity is not of direct interest to isolated systems and I shall restrict myself only to spatial and null regimes. (It is, however, of interest to cosmology. See Eardley and Sachs, 1973, and Porrill, 1982).

The first problem is that of introducing boundary conditions which the gravitational fields of isolated systems should satisfy in these two regimes. This is achieved by introducing the notion of *asymptotically flat space-times*. The idea is to require that the space-time curvature should die off - i.e. the space-time should resemble Minkowski space - at a suitable rate as one recedes from sources in space-like and null directions. The problem now reduces to that of making these ideas precise. For this, let us begin by recalling the situation in Minkowski space itself. Here, one can conformally rescale the space-time metric and imbed the resulting space-time isometrically in the Einstein static universe, which is topologically $S^3 \times R$ (see, e.g. Hawking and Ellis, 1973, p. 118-124). The imbedding sends Minkowski space into an open region of the Einstein cylinder, the boundary of which consists of the null cone of a certain point i^0 (together with future and past vertices of the null cone, i^+ and i^-). The boundary points are all at infinity w.r.t. the physical (i.e. unrescaled) Minkowskian metric, although they are only a finite distance away w.r.t. the conformal metric; the rescaling brings infinity to finite distance. The point i^0 is space-like related to all points of the physical space-time and is the only point in the Einstein cylinder with this property. It represents *spatial infinity* of Minkowski space. Its null cone, denoted by $\mathscr{I}$ represents *null infinity*. $\mathscr{I}$ has two disconnected components $\mathscr{I}^{\pm}$, the future and the past light cones, called, respectively, future and past null infinity. All space-like geodesics in Minkowski space terminate at i^0 while each null geodesic has two end points on $\mathscr{I}$, one on $\mathscr{I}^+$ and the other on $\mathscr{I}^-$. The idea is to require that asymptotically flat space--times should admit a similar conformal completion. We have[(4)] (Ashtekar, 1980):

Definition 1: A space-time (M, g_{ab}) will be said to be *asymptotically empty and flat at spatial infinity* if there exists a space-time $(\hat{M}, \hat{g}_{ab})$ and an imbedding of M into $\hat{M}$, with $\hat{g}_{ab}=\Omega^2 g_{ab}$ on M, such that

i) $\hat{M}$ admits a point i^0 which is space-like related to all points of M;

ii) At i^0, $\hat{g}_{ab}$ is $C^{>0}$, and, $\Omega=0$, $\hat{\nabla}_a\Omega=0$ and $\hat{\nabla}_a\hat{\nabla}_b\Omega=2\hat{g}_{ab}$; and,

iii) The Ricci tensor R_{ab} of g_{ab} is such that R_{ab} admits a (regular direction dependent) limit at i^0.

Let us examine the conditions one by one. Condition i) just serves to locate the point i^0 at spatial infinity. The conditions on the conformal factor in ii) ensure that Ω 'falls-off as $1/r^2$', i.e. has Minkowskian behavior at i^0 while iii) asks that the stress-energy of matter sources should 'fall-off as $1/r^4$' in the physical space-time. These requirements are the obvious ones. The tricky part is the differentiability condition on $\hat{g}_{ab}$ contained in ii). In Minkowski space, $\hat{g}_{ab}$ can be made C^∞ - or even analytic - at i^0. In a general space-time, however, we cannot even demand that $\hat{g}_{ab}$ be C^1: as shown in Appendix 2, this would imply that the total mass of the space-time is zero. Why then do we not simply require $\hat{g}_{ab}$ to be C^0 at i^0? In this case, the mass cannot even be defined in general! Thus, in order to incorporate space-times with finite mass - the minimum one should incorporate if the definition is to be useful - one is forced to impose a differentiability requirement which is 'in between' C^0 and C^1. We have essentially required that the Christoffel symbols should admit limits at i^0 but which may depend on the direction of approach. As shown in Appendix 2, this implies that the physical metric approaches a flat metric as $1/r$; it is the presence of $1/r$ terms which prevent $\hat{g}_{ab}$ from being C^1. The necessity of such a differentiability requirement, I believe, is the principal reason behind the 10-15 years gap between the exploration of structure of $\mathscr{I}$ and that of i^0.

How do these requirements compare with those imposed in the 3+1 framework of Arnowitt, Deser and Misner (1962)? Consider any space-like 3-flat in the tangent space at i^0. Then, there exists a space-like sub-manifold $\hat{\Sigma}$ in $\hat{M}$ tangential to the given flat at i^0 such that the Cauchy data induced on $\Sigma \equiv \hat{\Sigma} - i^0$ by the physical metric g_{ab} satisfy the ADM conditions. Thus, while the present framework is entirely 4-dimensional, one can recover the ADM framework from it by carrying out a 3+1 splitting. However, because it does not *require* a 3+1 decomposition in itself, it avoids many conceptual and technical problems faced by the ADM framework (for details, see Astekar and Hansen, 1978).

Given a definition, one is immediately led to ask if it is strong enough to lead to interesting structure, and, at the same time, weak enough to admit a wide class of examples. That Definition 1 leads to interesting structure will be clear from the discussion contained in the next two sections. Here, we summarize the status of examples. First of all, explicit construction (Ashtekar and Hansen, 1978) shows that the Kerr family satisfies the definition. The construction strongly suggests that any stationary solution which is asymptotically flat, say in the sense of ADM, would also be so in the sense of Definition 1. What about non-stationary space-times? Recently proved boost theorems show that non-stationary space-times whose initial data is 'close to' that of the Kerr family also satisfy the definition. Here, one argues in two steps. In the first (Choquet-Bruhat and Christodoulou, 1981), one shows that the initial value (i.e. constraint) equations on the gravitational Cauchy data admit a 'Sobolev space worth' of solutions whose deviations from the Cauchy data of the Kerr family are 'small', and, in the second (Christodoulou and O'Murchadha, 1981), one concludes by proving that, when evolved, these data yield space-times satisfying, in essence, Definition 1. Thus, the definition is quite satisfactory as far as examples are concerned.

Finally, let us consider null infinity. As the discussion of Minkowski space suggests, the light cone (in $\hat{M}$) of i^0 will be identified with null infinity. For the space-time to be asymptotically flat at null infinity, we have to require that this null cone has the appropriate structure.

Definition 2: A space-time (M, g_{ab}) satisfying Definition 1 will be said to be *asymptotically empty and flat at null and spatial infinity* (AEFANSI) if the following additional conditions hold:

iv) the boundary of the region in $\hat{M}$ causally connected to i^0 $(=:\mathscr{I})$ is topologically $S^2\times R$, with complete generators;

v) At $\mathscr{I}$, $\Omega=0$, $\hat{\nabla}_a\Omega\neq 0$; and,

vi) $\Omega^{-2}R_{ab}$ admits a (smooth) limit to $\mathscr{I}$.

The role of the last condition is obvious: v) demands that Ω should fall-off as $1/r$ at null infinity. (In Minkowski space, Ω falls-off as $1/uv$. Since one approaches $\mathscr{I}^+$ along u=const. (and $\mathscr{I}^-$ along v=const.) curves, Ω falls-off as $1/r$ at null infinity. At i^0, both u and v diverge as r, whence Ω falls off as $1/r^2$). iv) identifies $\mathscr{I}$ as the null cone of i^0, care being required in the definition of 'null cone' since $\hat{g}_{ab}$ is not C^1 at i^0. (One cannot, for example, simply say that $\mathscr{I}$ is formed by null geodesics emanating from i^0). The completeness of generators - more precisely, that of the vector field $\hat{n}^a\equiv\hat{\nabla}^a\Omega$, in the conformal frame in which $\mathscr{I}$ is divergence-free, i.e., in which $\hat{\nabla}_a\hat{n}^a=0$ on $\mathscr{I}$ - ensures that we have entire null infinity and not just a piece of it (Geroch and Horowitz, 1978). It is straightforward to verify that AEFANSI space-times are asymptotically flat at null infinity in the standard sense (Penrose 1963, 1965; Geroch and Horowitz, 1978).

What is the status of examples as far as null infinity is concerned? In the stationary case, there are lot of examples; one knows (Koul,1983a, 1983b) that any such space-time which is asymptotically flat at spatial infinity in the sense of 3+1 frameworks (more precisely, in the sense of Geroch, 1971, and Hansen 1974) is asymptotically flat at null infinity. In the non-stationary context, however, the situation is not as conclusive as that in the case of i^0 and in fact there is some controversy as to whether physically interesting space-times with $\mathscr{I}$ exist. This issue is discussed in Section 5. That the issue is still unsettled is, in any case, quite unsatisfactory since the main application of the null infinity framework to isolated systems is in problems involving radiation.

3. THE FIRST ORDER STRUCTURE AT INFINITY

By studying the leading order deviations of the structure at i^0 and $\mathscr{I}$ of an asymptotically flat space-time from that of Minkowski space-time, one can single out physical fields which, when used with asymptotic symmetries, yield the 4-momentum and the angular momentum content of the space-time. The higher order structure is more difficult to disentangle unambiguously and leads to higher multipoles.

3.1. Spatial infinity

Let us consider any space-time (M, g_{ab}) satisfying Definition 1 and consider therein diffeomorphisms which preserve the asymptotic conditions. That is, if $(\hat{M}, \hat{g}_{ab})$ is a completion of (M, g_{ab}) satisfying the definition, consider the group D of diffeomorphisms of $\hat{M}$ which leave i^0, the value of $\hat{g}_{ab}$ at i^0 and its $C^{>0}$ - character invariant. Take the quotient, $G=D/D_0$ of D by its subgroup D_0 consisting of diffeomorphisms which are asymptotically identity. (A diffeomorphism of compact support in M is, for example, in D_0). Then G is the asymptotic symmetry group at spatial infinity, *the spi group*. A detailed analysis (Ashtekar, 1980) shows that G has the following structure. It admits an infinite dimensional, Abelian normal subgroup $\mathscr{S}$, the subgroup of supertranslations and the quotient, $G/\mathscr{S}$ is isomorphic with the Lorentz group $\mathscr{L}$. In order to define $\mathscr{S}$, consider the hyperboloid $\mathscr{H}$ of unit space-like vectors in the tangent space at i^0. Then $\mathscr{S}$ is naturally isomorphic to the additive group of functions on $\mathscr{H}$. Thus, G is a semi-direct product of $\mathscr{S}$ with the Lorentz group $\mathscr{L}$-$\mathscr{L}$ has a natural action on $\mathscr{H}$, and hence, on $\mathscr{S}$. Thus, in its structure, G is like the Poincaré group. It is, however, infinite dimensional and, although it admits a preferred 4-parameter (normal) subgroup of translations[5], it does *not* admit a preferred Poincaré subgroup. Had the metric $\hat{g}_{ab}$ been, in addition, C^1 at i^0, the group *would have reduced to the Poincaré group*; it is the coulombic, 1/r- terms due to the presence of mass that introduce the 'supertranslation ambiguities'.

Next, let us consider the physical fields at i^0. The $C^{>0}$ differentiability of $\hat{g}_{ab}$ implies that its Riemann tensor $\hat{R}_{abcd}$ is such that $\Omega^{\frac{1}{2}}\hat{R}_{abcd}$ admits a limit at i^0 which can, however, depend on the direction of approach to i^0. The 'Weyl-part' of this limit gives rise to - and is coded in - two symmetric, trace-free tensor fields, E_{ab} and B_{ab} on the hyperboloid $\mathscr{H}$ at i^0. These fields - called the electric and magnetic parts of the asymptotic Weyl tensor - capture the $1/r^3$, i.e. the leading contribution of the physical space-time curvature. The Ricci part of $\hat{R}_{abcd}$ only provides certain potentials for these fields. Of particular interest is a tensorial potential K_{ab} for B_{ab} satisfying

$$B_{ab} = \epsilon_{mn(a} D^m K^n{}_{b)} \tag{1}$$

where ϵ_{amn} and D are the natural alternating tensor and the derivative operator induced by the natural metric h_{ab} on $\mathscr{H}$. Finally, the asymptotic field equations (condition iii) of Definition 1 imply:

$$D_{[a}E_{b]c} = 0 \quad \text{and} \quad D_{[a}B_{b]c} = 0. \tag{2}$$

By investigating the geodesic deviation equations governing the motion of physical test particles in the asymptotic regime, one can identify the 'mass-aspect' - the primary factor governing the deviation - in terms of the above first-order fields (Ashtekar 1980). This leads to the following definition of the total 4-momentum, P_a, of the isolated system under consideration: P_a is a 4-vector at i^0 given by

$$P_a V^a = \frac{1}{8\pi} \oint_C E_{ab} V^b dS^a \tag{3}$$

where V^a is any vector at i^0, and, C, any 2-sphere cross-section of i^0 (non-contractable to zero). Equation (2) implies that (3) is independent of the choice of C; P_a is absolutely conserved. There are a number of theorems concerning the behavior of P_a. First of all, one can show (Ashtekar and Magnon-Ashtekar, 1983) that P_a is the same as the ADM 4-momentum defined in terms of the initial data on a 3-surface. Secondly, one can show (Ashtekar and Magnon-Ashtekar, 1979) that, if (M, g_{ab}) is stationary, P_a is proportional to the asymptotic time translation defined by the Killing field, the factor of proportionality being the Komar integral (Komar, 1959). Finally, and more importantly, one has the positive energy theorems (Schoen and Yau, 1979, 1981; Witten, 1981): if (M, g_{ab}) admits a complete, space-like hypersurface on which the initial data induced by g_{ab} are asymptotically flat and the stress-energy tensor satisfies the dominant energy condition, P_a is future directed and time-like (Ashtekar and Horowitz, 1982) and vanishes iff (M, g_{ab}) is Minkowski space(6). This is by far the most important result in the field in recent years and the only reason why I have not devoted more space to it is that it - as well as its ramification - are discussed in themselves elsewhere in this volume (see Witten's article and reports of workshops A1, A5 and D2).

What happens if we replace E_{ab} by B_{ab} in (3)? We again obtain a conserved quantity, the 'magnetic' or the 'NUT' 4-momentum, which, however, vanishes in virtue of (1) and smoothness of K_{ab} guaranted by Definition 1. To obtain a non-zero 4-vector, we must allow K_{ab} to develop 'wire singularities' and this is not easy to incorporate in the present framework. In the ADM framework, it is straightforward to weaken the requirements to permit a non-zero NUT energy (Ashtekar and Magnon-Ashtekar, 1983) - one has to allow the extrinsic curvature to develop 'wire singularities' but *can* keep the 3-metric regular - but it is complicated to do the same for the NUT 3-momentum. For classical general relativity, of course, NUT 4-momentum is not of direct physical interest. However, it may be of considerable interest to quantum gravity.

Next, let us consider angular momentum. The problem is that, with the boundary conditions imposed so far, angular momentum cannot be defined. There are two difficulties:whereas angular momentum normally refers to Poincaré - or Lorentz - group, G is infinite dimensional and does not admit such a preferred subgroup, and, while one expects the angular momentum information to reside in the $1/r^4$-part of the physical space curvature, so far,we have access only to the $1/r^3$-part. That there is a problem may appear, at first, surprising since I have said that the present asymptotic conditions do imply the ADM conditions and many textbooks give the angular momentum formula using the ADM definition. To see where the problem lies, let us briefly recall the ADM definition. Let Σ be a 3-surface in (M, g_{ab}) with asymptotically flat initial data (h_{ab}, π^{cd}). To define angular momentum, one considers a flat 3-metric f_{ab} to which h_{ab} approaches as $1/r$ and a rotational Killing vector ξ^a of f_{ab}. Then, the component of angular momentum along the axis of ξ^a is claimed to be

$$J_\xi = \frac{1}{8\pi} \lim_{r_0 \to \infty} \oint_{r=r_0} \pi_{ab} \xi^a ds^b \tag{4}$$

where r is a radial coordinate of f_{ab}. The problem is that the expression is tied too rigidly to the choice of f_{ab}. Let $x_{\underset{\sim}{k}}, \underset{\sim}{k}=1,2,3$ be a set of cartesian coordinates of f_{ab}. Set $\bar{x}_{\underset{\sim}{k}} = x_{\underset{\sim}{k}} + \Lambda_{\underset{\sim}{k}}(\theta,\phi)$ and consider the flat metric $\bar{f}_{ab}$ for which $\bar{x}_{\underset{\sim}{k}}$ constitutes a cartesian chart. It is easy to verify that, because $\Lambda_{\underset{\sim}{k}}$ depend only on θ and ϕ, the difference $\bar{f}_{ab}-f_{ab}$ - and hence $h_{ab}-\bar{f}_{ab}$- falls-off as 1/r. Thus, $\bar{f}_{ab}$ is 'as good a flat metric' as f_{ab}. Let us therefore consider the rotational Killing field $\bar{\xi}^a$ of $\bar{f}_{ab}$ (with the same axis as ξ^a) and compute $J_{\bar{\xi}}$. Since $\bar{\xi}^a-\xi^a$ is a vector field involving $\Lambda_{\underset{\sim}{k}}$ and their derivatives, and since $\Lambda_{\underset{\sim}{k}}$ are arbitrary, $J_{\bar{\xi}}-J_\xi$ can be made to assume any value one pleases (except in special cases such as the one in which π_{ab} vanishes, or is pure-trace). The 'angle--dependent translations', $x_{\underset{\sim}{k}} \to \bar{x}_{\underset{\sim}{k}}$, are precisely the spi supertranslations and their presence is what makes the expression (4) ill-defined. To remedy this situation, one has to impose stronger boundary conditions which can enable one to pick a family of flat metrics which are related to each other only by *translations*, rather than supertranslations, and use the Killing vectors of this family in (4).

In the 4-dimensional spi framework, this elimination of the supertranslation freedom is achieved by imposing the condition[7]

$$B_{ab}=0 \quad \text{on } \mathscr{H}, \tag{5}$$

and requiring that the next-order - i.e. $1/r^4$-part - of the magnetic Weyl curvature admits a limit β_{ab} at i^0. (1) now implies that we can demand that K_{ab} itself must vanish. This last requirement picks out a preferred class of conformal metrics $\hat{g}_{ab}$ which are mapped among each other by translations but not supertranslations. If this class is now included in the list of structures to be preserved by asymptotic symmetries, one retains, from G, only a Poincaré subgroup thereof. Thus, under the stronger boundary condition (5), the asymptotic symmetry group at i^0 reduces to the Poincaré group. One, can now define a tensor M_{ab} at i^0 by:

$$M_{ab}F^{ab}=\frac{1}{8\pi}\oint_C \beta_{ab}\xi^a dS^b \tag{6}$$

where F_{ab} is any skew tensor at i^0 and $\xi^a=\frac{1}{2}\epsilon^{abcd}F_{cd}\eta_b$ where η^b is the unit vector field normal to $\mathscr{H}$ in the tangent space of i^0. M_{ab} is the required angular momentum. It transforms correctly under the action of translations (Ashtekar and Hansen, 1978; Ashtekar, 1980). One can show that the definition reduces to the familiar ones in axisymmetric as well as stationary space-times (Ashtekar and Magnon-Ashtekar, 1979).

What does this contruction correspond to in the physical space language? It follows from Appendix 2 that the preferred family of conformal metrics $\hat{g}_{ab}$ - i.e. of conformal factors Ω - leads to a preferred family of flat metrics in the physical space; the preferred conformal factors are given in terms of the radial coordinates of the flat metric by $\Omega=1/r^2$. Any two flat metrics in this family are related by asymptotic translations rather than supertranslations and this eliminates the supertranslation freedom. In the 3+1 framework, the extra boundary condition (5) is difficult to motivate, although, once it is imposed, the group-reduction has a nice geometrical interpretation (Ashtekar and Magnon-Ashtekar, 1983). The part of (5) relevant to the 3-surface Σ is:

$$\lim_{r\to\infty} r^3\mathscr{R}^{ab}(\nabla_a r)\gamma_{bc}=0 \tag{7}$$

where $\mathscr{R}_{ab}$ is the Ricci tensor of the 3-metric h_{ab}; r a radial coordinate of *any* flat metric to which h_{ab} approaches, and γ_{bc}, the projection operator into the 2-spheres r=constant. (Both (5) and (7) are supertranslation invariant). Thus, one requires that the radial-angular part of the 3-dimensional Ricci tensor should fall-off faster than expected. (Note that the radial-radial part cannot fall-off faster than $1/r^3$ unless the total mass, the norm of P_a, vanishes). If this condition is satisfied, one can require that the projection of K_{ab} into Σ should vanish. In the 3+1 language, this corresponds to requiring that the extrinsic curvature, ${}^2\pi_{ab}$, of the r=constant 2-spheres should satisfy[8]:

$$\lim_{r\to\infty} r^2({}^2\pi_{ab}-\frac{1}{r}\gamma_{ab})=0.$$

This is an implicit condition on the flat metric whose

radial coordinate is being used in the construction of 2-spheres. The condition is *not* satisfied by a general flat metric; it selects a preferred family of flat metrics which are related to each other by translations. Roughly speaking, the metric 2-spheres of this family have 'fewer wiggles' (as seen by the physical metric h_{ab}) than the metric 2-spheres of a generic flat metric. Now, if in the ADM formula (4) one uses for ξ^a a rotational Killing field of one of the preferred flat metrics, the spi expression (6) reduces to (4) (Ashtekar and Magnon-Ashtekar, 1983). The equality does *not* hold for other (supertranslated) choices of ξ^a. Thus, the ADM construction is not wrong; it is incomplete. Also, the ADM angular momentum considered so far is associated with spatial rotations within Σ and the reduced group obtained from (7) is an Euclidean subgroup of the Poincaré group at i^0. It seems very difficult to remove the supertranslation ambiguities from boosts - i.e. to get rid of supertranslations which are identity transformations on Σ - in the 3+1 framework.

Finally, note that the necessity of imposing (5) implies that the expressions (4) and (6) cannot be well-defined in presence of the NUT 4-momentum. This may have interesting ramifications to the notion of spin in quantum gravity (e.g. in the framework of Friedman and Sorkin, 1980).

3.2. Null infinity

Let us now suppose that (M, g_{ab}) is AEFANSI and examine the structure at $\mathscr{I}$. Since most of this structure was well-understood already in mid-sixties (Bondi et al., 1980; Sachs, 1962a, 1962b; Penrose, 1963,1965), the discussion here will be brief. (For detailed reviews, see Geroch, 1976; Newmann and Todd, 1980). $\mathscr{I}$ is a null 3-manifold with null normal $\hat{n}^a \equiv \hat{\nabla}^a \Omega$; if $\hat{q}_{ab}$ denotes the pull-back to $\mathscr{I}$ of $\hat{g}_{ab}$, then $\hat{q}_{ab}\hat{v}^b=0$ iff $\hat{v}^a$ is proportional to $\hat{n}^a$. The asymptotic symmetry group at null infinity - The Bondi-Metzner-Sachs group B - consists of diffeomorphisms on $\mathscr{I}$ which send the pair $(\hat{q}_{ab}, \hat{n}^a)$ to a rescaled pair $(\omega^2\hat{q}_{ab}, \omega^{-1}\hat{n}^a)$ for some function ω on $\mathscr{I}$ satisfying $\mathscr{L}_{\hat{n}}\omega=0$. (As in the case of i^0, the group B can be obtained by taking a quotient of the group of diffeomorphisms which preserve the asymptotic conditions by its subgroup consisting of diffeomorphisms which are asymptotically identity (Tamburino and Winicour, 1966)). B is again a semi-direct product of an infinite dimensional supertranslation group $\tilde{\mathcal{S}}$ and the Lorentz group $\mathcal{L}$. However, the structure of $\tilde{\mathcal{S}}$ is different from that of the analogous group $\mathcal{S}$ at i^0: $\tilde{\mathcal{S}}$ is the additive group of functions on a *2-sphere* with conformal weight 1. Furthermore, unlike at i^0, one cannot obtain a physically interesting Poincaré reduction B: imposition of an additional condition such as (5) does reduce B to the Poincaré group but it also implies absence of gravitational radiation! Thus, in presence of radiation, one is stuck with an infinite dimensional enlargement of the Poincaré group. The occurrence of this enlargement was puzzling at first. It now appears that it has a deep significance: the enlargement seems to be the imprint left on classical general relativity by the infrared behavior of the quantum gravitational field (Ashtekar, 1981b,1981c). However, a complete understanding of this issue must await a detailed treatment of asymptotic quantum dynamics.

The leading order physical fields at $\mathcal{I}$ are the following. First, the connection $\hat{\nabla}$ of $\hat{g}_{ab}$ induces a connection $\hat{D}$ on $\mathcal{I}$ and by considering various permissible conformal metrics, one gets an equivalence class, $\{\hat{D}\}$, of such connections. The two radiative modes of the gravitational field are coded in this equivalence class. The curvature tensor of $\{\hat{D}\}$ - i.e. the conformally invariant part of the curvature of $\hat{D}$ - defines a symmetric, trace-free tensor field N_{ab} on $\mathcal{I}$ satisfying $N_{ab}\hat{n}^a=0$. This is the *Bondi news tensor*. Its square, $N_{ab}N^{ab}$, gives the flux of energy-density carried away by gravitational waves. The algebraic properties of N_{ab} imply that the flux is manifestly positive. Finally, one can show (Penrose 1963, 1965) that the Weyl tensor $\hat{C}_{abcd}$ of $\hat{g}_{ab}$ vanishes on $\mathcal{I}$ and that $\Omega^{-1}\hat{C}_{abcd} = {:}\hat{K}_{abcd}$ admits a smooth limit thereon. (In the Newman--Penrose (1966, 1968) notation, the ten components of $\hat{K}_{abcd}$ are denoted by the five complex quantities, $\Psi_4^0, \Psi_3^0, \ldots \Psi_0^0$). The triplet ($\{D\}$, N_{ab}, $\hat{K}_{abcd}$) represents the required leading order fields. (For details, see Geroch, 1976 and Ashtekar, 1981).

Finally, let us define the conserved quantities. First of all, as at i^0, the asymptotic symmetry group B admits a 4-parameter subgroup of translations. (Unlike at i^0, this subgroup is the *only* 4-dimensional, Abelian normal subgroup of B (Sachs, 1962a, 1962b)). These translations are generated by vector fields of the type $\hat{\alpha}\hat{n}^a$ where $\hat{\alpha}$ belong to a certain 4-dimensional vector space of (conformally weighted) functions on $\mathcal{I}^{(9)}$. The first conserved quantity is the Bondi 4-momentum (Bondi et al., 1962; Sachs, 1962a, 1962b; Geroch, 1976) associated with any cross-section C of

$\mathscr{I}$. This is a 4-vector which lies in the space dual to that of translations and represents the 4-momentum 'left-over' at the retarded time instant defined by C. Its component along a translation $\hat{\alpha}\hat{n}^a$ is given by Geroch (1976):

$$\frac{1}{8\pi}\oint_C \left\{\frac{1}{4}\hat{\alpha}\hat{K}^{ambn}\hat{n}_m\hat{n}_n\hat{l}_b + (\hat{\alpha}\hat{D}_m\hat{l}_b + \hat{l}_m\hat{D}_b\hat{\alpha})\hat{N}_{cd}\hat{q}^{bd}\hat{q}^{c\,[a}\,\hat{n}^{m]}\right\} d\hat{S}_a, \quad (8)$$

where $\hat{l}_n$ is any covector on $\mathscr{I}$ satisfying $\hat{l}_m\hat{n}^m=1$ and $\hat{q}^{ab}$ is any 'inverse' of $\hat{q}_{ab}$ (i.e. satisfies $\hat{q}_{am}\hat{q}_{an}\hat{q}^{mn}=\hat{q}_{ab}$). Unlike the 4-momentum at i^0, the Bondi 4-momentum is *not* absolutely conserved. Instead it satisfies a 'balance equation': the difference between the 4-momentum associated with 2 cross-sections is the 4-momentum radiated away in the form of gravitational waves. Thus, the Bondi-energy (i.e. any time-component of the 4-momentum) is a non-increasing function of retarded time (Bondi et al., 1962; Sachs 1962a, 1962b), the flux of energy density being given by the square of the news tensor as noted above. This leads one to expect that a simple relation should exist between the Bondi 4-momentum and the ADM 4-momentum at i^0. This expectation is indeed correct: if the energy radiated up to a retarded instant of time is finite, the Bondi 4-momentum at that instant equals the difference between the ADM 4-momentum at i^0 and the 4-momentum radiated away until that retarded instant (Ashtekar and Magnon-Ashtekar, 1979a). Finally, positive energy theorems have also been proved at null infinity (in fact by an extension of the arguments used at spatial infinity, Shoen and Yau, 1982; Horowitz and Perry, 1982; Ludvigsen and Vickers, 1982). We have: the Bondi 4-momentum at a cross-section C is either time-like (Ashtekar and Horowitz, 1982) and future directed, or zero, provided there exists a complete space-like 3-surface with boundary C on which the matter-sources satisfy the dominant energy condition and, the 4-momentum vanishes if and only if (M, g_{ab}) is flat in a neighborhoud of Σ. Thus, as at i^0, the intuitive content of the result is that the energy at infinity is positive provided the local energy-density of matter is positive. In the light of the relation between the ADM and the Bondi 4-momenta, the positive energy theorem at null infinity is clearly stronger than that at spatial infinity.

One can define the 'magnetic' analogue of the Bondi 4-momentum - the NUT 4-momentum-at a cross-section C of $\mathscr{I}$ by replacing $\hat{K}^{ambn}$ by its dual, ${}^*\hat{K}^{ambn}=\frac{1}{2}\hat{\epsilon}^{amcd}\hat{K}_{cd}{}^{bn}$, in the expression (8), although, as at i^0, the quantity vanishes identically under present boundary conditions. However, interestingly, one *can* weaken the asymptotic conditions at $\mathscr{I}$ in such a way that the 'potential' $\{D\}$ is allowed to develop 'wire singularities' keeping the fields N_{ab} and the Weyl curvature smooth. When this is done, the NUT 4-momentum can be non-zero. Furthermore, it is *absolutely conserved*, even when the Bondi 4-momentum is radiated away! (Ramaswami and Sen, 1981; Ashtekar and Sen 1983). Again, this conserved quantity is of interest only in the quantum regime.

Finally, one may attempt to define angular momentum at $\mathscr{I}$. However because the BMS group does not admit a natural Poincaré reduction in presence of radiation, the program runs into difficulties. Since the issues involved have not been completely resolved yet, the discussion will be postponed to Section 5.

4. HIGHER MULTIPOLES

So far we have dealt only with the lowest multipoles. The higher order multipoles have been investigated in detail only in the case of stationary space--times. Hence, in this section, we shall restrict ourselves to the stationary case; the situation in the general case will be summarized in the next section. Also, in all the available investigations, one works at spatial infinity rather than null. However, one knows that, for stationary space-times, asymptotic flatness in one regime implies that in the other (Koul 1983a, 1983b) and one should be able to translate the multipole framework at spatial infinity to the one which uses fields (say various "radial" derivatives of the Weyl tensor) at null infinity. Such a translation would be very useful in extending the definition of multipoles to non-stationary space-times and in investigating the effect of gravitational radiation on the multipoles.

Let (M, g_{ab}) be, then, a stationary space-time. Denote the stationay Killing field by t^a. Since 'nothing happens' along the orbits of t^a, all information about the space-time is coded in the manifold Σ of orbits of t^a. (Σ is a 3-manifold each point of which represents an integral curve of t^a. If (M, g_{ab}) is *static*, i.e. if t^a is hypersurface orthogonal, Σ is naturally isomorphic to the static slices 't=const.'). Indeed, all the information in the 4-metric is registered in the triplet $(h_{ab}, \omega_a, \lambda)$ where $\lambda=-t_a t^a$ is the norm of the Killing field, $\omega_a=\epsilon_{abcd}t^b\nabla^c t^d$ is the twist, and where $h_{ab}=g_{ab}-(\lambda^{-1})t_at_b$ is the natural metric on Σ. In the vacuum region, ω_a is curl-free, and, if the NUT 4-momentum of the space-time is zero, can be expressed as the gradient, $\omega_a=\nabla_a\omega$, of a function ω. Thus, in the stationary case, one can replace (M, g_{ab}) by $(\Sigma, h_{ab}, \lambda,\omega)$. This replacement is particularly convenient for the following reason. Let us suppose that (M, g_{ab}) satisfies Definition 1. This implies certain fall-off of the fields h_{ab},λ and ω (Ashtekar and Magnon-Ashtekar 1979b). Using this fall-off, one can show that (Σ, h_{ab}) admits a conformal completion $(\hat{\Sigma}=\Sigma\cup\Lambda, \hat{h}_{ab}=\Omega^2h_{ab})$ in which $\hat{h}_{ab}$ is *analytic* at the point Λ at infinity. (Λ, the point at spatial infinity of Σ, can be thought of as the 'projection' of i^0 under the mapping which sends an integral curve of t^a to a point of Σ). The proof of this assertion is not straightforward; it requires one to combine the elaborate techniques developed carefully by Beig and Simon (1980, 1982). The basic reason behind this tremendous gain of differentiability in the passage from the four to the three dimensional scenario is that the fields $(h_{ab}, \lambda, \omega)$ satisfy elliptic differential equations on Σ. Finally, one can show that the combinations

$$\Phi^{(1)}=\frac{1}{4}\lambda^{-1}(\lambda^2+\omega^2-1)\Omega^{\frac{1}{2}} \text{ and } \Phi^{(2)}=\frac{1}{2}\lambda^{-1}\omega\Omega^{\frac{1}{2}} \tag{9}$$

of the norm and the twist are also analytic at Λ (Beig and Simon 1981; Kundu 1981b). (Here, we are assuming that t^a is so normalized that $\lambda\to 1$ at infinity and that the integration constant in ω is so chosen that $\omega\to 0$ at infinity). These combinations are called the Hansen (1974) potentials and their expansions around Λ lead to the desired multipoles (Geroch 1970; Hansen 1974). Thus, there are two sets of multipoles - one for mass and the other for angular momentum - which are analogous to the electric and magnetic multipoles in electromagnetism. They are defined by

$$P^{(A)}=\Phi^{(A)}\Big|_{\Lambda}, \text{ and the recursion relation,}$$

$$P^{(A)}_{i_1 i_2\ldots i_{n+1}}=\mathscr{S}\Big[\hat{D}_{i_1}P^{(A)}_{i_2\ldots i_{n+1}}-\frac{1}{2}n(2n-1)\hat{\mathscr{R}}_{i_1 i_2}P^{(A)}_{i_3\ldots i_{n+1}}\Big]\Big|_{\Lambda}. \tag{10}$$

Here A runs over 1, 2; $\hat{D}$ and $\hat{\mathscr{R}}_{ab}$ are the derivative operator and the Ricci tensor of $\hat{h}_{ab}$ on $\hat{\Sigma}$; and, $\mathscr{S}$

stands for the operation of taking symmetric trace--free parts. Thus, multipoles are two sets of symmetric trace-free tensors at Λ; $\Phi^{(1)}$ yields mass multipoles and $\Phi^{(2)}$ yields angular momentum multipoles. (Under the action of asymptotic translations, they transform as in Newtonian theory (Geroch, 1970; Hansen, 1974)). Now, in Newtonian theory, multipoles provide a powerful tool because they *determine* the gravitational field. Is this true in general relativity? A major development in 'asymptotics' is the recent proof (Beig and Simon, 1981; Kundu, 1981b) that this is so. More precisely, the statement is the following: if two asymptotically flat stationary space-times have the same multipoles, they are isometric. As one might expect, it is the analyticity of various fields which lies at the heart of the proof. (The analogous result in the simpler static case was proved in Beig and Simon, 1980, and, under somewhat stronger assumptions at Λ, in Kundu, 1981a. For a review, see Beig, 1981a).

There is another approach to multipoles in which one works in the physical space-time itself, without conformal completion. The idea here is to identify multipoles as coefficients, in the 1/r-asymptotic expansions, of various physical fields. Thus, for example, one can take the 4-metric g_{ab} itself and expand its components in an asymptotically cartesian chart as a power series in 1/r. The problem of course is that of eliminating spurious degrees of freedom. This is achieved by imposing suitable coordinate conditions. For example, if one requires that the coordinate system by asymptotically cartesian, stationary (i.e. metric components are manifestly time independent), *and* satisfy de Donder condition, one can eliminate gauge degrees and 'read-off' multipoles from the power series expansions of the metric components: expansion of g_{00} gives mass-moments and that of g_{0i} gives angular momentum moments (Thorne, 1980). Furthermore, in practice, one can simplify computations by avoiding to go to the de Donder gauge if one is interested only in a finite number of moments. Finally, one can show (Gürsel, 1983) that the moments so defined are the same as those defined in the conformal picture using (9). Instead of expanding the components of the 4-metric, one can expand the Hansen potentials and again read-off multipoles (Beig and Simon, 1982). Again, the resulting moments are the same as those obtained in the conformal picture. Furthermore, in this approach one can show that the moments are 'algebraically independent' in the following sense: one can choose an arbitrary list of moments, construct therefrom a power series expansion (in 1/r) for a stationary 4-metric which reproduces these moments and show that it satisfies Einstein's equation order by order. This strongly suggests that 'there are as many stationary, asymptotically flat and empty solutions of Einstein's equation as there are lists of multipoles'. To conclude the analysis, one would like to show that these power series converge. This seems to be a very difficult task because the 1/r-expansions involve quantities which do not have an intrinsic geometric meaning and this makes it difficult to make the necessary estimates. Thus, for example, it is not known if the power series converges even in the case in which the list contains only a finite number of non-zero multipoles. (For details see Beig and Simon, 1982 and Simon, 1980).

Finally, although the above analyses use the Hansen potentials (9) for convenience, these are by no means 'unique' or 'canonical': because of field equations, distinct potentials can give rise to the *same* moments. Thus, for example, although the Hansen potentials do not reduce to the one used by Geroch (1970) in the static case, the resulting multipoles by Hansen (1974) and Geroch (1970) are the same. (For details, see Beig and Simon, 1982, Beig, 1981a and Simon, 1980).

5. OPEN PROBLEMS

In this section, I shall list some of the major open problems in the field. The list is of course not exhaustive. However, I feel that it does reflect the general consensus of the community as far as desirable directions for the evolution of the field are concerned.

5.1. Stationary space-times.

i) The problem of 'algebraic independence of multipoles'. This was partially discussed in Section 4. The problem is that of finding conditions on the lists of multipoles which yield a stationary metric - i.e. for which the series in question converge - in a neighborhood of Λ.

ii) A more difficult problem is the relation of multipoles to sources. In Newtonian theory, the field multipoles are the *same* as the source multipoles. In general relativity, it is difficult to even define source multipoles since the curvature is non-negligible in the source region. However, there does exist a treatment of source multipoles in the literature (Dixon, 1977) and first steps towards relating them to field multipoles have been taken (Schattner and Streubel, 1980). Nonetheless, I feel that it would be more interesting to try to develop approximation methods - say along the lines of Thorne (1980) - which can carry over to non-stationary contexts. One possibility would be to define multipoles on $\mathscr{I}$ and see how they are affected by radiation. In particular, it should be possible to define quadrupole on cross-sections of $\mathscr{I}$ and check whether its third time derivative gives mass--loss in the exact theory. Approximation methods can then be developed.

5.2. General space-times: i^0

i) Perhaps the most basic problem here is that of uniqueness of the completion leading to i^0 (10). Ideally, one would have liked to show that if $(\hat{M}, \hat{g}_{ab})$ and $(\bar{M}, \bar{g}_{ab})$ are two completions of (M, g_{ab}) satisfying Definition 1, then there is a mapping from $\hat{M}$ to $\bar{M}$ which maps $M \not\subset \hat{M}$ to $M \not\subset \bar{M}$, $\hat{i}^0$ of $\hat{M}$ to $\bar{i}^0$ of $\bar{M}$, $\hat{g}_{ab}$ at $\hat{i}^0$ to $\bar{g}_{ab}$ at $\bar{i}^0$ *and is* $C^{>1}$ at i^0.
Unfortunately, this is false: one can obtain inequivalent completions already for Minkowski space (Beig, 1981b, 1983; Beig and Schmidt, 1982). However, the particular ambiguity pointed out in these references does *not* affect the definition of asymptotic symmetries and of conserved quantities at i^0 (Appendix 1). The question is: are there other ambiguities? Note that ambiguities in the conformal completion show up also in physical space treatments (e.g. the ADM framework) in a different disguise and cannot be avoided by a simple change of strategy.
ii) Multipoles at i^0. One would like to extend the multipole framework to non-stationary space-times. So far, even the definitions are not available. To obtain these, there exist two avenues. The first involves casting the i^0 framework in the physical space language thereby obtaining a treatment of i^0 which is analogous to the Bondi treatment of $\mathscr{I}$, and analyzing the structure of the coefficients of various powers of $1/r$ in the expansion (Beig and Schmidt, 1982; Beig, 1983). The second avenue is to use the Witten equation (Witten, 1981) which featured in a proof of the positive energy theorem. This equation enables one to transport constant spinors at infinity to the interior of a

3-surface. However, a normalized spin-dyad at infinity does *not* yield a normalized dyad after transport; a complex scale factor is involved. In the static case, it appears that this scale factor can be used in place of the Hansen potential to extract multipoles. The complex scalar, however, exists in absence of any Killing field and could therefore be used to define multipoles in the general case. Definition of multipoles is of course just the first step. The next - and more important - step would be to extract information about dynamics in the interior from observing the evolution of these multipoles. Even though one would be working at i^0, this program *can* work at least in principle (Schmidt, 1983).

5.3. General space-times: $\mathscr{I}$

i) The most outstanding problem here is that of existence of space-times which admit $\mathscr{I}$ *with* gravitational radiation. A number of space-times are known (Schmidt, 1981; Ashtekar and Dray, 1981; Bičak et al., 1983) for which $\mathscr{I}$ is topologically $S^2 \times R$ and radiation is present; however, the generators of $\mathscr{I}$ are not complete. What is worse, these space-times admit a boost-rotation symmetry, whence they have *zero* ADM 4-momentum. On the positive side, significant advances have been made on the characteristic initial value problem on $\mathscr{I}$ (Friedrich, 1982, 1983) and these show that a large class of space-times with $\mathscr{I}^+$ or $\mathscr{I}^-$ exist. However, the conformal singularity at i^0 has prevented one from showing that both $\mathscr{I}^+$ *and* $\mathscr{I}^-$ exist in physically interesting non-stationary situations. In fact, perturbation calculations have been done off Schwarzschild space-time to argue that the conformal singularity is such that both $\mathscr{I}^+$ and $\mathscr{I}^-$ will not exist (Porrill and Stewart, 1981). However, it is also known that a perturbation off any-space-time with $\mathscr{I}$ - and this, of course, includes the Schwarzschild space-time - will be well-behaved on $\mathscr{I}$ if the initial data of this perturbation is of compact support on some Cauchy surface (Geroch and Xantopoulos, 1978); asymptotic flatness at $\mathscr{I}$ is 'stable' under such perturbations. In perturbative calculations, one ignores most non-linear effects, and hence, if, as in the present case, they suggest conflicting results for the full theory, one cannot a priori know which of the scenarios used in the calculations is more realistic. Finally, note that 'boost theorems' (Choquet-Bruhat and Christodoulou, 1981; Christodoulou and O'Murchadha, 1981) referred to in Section 1 do not shed light on the existence of $\mathscr{I}$ because one has to make an infinite boost to go from a Cauchy surface to $\mathscr{I}$ and the theorems can handle only finite boosts.

ii) The next outstanding problem is that of the definition of angular momentum at $\mathscr{I}$. One may adopt the view that the impossibility of reduction of the BMS group to the Poincaré group has a physical origin and that one simply has to live with the supertranslation ambiguities. This is a viable viewpoint at null infinity because supermomenta are well-defined there (Geroch, 1976, Ashtekar and Streubel, 1981). (At i^0, supermomenta are *not* well-defined (Ashtekar and Hansen, 1978)). However, one is still faced with two problems: what physical effects do the supertranslation ambiguities have and what is the formula for the (generalized or BMS) angular momentum? Nothing is known in response to the first question. As for the second, several expressions are available (Winicour, 1968; Bramson, 1975; Prior, 1977; Streubel, 1978; Geroch and Winicour, 1981) none of which is fully satisfactory. (For summary of what is needed, see, e.g. Ashtekar and Winicour, 1983). At i^0, one can get the 'correct' angular momentum by Hamiltonian methods and, I feel, a suitable Hamiltonian framework based on partial Cauchy surfaces which intersect $\mathscr{I}$ may settle the issue at $\mathscr{I}$. A completely new approach is suggested by twistor methods (Penrose, 1982) although the viewpoint towards the resulting definition of angular momentum is not yet settled (Shaw, 1983b).

APPENDIX 1: UNIQUENESS OF THE i^0 COMPLETION

Let (M, η_{ab}) be Minkowski space. The easiest way to complete (M, η_{ab}) to satisfy Definition 1 is the following. Let $x^{\underset{\sim}{a}}$, $\underset{\sim}{a}=0, 1, 2, 3$, be a cartesian chart of η_{ab}. In the region space-like related to the origin of this chart, set $\hat{x}^{\underset{\sim}{a}}=x^{a}/\rho^2$, with $\rho^2=-t^2+r^2$ and set $\hat{\eta}_{ab}=\hat{\Omega}^2\eta_{ab}$ with $\hat{\Omega}=1/\rho^2=\hat{\rho}^2$. The metric $\hat{\eta}_{ab}$ is again flat and $\hat{x}^{\underset{\sim}{a}}$ is a cartesian chart for it. Let $\hat{M}$ be the manifold obtained by adding to M all points covered by the chart $\hat{x}^{\underset{\sim}{a}}$. The origin of this chart is the point i^0. Since $\hat{\eta}_{ab}$ is flat, $(\hat{M}, \hat{\eta}_{ab})$ obviously satisfies the necessary conditions near i^0. (As defined, $\hat{\eta}_{ab}$ is badly defined on the null cone of the origin of $x^{\underset{\sim}{a}}$. This can be easily remedied, however, by smoothly changing Ω outside a neighborhood of i^0). The differential structure near i^0 is given by the hatted chart.

There is, however, another inequivalent completion (Beig, 1981b, 1983; Beig and Schmidt, 1982). This is obtained by using, in place of $\hat{x}^a$, coordinates $\bar{x}^a$ given by

$$\hat{x}^a = \bar{x}^a(1+2\bar{x}_b c^b \ln\bar{\rho}) - \bar{\rho}^2 c^a \ln\bar{\rho}$$

where c^a are arbitrary constants and $\bar{\rho}^2 = -\bar{t}^2 + \bar{r}^2$. One can then set $\bar{\eta}_{ab} = \omega^2 \hat{\eta}_{ab}$ with $\omega = (1-2c^a \hat{x}_a \ln\hat{\rho})$ and obtain a completion $(\bar{M}, \bar{\eta}_{ab})$ of (M, η_{ab}) by adding to M points covered by the chart $\bar{x}^a$. One can explicitly check that $\bar{\eta}_{ab}$ is (not flat but) $C^{>0}$ at i^0, $\bar{\Omega} \equiv \omega\hat{\Omega}$ is C^2 (in fact $C^{>2}$) at i^0 and that $\bar{\nabla}_a \bar{\nabla}_b \bar{\Omega}$ equals $2\bar{\eta}_{ab}$ there. Thus conditions of Definition 1 are satisfied.

How are the two completions related? One can check that ω is C^0 but *not* $C^{>0}$ at i^0. How can $\bar{\eta}_{ab}$ be $C^{>0}$ at i^0 then? The answer lies in the fact that the $C^{>1}$ differentiable structure at $\hat{i}^0$ of $\hat{M}$ is C^1 related, but *not* $C^{>1}$ - related to that at $\bar{i}^0$ of $\bar{M}$. That is, $\partial\hat{x}^a/\partial\bar{x}^b$ admit continuous limit at $\hat{x}^a=0$ (or $\bar{x}^a=0$) but $\partial^2\hat{x}^a/\partial\bar{x}^b\partial\bar{x}^c$ do not admit even a direction-dependent limit there. Thus, the incompatibility of the differential structures just compensates for the bad behaviour of ω to make $\bar{\eta}_{ab}$ $C^{>0}$ at i^0.

The natural question is whether this ambiguity affects the definitions of physical quantities. Let (M, g_{ab}) be a general space-time which admits two completions, $(\hat{M}, \hat{g}_{ab})$ and $(\bar{M}, \bar{g}_{ab})$, satisfying Definition 1. On M, we know the differentiable structure, i.e., smooth charts. But these charts are bad at i^0 and the completion involves the introduction of new differential structure there. Let $\hat{x}^a$ and $\bar{x}^a$ be the charts which define the manifold structures of $\hat{M}$ and $\bar{M}$ respectively at infinity. Using the fact that $\hat{g}_{ab}$ and $\bar{g}_{ab}$ are C^0 at i^0 and that $\hat{g}_{ab}$ and $\bar{g}_{ab}$ have the same value there, one can argue that $\hat{M}$ and $\bar{M}$ are identical as far as C^1-structure is concerned; $\hat{x}^a$ and $\bar{x}^a$ are C^1-related. Thus, we have only one C^1-manifold and only one class of C^0 conformally related metrics at i^0. Now, suppose $\bar{g}_{ab} = \omega^2 \hat{g}_{ab}$. The problem is the following: although $\hat{g}_{ab}$ and $\bar{g}_{ab}$ are both $C^{>0}$ (w.r.t. $\hat{x}^a$ and $\bar{x}^a$ charts, resp.), ω need not be $C^{>0}$ at i^0 because $\hat{x}^a$ and $\bar{x}^a$ need not be $C^{>1}$-related. Let us now suppose that $\omega = 1-2\hat{x}_a c^a \ln\hat{\rho}$

where $c^{\underset{\sim}{a}}$ are constants (no explicit assumption on the relation between $\hat{x}^{\underset{\sim}{a}}$ and $\bar{x}^{\underset{\sim}{a}}$ will be made). How are structures at $\hat{i}^0$ and $\bar{i}^0$ related?

First, one can show that the asymptotic symmetry groups are the same: a vector field ξ^a on M satisfies the requirements (Ashtekar, 1980) of a generator of a spi-symmetry in $\hat{M}$ if and only if it does so in $\bar{M}$. Next, since E_{ab} and B_{ab} are constructed from Weyl tensor, since $\hat{M}$ and $\bar{M}$ endow i^0 with the same tangent space, and since $\omega=1$ at i^0, we have: $\hat{E}_{ab}=\bar{E}_{ab}$ and $\hat{B}_{ab}=\bar{B}_{ab}$.

Hence, the 4-momentum defined using $\hat{M}$ is the *same* as that defined using $\bar{M}$. The potentials for E_{ab} and B_{ab} are, however, different: in the terminology of Ashtekar (1980),

$$\bar{E}=\hat{E}+4c^{\underset{\sim}{a}}\eta_{\underset{\sim}{a}}$$

$$\bar{U}_{ab}=\hat{U}_{ab}+4(c^{\underset{\sim}{m}}\eta_{\underset{\sim}{m}})h_{ab}$$

whence,

$$\bar{K}_{ab}=\hat{K}_{ab}.$$

But the difference does not affect physical structure. For, since $\bar{K}_{ab}=\hat{K}_{ab}$, the Poincaré group obtained (by imposing $B_{ab}=0$) in $\hat{M}$ is the same as that obtained in $\bar{M}$. Finally, a direct calculation shows that $\hat{\beta}_{ab}$ equals $\bar{\beta}_{ab}$ at i^0, whence the angular momenta are the same. Thus, although $(\hat{M}, \hat{g}_{ab})$ and $(\bar{M}, \bar{g}_{ab})$ are inequivalent completions, they provide the same physical information about (M, g_{ab}) (at least to the leading asymptotic order).

APPENDIX 2: PHYSICAL SPACE PICTURE

Let (M, g_{ab}) satisfy Definition 1 with a completion $(\hat{M}, \hat{g}_{ab})$. Let $\hat{\Sigma}$ be a $C^{>1}$, space-like, 3-dimensional sub-manifold of $\hat{M}$, passing through i^0. Then, as shown in (Ashtekar and Hansen, 1978), the intrinsic 3-metric $\hat{h}_{ab}$ and the extrinsic curvature $\hat{\pi}^{ab}$ of $\hat{\Sigma}$ satisfy the Geroch (1972) version of the ADM conditions. In this section, we consider the initial data h_{ab} and π_{ab} induced by g_{ab} (rather than $\hat{g}_{ab}$) on $\Sigma\equiv\hat{\Sigma}-i^0$ and show that this satisfies the original ADM conditions.

Introduce on $\hat{\Sigma}$ a $C^{>1}$-chart $\hat{x}^{\underset{\sim}{k}}$, $\underset{\sim}{k}=1,2,3$ which is centered at i^0 and which is such that the components of $\hat{h}_{ab}$, *at* i^0, are $\delta_{\underset{\sim}{ab}}$. Consider the function

$\hat{r}^2 = \hat{x}_{\underset{\sim}{k}} \hat{x}^{\underset{\sim}{k}}$. It is easy to verify that $\hat{r}^2$ is as good a conformal factor as Ω. Let us therefore assume $\Omega = \hat{r}^2$. (This requires transition to a new conformal frame. From now on, we use hatted tensor fields to refer to *that* frame). Next, set $x^{\underset{\sim}{k}} = \hat{x}^{\underset{\sim}{k}}/\hat{r}^2$. Then, $x^{\underset{\sim}{k}}$ may be used as a chart on $\Sigma \equiv \hat{\Sigma} - i^0$. Let f_{ab} be the flat metric for which $x^{\underset{\sim}{k}}$ are cartesian co-ordinates. Then, using the fact that $\hat{h}_{ab}$ is $C^{>0}$ at i^0, one can readily show that,

$$(h_{\underset{\sim}{a}\underset{\sim}{b}} - f_{\underset{\sim}{a}\underset{\sim}{b}}) = O(1/r);\ h_{\underset{\sim}{a}\underset{\sim}{b},\underset{\sim}{c}} = O(1/r^2),\ h_{\underset{\sim}{a}\underset{\sim}{b},\underset{\sim}{c},\underset{\sim}{d}} = O(1/r^3), \quad \text{etc.etc.}$$

where indices with tilde denote components in the $x^{\underset{\sim}{k}}$ chart, and, comma denotes the derivative operator of f_{ab}. (This holds to all orders because in the definition of $C^{>k}$-structure one requires that, if a function f is $C^{>0}$, $\mathrm{Lim}\,\Omega^{\frac{1}{2}}\hat{\nabla}_{a_1} \ldots \Omega^{\frac{1}{2}}\hat{\nabla}_{a_n} f = \partial_{\underline{a}_1} \ldots \partial_{\underline{a}_n} f$ for *all* n (Ashtekar, 1980). One can, of course, weaken this requirement).

If $\hat{g}_{ab}$ is, in addition, C^1 at i^0, we could require that $\hat{\Sigma}$ be C^2 and then $\hat{h}_{ab}$ would again be C^1. In this case, one can always choose a chart $\hat{x}^{\underset{\sim}{k}}$ such that, in the resulting $x^{\underset{\sim}{k}}$-chart, $h_{\underset{\sim}{a}\underset{\sim}{b},\underset{\sim}{c}} = O(1/r^3)$. Also, in this case $\hat{\pi}_{ab}$ would be C^0 at i^0, whence $\pi_{\underset{\sim}{a}\underset{\sim}{b}} = O(1/r^3)$. From the ADM expression of energy-momentum, it now follows that the energy-momentum must vanish. Also, in this case, the Ricci tensor $\mathscr{R}_{ab}$ of h_{ab} automatically satisfies (7) - in fact, $\underset{r\to\infty}{\mathrm{Lim}}\, r^3 \mathscr{R}_{ab} = 0$ - whence the asymptotic symmetry group reduces to the Euclidean group as shown in Section 3.

Finally, note that, although the conditions in Definition 1 imply the finiteness of energy-momentum, they are *not required* for finiteness: as has been emphasized by O'Murchadha, because of constraint equations, the ADM integrals can admit limits as $r\to\infty$ although the individual terms in the integrand may actually diverge. Thus, the asymptotic conditions in Definition 1 are by no means the weakest possible ones for spatial infinity. However, it would be difficult - if not impossible - to treat $\mathscr{I}$ and i^0 in a unified way with weaker conditions.

NOTES

* Alfred P.Sloan Research Fellow. Work supported in part by the NSF contract PHY 80 08155 with Syracuse University.

(1) See, however, Thorne (1980) where first steps are taken towards connecting approximation methods with the geometrical framework at spatial infinity.

(2) See, however, T. Damour's article in this volume. I feel optimistic that the work of the Paris group will eventually bridge the gap between the geometrical framework at null infinity and approximation methods in equations of motion.

(3) Examples are: characteristic initial-value problem and S-matrix theory of linear zero rest mass fields in Minkowski space, due to Penrose; solutions to and the S-matrix theory of classical, self-dual, Yang-Mills and Einstein's equations, due to Newman and collaborators; the S-matrix theory of zero-charge, real, Yang-Mills fields due to Christodoulou and Choquet-Bruhat; the classification of the infrared sectors of QED, due to Ashtekar and Narain; the calculation of instanton effects in the quantum S-matrix theory of Yang-Mills-Higgs system, due to Hawking and Pope; and, the original derivation of the Hawking effect. In all these analyses $\mathscr{I}$-framework plays a crucial role.

(4) Unless otherwise specified, all fields will be assumed to be C^∞ for simplicity.

(5) In the physical space language, in an asymptotically cartesian chart x^a (See Appendix 2), supertranslations correspond to transformations

$x^a \to x^a + \Lambda^a(\theta,\phi)$ while translations correspond to the case when Λ^a are constants, independent of θ and ϕ.

(6) In Gibbons et al. (1983), this result is extended to situations with trapped surfaces, where the energy conditions on matter are imposed only *outside* these surfaces. However, the proof assumes the existence of 3-surfaces whose conformal

completion is C^∞, an assumption which is yet to be justified in non-stationary contexts.

(7) The 'boost-theorems' (Choquet-Bruhat and Christodoulou, 1981; Christodoulou and O'Murchadha, 1981) referred to in Section 1 imply that (5) is also satisfied in a wide class of examples. It is curious to note that (5) is precisely the condition needed to ensure the existence of an asymptotic twistor space and therefore features in other discussions as well (Shaw, 1983; Hawking, 1983).

(8) If h_{ab} itself were flat, ${}^2\pi_{ab}$ would have been exactly equal to $(1/r)\gamma_{ab}$.

(9) In the physical space picture, translations just correspond to $x_{\underset{\sim}{k}} \to x_{\underset{\sim}{k}} + \Lambda_{\underset{\sim}{k}}$ where $x_{\underset{\sim}{k}}$ are asymptotically cartesian and $\Lambda_{\underset{\sim}{k}}$ are constants.

(10) The analogous result *is* true for $\mathscr{I}$ (Geroch, 1976).

NOTE ADDED IN PROOF

Since this report was completed, substantial progress was made on two issues mentioned in Section 5. Using Penrose's (1982) quasi-local quantities, William Shaw has 'integrated' the flux expressions of Ashtekar and Streubel (1981) at null infinity. The result is a new formula for the BMS angular momentum at $\mathscr{I}$ which seems to be free of the drawbacks of the earlier formulae. (However, this is still the BMS angular momentum and not Poincaré; the supertranslation ambiguities are not eliminated). The second issue concerns the uniqueness of the i^0 completion. A systematic procedure has been found to eliminate the logarithmic-translation ambiguities. The main idea here is based on Robert Bartnik's theorem on the existence of maximal slices in asymptotically flat space-times. For details, see the articles by Shaw, Bartnik and Ashtekar in the Proceedings of the 1983 Oregon Conference on Mass and Asymptotic Structure of Space-time (edited by F. Flaherty, Springer-Verlag, to appear).

REFERENCES

Arnowitt, R., Deser, S., and Misner, C.W.: 1962, "The Dynamics of General Relativity", in L.Witten (ed.), *Gravitation, an Introduction to current research*, Wiley, New York, p. 227.

Ashtekar, A.: 1980, "Asymptotic Structure of the Gravitational Field and Spatial Infinity", in A. Held (ed.), *General Relativity and Gravitation, One Hundred Years After the Birth of Albert Einstein*, Plenum, New York, p. 37.

Ashtekar, A.: 1981a, "Radiative Degrees of Freedom of the Gravitational Field in Exact General Relativity", *J.Math. Phys.*, 22, 2885.

Ashtekar, A.: 1981b, "Asymptotic Quantization of the Gravitational Field", *Phys. Rev. Lett.*, 46, 573.

Ashtekar, A.: 1981c, "Quantization of the radiative modes of the gravitational field", in C.J.Isham, R.Penrose, and D.W.Sciama (eds), *Quantum Gravity 2*, Oxford University Press, London, p. 416.

Ashtekar, A., and Dray, T.: 1981, "On the Existence of Solution to Einstein's Equation with Non-Zero Bondi News", *Commun. Math. Phys.*, 79, 581.

Ashtekar, A., and Hansen, R.O.: 1978, "A unified treatment of null and spatial infinity in general relativity. I. Universal structure, asymptotic symmetries, and conserved quantities at spatial infinity", *J.Math. Phys.*, 19, 1542.

Ashtekar, A., and Horowitz, G.T.: 1982, "Energy-Momentum of isolated systems cannot be null", *Phys. Lett.*, 89A, 181.

Ashtekar, A. and Magnon-Ashtekar, A.: 1979a, "Energy-Momentum in General Relativity", *Phys. Rev. Lett.*, 43, 181.

Ashtekar, A., and Magnon-Ashtekar, A.: 1979b, "On conserved quantities in general relativity", *J. Math. Phys.*, 20, 793.

Ashtekar, A.: and Magnon-Ashtekar, A.: 1983, "From i^0 to 3+1 description of spatial infinity", preprint.

Ashtekar, A., and Sen, A.: 1982, "NUT 4-momenta are forever", *J.Math. Phys.*, 23, 2168.

Ashtekar, A., and Streubel, A.: 1981, "Symplectic geometry of radiative modes and conserved quantities at null infinity", *Proc. R.Soc. London*, A376,585.

Ashtekar, A.,and Winicour, J.: 1982, "Linkages and Hamiltonians at null infinity", *J.Math. Phys.*, 23, 2410.

Beig, R.: 1981a, "The multipole Expansion in General Relativity", *Acta Phys.Austriaca*, 53, 249.

Beig, R.: 1981b, Private Communication.

Beig, R.: 1983, "Integration of Einstein's equations near spatial infinity", preprint-UWThPh 1983-10.

Beig, R., and Schmidt, B.G.: 1982, "Einstein's Equations near Spatial Infinity", *Commun. Math. Phys.*, 87, 65.

Beig, R., and Simon, W.: 1980, "Proof of a Multipole Conjecture due to Geroch", *Commun. Math. Phys.*, 78, 75.

Beig, R., and Simon, W.: 1981, "On the multipole expansion for stationary space-times", *Proc. R.Soc. London*, A376, 333.

Beig, R., and Simon, W.: 1982, "The multipole structure of stationary space-times", preprint UWThPh 1982-7.

Bičak, J., Hoenselaers, C., and Schmidt, B.G.: 1983, "The solutions of the Einstein equations for uniformly accelerated particles without nodal singularities", Max Planck Institute preprint, to appear in *J. Math. Phys.*

Bondi, H., Van der Burg, M.J.G., and Metzner, A.W.K.: 1962, "Gravitational waves in general relativity VII. Waves from axi-symmetric isolated systems", *Proc. R.Soc. London*, A269, 21.

Bramson, B.D.: 1975, "Relativistic angular momentum for asymptotically flat Einstein-Maxwell manyfolds", *Proc. R.Soc. London*, A341, 463.

Choquet-Bruhat, Y. and Christodoulou, D.C.: 1981, "The

Cauchy problem I", *Acta Math.*, 146, 129.

Christodoulou, D.C., and O'Murchadha, N.: 1981, "The Boost Problem in General Relativity", *Commun. Math. Phys.*, 80, 271.

Dixon, W.G.: 1977, in J.Ehlers (ed.) "Extended bodies in general relativity, their description and motion", *Isolated gravitating systems in general relativity*, North-Holland, Amsterdam.

Eardley, D. and Sachs, R.K.: 1973, "Space-Times with a future projective infinity", *J.Math. Phys.*, 14, 209.

Friedrich, H.: 1982, "On the existence of asymptotically flat and empty space-times", in N.Deruelle, and T.Piran (eds), *Gravitational Radiation*, North-Holland, Amsterdam.

Friedrich, H.: 1983, "The Hyperboloidal Cauchy Problem in General Relativity", in B.Bertotti, F. de Felice, A.Pascolini (eds), *10th International Conference on General Relativity and Gravitation. Contributed Papers*, CNR Roma, vol.1,p. 401.

Friedman, J.L., and Sorkin, R.: 1980, "Spin ½ from Gravity", *Phys.Rev. Lett.*, 44, 1100.

Geroch, R.: 1970, "Multipole Moments. II. Curved Space", *J.Math. Phys.*, 11, 2580.

Geroch, R.: 1971, "A Method for Generating Solutions of Einstein's Equations", *J.Math. Phys.*, 12,918.

Geroch, R.: 1972, "Structure of the Gravitational Fields at Spatial Infinity", *J.Math. Phys.*, 13, 956.

Geroch, R.: 1976, "Asymptotic structure of space-time", in P.Esposito,and L.Witten (eds), *Asymptotic structure of space-time*, Plenum, New York, p. 1.

Geroch, R., and Horowitz, G.T.: 1978, "Asymptotically Simple Does Not Imply Asymptotically Minkowskian", *Phys. Rev. Lett.*, 40, 203.

Geroch, R., and Winicour, J.: 1981, "Linkages in general relativity", *J.Math. Phys.*, 22, 803.

Geroch, R., and Xanthopoulos, B.C.: 1978 "Asymptotic simplicity is stable", *J.Math. Phys.*, 19, 714.

Gibbons, G.W., Hawking, S.W., Horowitz, G.T., and Perry, M.T.: 1983, "Positive Mass Theorem for Black Holes", *Commun. Math. Phys.*, 88, 2915.

Gürsel, Y.: 1983, "Multipole Moments for Stationary Systems: The equivalence of the Geroch-Hansen Formulation and the Thorne Formulation", *Gen.Rel. & Grav.*, 15, 737.

Hansen, R.O.: 1974, "Multipole moments of stationary space-times", *J.Math. Phys.*, 15, 46.

Hawking, S.W.: 1983, "The boundary conditions for gauged supergravity", University of Cambridge Preprint.

Hawking, S.W., and Ellis, G.F.R.: 1973, *The large scale structure of space-time*, Cambridge University Press, Cambridge.

Horowitz, G.T., and Perry, M.J.: 1982, "Gravitational Energy Cannot Become Negative", *Phys. Rev. Lett.*, 48, 371.

Komar, A.: 1959, "Covariant Conservation Laws in General Relativity", *Phys. Rev.*, 113, 934.

Koul, R.: 1983a, "Relation between asymptotic flatness at null infinity and spatial infinity for stationary space-times", in B.Bertotti, F. de Felice, and A. Pacolini (eds), *10th International Conference on General Relativity and Gravitation, Contributed Papers*, CNR, Roma, vol. 1, p. 82.

Koul, R.: 1983b, private communication.

Kundu, P.: 1981a, "Multipole expansion of stationary asymptotically flat vacuum metrics in general relativity", *J.Math. Phys.*, 22, 1236.

Kundu, P.: 1981b, "On the analyticity of stationary gravitational fields at spatial infinity", *J.Math. Phys.*, 22, 2006.

Ludvigsen, M., and Vickers, J.A.G.: 1982, "A simple proof of the positivity of the Bondi mass", *J. Phys.*, A15, 167.

Newman, E.T., and Penrose, R.: 1966, "Note on the Bondi-Metzner-Sachs Groups", *J.Math. Phys.*, 7,863.

Newman, E.T., and Penrose, R.: 1968, "New conservation laws for zero rest-mass fields in asymptotically flat space-time", *Proc. R.Soc. London*, A305, 175.

Newman, E.T. and Tod, P.: 1980, "Asymptotically Flat Space-Times", in A.Held (ed.), *General Relativity and Gravitation, one hundred years after the birth of Albert Einstein*, Plenum, New York.

Penrose, R.: 1963, "Asymptotic properties of fields and space-times", *Phys. Rev. Lett.*, 10, 66.

Penrose, R.: 1965, "Zero rest-mass fields including gravitational: asymptotic behavior", *Proc. R.Soc. London*, A284, 159.

Penrose, R.: 1982, "Quasi-local mass and angular momentum in general relativity", *Proc. R.Soc. London*, A381, 53.

Porrill, J., and Stewart, J.M.: 1981, "Electromagnetic and gravitational fields in a Schwarzschild space-time", *Proc. R.Soc. London*, A376, 451.

Porrill, J.: 1982, "The structure of timelike infinity for isolated systems", *Proc. R.Soc. London*, A381, 323.

Prior, C.R.: 1977, "Angular momentum in general relativity I. Definition and asymptotic behavior", *Proc. R.Soc. London*, A354, 379.

Ramaswami, S., and Sen, A.: 1981, "Dual-mass in general relativity", *J.Math. Phys.*, 22, 2612.

Sachs, R.: 1962a, "Gravitational waves in general relativity VIII. Waves in asymptotically flat space-time", *Proc. R.Soc. London*, A270, 103.

Sachs, R.: 1962b, "Asymptotic Symmetries in Gravitational Theory", *Phys. Rev.*, 128, 2851.

Schatner, R., and Streubel, M.: 1981, "The connection between local and asymptotically structures for isolated gravitating systems with isometries", *Ann. Inst. Henri Poincaré*, Sect. A., 34, 117.

Schmidt, B.G.: 1981, "The decay of the Gravitational Field", *Commun. Math. Phys.*, 78, 447.

Schmidt, B.G.: 1983, Private Communication.

Schoen, R., and Yau, S.T.: 1979, "On the Proof of the Positive Mass Conjecture in General Relativity", *Commun. Math. Phys.*, 65, 45.

Schoen, R., and Yau, S.T.: 1981, "Proof of the Positive Mass Theorem II", *Commun. Math. Phys.*, 79, 231.

Schoen, R., and Yau, S.T.: 1982, "Proof that the Bondi Mass is Positive", *Phys. Rev. Lett.*, 48, 369.

Shaw, W.T.: 1983a, "Twistor theory and the energy-momentum and angular momentum of the gravitational field at spatial infinity",preprint.

Shaw, W.T.: 1983b, "On the definition and radiation of angular momentum and energy-momentum at null infinity",preprint.

Simon, W.: 1980, Thesis, Universität Wien, Wien.

Streubel, M.: 1978, " "Conserved" Quantities for Isolated Gravitational Systems", *Gen. Rel. and Grav.*, 9, 551.

Tamburino, L.A., and Winicour, J.H.: 1966, "Gravitational Fields in Finite and Conformal Bondi Frames", *Phys. Rev.*, 150, 1039.

Thorne, K.S.: 1980, "Multipole expansions of gravitational radiation", *Rev. Mod. Phys.*, 52, 299.

Winicour, J.: 1968, "Some Total Invariants of Asymptotically Flat Space-Times", *J.Math. Phys.*, 9, 861.

Witten, E.: 1981, "A New Proof of the Positive Energy Theorem", *Commun. Math. Phys.*, 80, 381.

EXACT SOLUTIONS AND SINGULARITIES
REPORT OF WORKSHOP A2

M.A.H. MacCallum

Department of Applied Mathematics
Queen Mary College
Mile End Road, London E1 4NS, UK

1. INTRODUCTION

Reviews by Kramer et al. (1980) and Tipler at al. (1980) of both the major areas involved in the workshop appeared about the time of the 9th Conference on General Relativity and Gravitation (GR9) in Jena, giving a starting point for discussion of progress in the following three years. The workshop attracted more papers (86 final abstracts and 4 late papers) than any other at GR10 (Bertotti at al., 1983): of these 71 concerned exact solutions of Einstein's equations, 26 of them being about solutions with two commuting Killing vectors, 10 concerned solutions of theories other than general relativity, and only 9 concerned topology or singularities. As far as I know, authors of only 47 of the papers actually attended GR10.

In this account I first summarise the sessions and then give a personal (biassed) view of the progress and prospects for work on exact solutions. I have not attempted to review the progress on singularities where the relatively few contributions perhaps reflect the difficulty of an area in which rather advanced techniques are required to achieve useful results: most of the unresolved problems mentioned in Tipler et al. (1980) remain unresolved and many of the best-known experts in this area turned their attention to other matters during 1980-83.

The workshop was allocated 2 sessions of 2 hours and one of 1.5 hours. In view of the number of papers they were divided into groups, from most of which one

B. Bertotti et al. (eds.), General Relativity and Gravitation, 69–81.

was selected in a semi-random manner for verbal presentation (the choice did reflect my expectations of whether speakers would be present in Padova, and attempted some balance of subject and geographical origin, but certainly was not intended to represent a judgement on the relative merits of the papers submitted).

Each group of papers was discussed as follows. All authors who were present were asked to identify themselves to the audience. Then those asked to speak were each given 15 minutes, with time for questions afterwards. Then I briefly reviewed the remaining papers in the group and opened them for discussion. The review below roughly follows my summaries, with slightly longer comments on the speakers' papers. About one-third of the total time was allocated to the general discussion and questions.

Although I have attempted in the review of the sessions to simply represent the authors' view without editorial comment, reserving the expression of my prejudices for the concluding commentary, it is unlikely that I have succeeded in accurately reporting every one of the papers and comments, despite the assistance of the workshop secretaries, Drs. M. Carfora and F. Salmistraro, and written versions of remarks by Drs. W. Dietz, C. Hoenselaers and D. Kramer and Prof. W.B. Bonnor, for all of which I am most grateful. I therefore apologise in advance for any omission or inaccuracy in my account, and hope the victims will forgive the faults as inevitable in such a task. In addition to those just named I must thank again the speakers, who all kept admirably to time, and the authors who agreed to act as reserve speakers and were never called.

2. FIRST SESSION

2.1 General Methods

The first speaker was Dr. G.S. Hall. His paper proved that given the coordinate components of the curvature tensor the metric is not uniquely determined if the curvature, as a map on bivector space, has rank $r < 3$, may be if $r = 3$, and is if $r > 3$. Secondly, he established results linking sectional curvature to the metric and the algebraic type of the Ricci tensor. The results can be translated into statements about curvature collineations and similar symmetries. The paper provoked considerable interest and discussion, e.g. on the dependence on the dimension of the manifold, the rank of the Riemann tensor and the signature of the metric.

Dr. A. Krasinski next presented his paper on symmetry transformations preserving the tensor $R^{ab}{}_{cd}$. He showed that any vector field in a space of constant curvature or in subspaces of constant curvature formed as orbits of isometry groups had this property, and gave some examples. He had found 15 spherically symmetric metrics satisfying the necessary algebraic conditions but had not yet checked the derivation of curvatures from metrics. In discussion, Dr. C.B.G. McIntosh noted that the tensor $R^a{}_{bcd}$ was the one which arose most naturally.

There were two papers by A.V. Aminova concerned with symmetries preserving geodesics whose initial tangent lay in special planes or cones. A.M. Baranov discussed the algebraic classification of Ricci and Weyl tensors using catastrophe theory. I. Gottlieb and L. Alexandru wrote on transformations between different "3+1" splittings of space-time. E. Leibowitz studied vector fields such that invariance at p under rotation about p implied invariance of the vector field at all points under that rotation. N.V. Mitskievich discussed heuristic generation of metrics from Minkowski space by Lorentz transformations and scalings, giving the Kerr solution as an example and also, with N.F. Dandach re-considered the junction conditions between Schwarzschild and Friedmann regions and the desirability of a more physical "buffer-zone" in place of the jump discontinuities.

My own paper on a proposed coding of known solutions of Einstein's equations for a database provoked some suggestions for additional data to be included, e.g. the authors' names, asymptotic and singularity properties, and relation to more general families as specialisations or limits.

2.2 Symmetry Inheritance and Maxwell Fields

These papers were about the relation of symmetries of the metric and symmetry of the source fields. Dr. F. Salmistraro spoke about his work with Drs. R. Catenacci and A. Marzuoli in which, using null tetrad methods, they proved the long-conjectured result that in a static metric solving the (source-free) Einstein-Maxwell equations, the Maxwell field must also be static. M. Henneaux noted that for Maxwell fields the problem could be phrased in terms of maps from the symmetry group of the metric to the group SO(2) of duality rotations, while L. Herrera and L. Carigi rewrote it in Newman-Penrose notation and also considered the determination of the metric from symmetries preserving the Maxwell field and energy-momentum. G.G. Ivanov submitted a paper considering like issues for non-linear scalar fields, short-range vector fields, and Einstein-Vlasov solutions, while C.A. Kolassis similarly treated Dirac fields.

The paper by K. Duggal and R. Sharma was included with this group because it also concerned, in a certain sense, symmetry and the Maxwell field; it defined a 'Riemannian-Maxwellian singular structure' in terms of projection into a two-dimensional subspace and an associated two-form, and studied the resulting metrics.

2.3 Cosmological Solutions

Dr. Ch. Charach spoke about his work, with M. Carmeli and A. Feinstein, on the generalisation of the homogeneous anisotropic Bianchi IX cosmologies with rotational symmetry to solutions with only two Killing vectors covered by four coordinate patches each of Einstein-Rosen form. The known solution for the Bianchi IX case with scalar field was reviewed and the inhomogeneous models, in which, unlike the toroidal case, the (spherical) topology puts constraints on the modes of the scalar and gravitational perturbations, were examined. The questions concerned the relation between acoustic and gravitational waves in the perturbed solutions.

The other papers in this group had little direct interrelation. A.A. Coley and B.O.J. Tupper showed that the energy-momentum content of a Robertson-Walker solution could be provided by a viscous magnetohydrodynamic fluid rather that the conventional perfect fluid. R.T. Jantzen reviewed his work using techniques of modern dynamics to discuss homogeneous cosmologies, and its relation to the oscillatory singularities which Prof. E.M. Lifschitz had discussed in his plenary lecture. V.N. Melnikov and V.K. Shigolev investigated SU(2) gauge fields in the Einstein static universe, and M.A. Melvin considered homogeneous metrics with diagonal metric in which the strain tensor was axisymmetric (the "Taub submanifolds" of Jantzen's treatment), while I.S. Shikin discussed test particles following the dynamics of Bianchi V models in a Minkowski space.

2.4 Spherically Symmetric Metrics

L.Ya. Arifov and H-J. Schmidt used shell sources, Arifov's having a negative central mass and being composed of perfect fluid. H. Knutsen, P.I. Pronin and Yu. N. Vlasenko, D.C. Srivastava and S.S. Prasad, and G.C. McVittie also took perfect fluid sources, the last two assuming absence of shear. McVittie's paper listed the explicitly known solutions of this type and identified published versions (a paper published elsewhere by H. Stephani has shown the list to be exhaustive). Pronin and Vlasenko's fluid solutions were found by assuming forms for the metric coefficients. K.A. Bronnikov considered scalar and vector field sources, M.P. Korkina a non-linear scalar field, V.J. Dubey a null Maxwell field and S. Kitamura a "Lichnerowicz" magnetohydrodynamic source. L. Herrera and J. Jimenez considered junction conditions between the Vaidya rediating solution and a non-static interior.

3. SECOND SESSION

3.1 Algebraically Special Solutions

The first talk was by Prof. R.G. McLenaghan on his work with R. Debever and N. Kamran in which they derived a single metric containing all Petrov type D solutions with an aligned non-null Maxwell field and cosmological constant as sources. It makes the symmetry (all cases admit at least a G_2, which has orthogonal transitivity) and separability properties manifest, and contains 13 constants connected by 6 relations in general. The previously known particular cases are recovered without the use of singular limits. Questions to Dr. Mc.Lenaghan mainly concerned the physical interpretation of the parameters and the relation to the Plebanski-Demianski solution.

Dr. C.B.G. McIntosh spoke about his work on algebraically special complex solutions obtained by combining solutions of Petrov type N. The general form of pp waves involves complex functions constant on complex null 2-surfaces ('leaves') from which further solutions can be obtained using the 'Sachs duality' which in real space-time exchanges spacelike and timelike two-surfaces. By linearly combining these complex metrics, various complex solutions, some of them having real slices, can be obtained. In discussion, J.D. Finley questioned whether it would prove possible to similarly combine non-plane waves in a general case.

The remaining papers in this group, except for that by M. Hickman, were interrelated. Hickman discussed the existence of foliations by two-surfaces defined by self-dual bivectors and introduced the related (but not identical) concept of quarter flat spaces; these ideas suggest extensions of McIntosh's methods that could prove useful in summing more general spaces.

The links between the other papers were their concern with Petrov type N and/or Kerr-Schild forms, a Kerr-Schild metric being of the type $g_{ab} + Vk_a k_b$, where k_a is a null vector in the Minkowski metric g_{ab}: this idea can be generalised by allowing more general g_{ab}.

A.M. Bilge and M. Gürses investigated the general properties of this generalisation, while G.A. Gavrilina considered its restrictions to the case where initial and final metrics were both electrovacs (Einstein- Maxwell fields). S.I. Tertyczny and V.N. Ponomariov studied the singularity of the special radiating Kerr-Schild electrovac of Kundt's class, considering its black-hole-like properties. B.C. Xanthopoulos used the generalised Kerr-Schild ansatz in an (incomplete) attempt to arrive at twisting type N solutions. General results concerning such solutions had been presented by several authors at GR8 at Waterloo, but other than the Hauser solution, exact solutions are not known. A second attempt to find them was presented at GR10 by Finley, who used complex 'heavens'. A.M. Baranov wrote about type N pure radiation solutions as limits of Taub-NUT like solutions, and E.G.L.R. Vaz gave some results on curvature collineations in Robinson-Trautman (diverging non-twisting) type N solutions.

3.2 Other Solutions of Einstein's Equations

P.J. Adams, R.W. Hellings and R.L. Zimmerman announced a metric representing circularly-polarised gravitational waves, and D.M. Song, D.Y. Li, and Z.-J. Zhang wrote on conformally-flat perfect fluid metrics. A paper by V.G. Bagrov, V.V. Obukhov and A.V. Shapovalov, which appeared in workshop A1, about electrovacs in which the Hamilton-Jacobi equation separates was noted here.

3.3 Solutions of Gravity Theories other than General Relativity

To the 10 papers in this group in A2 were added 8 such papers sent to workshop A5. I briefly summarised all these. Three (Chatterjee, Singh and Singh, and Van den Bergh) concerned static or stationary axisymmetric fields in a generalised scalar-tensor theory. V.N. Melnikov and A.G. Radynov in two papers (A5) also considered theories involving scalar fields, in their case conformally-invariant. Two (Ponomariev and

Zhytnikov, and Müller-Hoissen and Nitsch (A5)) considered Poincaré gauge theory, the former showing when it was equivalent to Einstein's theory, and the latter also deriving cosmological solutions. Three (Minkevich (A5), Minkevich and Chuong (A5) and Staniukovich and Borisova) did the same for theories obtained with generalisations of the Lagrangian, which can be considered as gauge theories. Three, all in A5, (Bedran and de Andrade, Duarte, Teixeira and Tiomno, and Tsoubelis) considered static or stationary cylindrically-symmetric solutions with various energy-momenta in Einstein-Cartan theory. I.A. Adamuti discussed black hole and wave solutions of his own generalised Newtonian theory, G. Clement gave axisymmetric solutions in 5-dimensional relativity, D. Lorenz gave the homogeneous (Bianchi model) solutions of the Einstein theory for metrics of Euclidean signature, H.D. Pande found spherical wave solutions of the Kilmister-Newman theory, and R.D. Sorkin discussed Taub-NUT solutions in (5-dimensional) Kaluza-Klein theory. This last paper also considered the relation to spacetime topology, and could be regarded as a simplified model for the higher-dimensional Kaluza-Klein treatment arising in supergravity (workshop D2).

3.4 Topology and Singularities

Dr. C.J.S. Clarke spoke about the recent results on cosmic censorship, reviewing the various disagreements. Christodoulou had found exact counter-examples, but it could be argued these were not 'real', that one should consider only cases with curvature singularities. Two approaches had been followed. A. Krolak (who also submitted a paper to A2) had applied a physically reasonable condition due to Tipler and proved that in that sense singularities could not be naked, provided a certain "marginally outgoing" condition was satisfied. R.P.A.C. Newman had instead used a lower bound condition on the curvature near the singularity and found no such singularity could be naked. Clarke's work with de Felice had followed Newman's general approach and shown that any counter-example must be (highly) non-causal.

Dr. D.A. Konkowski spoke about her work with T.M. Helliwell and L.C. Shepley which studied the effect of test waves on quasi-regular singularities (at which the curvature is well-behaved) in Taub-NUT-like models. They had found that in four dimensions (in contrast to two) the waves showed divergences at the Cauchy horizons, implying instability of the singularity and suggesting it becomes a curvature singularity.

D. Blackmore considered a flow (vector field) on a singular spacetime from the point of view of dynamical systems (recurrence, strange attractors, etc,). G.J. Galloway proved that (under certain technical conditions) a rigidly rotating perfect fluid body with mean convex boundary had to be a handlebody, and that if a contracting body in a spacetime satisfying the usual generic, chronological and convergence conditions was not a handlebody the spacetime must be geodesically incomplete. J.I. Goldman re-examined the old question of the nature of the Schwarzschild radius by studying a particular interior solution in particular coordinates. G. Sardanashvili re-interpreted the singularity theorems in terms of singularities of foliations of space-time, and P.M. Williams considered the reformulation of geodesic incompleteness in terms of the behaviour of the geodesic spray, studying its stability by means of the differential geometry of jets.

4. THIRD SESSION

4.1 Generating techniques

Prof. B.K. Harrison spoke about the Backlund transformation methods for generating solutions with two Killing vectors in terms of the pseudo-potentials and prolongation structures of Estabrook and Wahlquist. 16 Backlund transformations, forming a group, had been found. Recently Harrison had used a different initial transformation to generate 12 more, and although these had been found to be combinations of previously known Backlund transformations with an Ehlers transformation they led to new solutions. Particular transformations, after change of variables, were shown to be those for which F.J. Chinea, in his contribution, had found a superposition principle.

Other ways of presenting generation techniques are the inverse scattering method of Belinskii and Zakharov (BZ) and the Hauser-Ernst (HE) integral equation form. G.A. Aleksejev presented a BZ derivation of N soliton solutions for electrovac and neutrino fields, the latter being treated by the HE method by N.R. Sibgatullin in his paper, while A. Eris, M. Gürses and A. Kasaru gave 2N soliton solutions from the BZ method reformulated in terms of harmonic maps.
J. Carminati and F.J. Cooperstock introduced a new method for constructing solutions, making use of the

possibility of adapting the coordinates to the electromagnetic equipotentials, and gave some new solutions. N. Sanchez discussed a broader class of solutions of the Einstein equations without symmetry, relating their treatment to methods for self-dual Yang-Mills fields, non-linear sigma models, and other non-linear systems in physics; in general the resulting solutions are complex and have Euclidean real slices, but (real) gravitational plane waves are also included.

4.2 Stationary Axisymmetric Solutions

Three papers in this group (two of them late contributions) were presented verbally. Dr. T. Ohta spoke about his work with T. Kimura on the equilibrium of two charged masses. By three methods (finding the many particle Hamiltonian up to PPN order, directly solving the field equations with sources to PPN order, and the $h \rightarrow 0$ limit of an S matrix quantum treatment) they had obtained the result that the charge to mass ratios should be equal. In discussion, W.B. Bonnor pointed out that this result depended on the choice of solution of the Poisson equation (i.e. on boundary conditions) and on the assumption that coefficients of terms in the interparticle separation should vanish identically, rather than the balance explicitly depending on the separation. In view of these points, Bonnor argued that the final balance condition was still unclear.

Four papers concerned the balance of spinning particles. M. Kihara, K. Oohara, H. Sato and A. Tomimatsu discussed the double Kerr solution, while A. Koppel, A. Ihermann and T. Lember considered its interpretation in a Newtonian framework, and M. Yamazaki studied the case of N Kerr particles. Unfortunately none of these authors were available to speak at A2 in Padova, but C. Hoenselaers and W. Dietz submitted a late paper which they spoke about. In this they generated the double Kerr solution, then made special choices of parameters to give asymptotic flatness and defined the mass and angular momentum of the particles in terms of Komar integrals. The general expressions for balance (i.e. no 'strut', or elementary flatness, at the axis) were too complicated for solution, but it was shown that two black holes with equal masses and angular momenta could not be in equilibrium, although two rotating discs could.

There were three more papers concerned with the sources of metrics. W.B. Bonnor studied the behaviour at the axis of the C-metric, interpreting it in Weyl coordinates as a spherical mass plus a semi-infinite rod, separated by a strut; this could be transformed to a non-static form in which the sources were interpreted as two spheres uniformly accelerated by a spring. G.J.G. Junevicus considered shell sources for Weyl metrics. D. Kramer contributed a late paper in which an interior perfect fluid solution was found; it is of Petrov type D, everywhere regular, has vanishing pressure at a closed boundary, rotates rigidly, has equation of state $\mu + 3p = \text{const.}$, and is a special case of the Petrov D fluid solutions of Wainwright.

Also in this group, T.I. Gutsunaev and V.A. Chernaiev found an ansatz yielding new asymptotically flat solutions, N.R. Sibgatullin, L.Yu. Blagenova-Miculich, A.T. Ilichev and Ju.R. Musin gave solutions with neutrino sources, and B. Xanthopoulos sought solutions with regular horizons.

The paper by J. Bicak and B.G. Schmidt concerned solutions with two Killing vectors acting on spacelike surfaces in one region and timelike surfaces in another. This arose from the fact that two commuting Killing vectors can be taken to be a 'boost' and a rotation, rather than as axial or time translation and rotation; in particular this leads to radiating solutions with two commuting Killing vectors, and the same authors wrote a paper on that aspect in workshop A1. Their paper formed a bridge between this group and the next, which concerned the solutions with two spacelike Killing vectors, rather than two Killing vectors on a timelike surface.

4.3 "Cylindrically Symmetric" Solutions

This title covers two cases, which differ in whether the gradient of the determinant of the metric in the group orbits is spacelike or timelike, the latter case giving cosmological solutions.

Dr. A. Curir spoke about her work with M. Francaviglia. They had studied the asymptotic behaviour in the solutions with solitons found by Belinskii and Francaviglia starting from a vacuum Bianchi II cosmology. These solutions were the first non-diagonal metrics for which the BZ inverse scattering method had been carried out, but initial investigation showed a drastic change in asymptotic structure at a certain parameter value. Curir and Francaviglia investigated this change using numerical and graphical computing and gave details of the behaviour near the discontinuity.

B.R. Iyer and C.V. Vishveshwara had considered

the solutions generated by 'magnetising' (applying Harrison transformations to) the stationary cylindrically symmetric vacua. G. Mohanty had found a solution with stiff fluid ($\mu = p$) and in a second paper, with Tiwari and Rao, had considered solutions with stiff fluid plus zero-mass scalar field and Maxwell fields. R.N. Tiwari and J.R. Rao, had considered solutions with stiff fluid plus zero-mass scalar field and Maxwell fields. G.N. Shikin had found a plane solitonic solution with interacting complex and real scalar fields. E.P. Vasconcellos and M. Som had studied cylindrically static metrics with Maxwell plus scalar field sources. including a general family dependent on the third Painleve transcendental function, and similar metrics in which the metric symmetry was not inherited. I announced that I had completed this work by showing that there were just two analogous non-inheriting classes, and finding all solutions of the inheriting type either explicitly or in terms of the Painleve transcendent.

5. COMMENTARY

The study of exact solutions is an area of classical relativity which often attracts the criticism that it is unrelated to physical problems, and that many papers merely write down "new" metrics of no real interest. I personally would agree that there are still too many papers in which (a) a complicated energy-momentum is used with simple geometry or (b) a (slight) variation on Einstein's equations is used or (c) the generating techniques are merely applied routinely. In all these cases, the work can only be justified, I believe, if very strong physical motivation and careful interpretation of the result is given. The papers at GR10 did show some move in this direction especially in the area of solutions with two commuting Killing vectors, where many of the papers were interpretative.

Another problem, which was one motivation for Kramer et al.(1980) was the frequent unwitting rediscovery of known solutions or known results, e.g. the conformally flat perfect fluid solutions were rather fully analysed by Stephani (see Kramer et al., 1980) but have been rediscovered in at least three papers since 1979. To avoid such rediscoveries one should always use invariant characterisations of the solutions; simply giving metric coefficients in specific coordinates is inadequate, since a complicated coordinate change can obscure the identity of the metric. It is for this reason that I am myself anxious to promote the use of invariant classifications of metrics of which I believe the technique developed by Cartan, Brans and Karlhede (for which see the plenary lecture by Cohen, Frick and Åman) to be the most effective, and to produce a database of known solutions using this technique, a project in which I am fortunate to have d'Inverno, Åman, Joly and others as collaborators already.

The effect of such moves, if, as is unlikely, everybody entirely agreed with my prejudices and followed the above advice, would be to substantially decrease the number and increase the mathematical complexity of papers on exact solutions. In fact, in seminars I like to say that my objective is to kill off the rather useless game of finding new exact solutions entirely; of course I then follow this provocative remark by giving reasons for reviving part of this art (a necessary reversal if only in justification of my own activities!). An immediate one is that knowledge of specific solutions can lead to a deeper understanding of the basic Einstein equations, as they have in many cases in the past (see Kramer at al., 1980).

Indeed, there are still some classical areas of investigation where there are surprising lacunae, and some where the current boundaries of knowledge can be pushed back quite a bit. There are also areas where tidying up of the diverse literature is still incomplete (the authors of Kramer at al.(1980) joke that they are only now realising what they should have written). These aspects were quite well illuminated at the workshop (as random examples, in the papers of Catenacci at al., McVittie, Debever et al., and Van den Bergh et al.).

Secondly there are still ways of obtaining general theorems about classes of solutions (e.g. papers by Hall and Harrison) or of aiming at general solutions. Of the latter the complex methods, represented in A2 by the papers of Finley, Hickman and McIntosh, but also covered in the language of twistors and 'heavens' in workshop A5, still look promising, as do the various generating transformation techniques, which are not entirely limited to solutions with two Killing vectors (see e.g. the paper of Sanchez). The latter methods, unfortunately, have so far been limited to combinations of fields all of which propagate along null geodesics, i.e. gravitation, source-free Maxwell fields, massless scalar fields (or the re-interpretation of these as stiff fluid) and neutrino fields. Little progress has been made with more realistic fluids. In this context the interior solution of Kramer (though still unrealistic) is a welcome addition.

It should also be noted that much of the work on Backlund transformations, inverse scattering, etc., can be transferred to (and from) other contexts, quite distinct from relativity, in which similar sets of equations arise.

One further development which could be of great interest is the use of extensions of the known techniques for finding exact solutions to the context of the higher dimensional field theories which are now of major importance in some approaches to quantising gravity, expecially supergravity. One or two of the contributions to the workshop could be seen as steps in the direction.

To sum up, I felt the workshop was fairly representative of current activity, with enough useful pointers to valuable future developments to outweigh any criticism that the field is one which lacks physical importance and value.

REFERENCES

Bertotti, B., de Felice, F., and Pascolini, A. (eds): 1983, *10th International Conference on General Relativity and Gravitation. Contributed Papers*, Vols. 1-2, CNR, Roma.

Kramer, D., Stephani, H., MacCallum, M. and Herlt, E.: 1980, *Exact Solutions of Einstein's Field Equations*, Deutscher Verlag der Wissenschaften, Berlin, and Cambridge University Press, Cambridge, (Russian edition, Energoizdat, Moscow, 1982).

Tipler, F.J., Clarke, C.J.S., and Ellis, G.F.R.: 1980, "Singularities and horizons. A review article" in A. Held (ed.) *General Relativity and Gravitation*, vol. 2, Plenum Press, New York.

INITIAL VALUE PROBLEM
REPORT OF WORKSHOP A3

Karel V. Kuchař

The University of Utah
Salt Lake City
Utah 84112, USA *

The logical distinction between intensive and extensive definitions finds an amusing reflection in the rather different ways in which contributions are solicited for a scientific meeting and in which the outcome of the meeting is reported in conference proceedings. While the instructions to participants describe what they ought to talk about, the proceedings summarize what they actually did talk about. In this spirit, the present report is purely extensive. It is not intended to impose on the reader my ideas about what is the contemporary worldwide status of the initial value problem and Hamiltonian formulation of the general theory of relativity, and even less my ideas about what this status ought to be. It is simply a brief summary of what happened in the two afternoon sessions devoted to the workshop. Anyone interested in details should go to the original papers by the participants.

It was arranged that about twenty people would have spoken about their work. In the end, through their own decision or by intervention of a *vis maior* (a convenient euphemism), this number was cut almost into a half. This fact in itself represented the first step from the intended intensive definition of the workshop to an extensive one.

In the sessions, the traditional initial value problem on spacelike hypersurfaces received relatively little attention, and the traditional Hamiltonian formulation of Einstein's theory even less. An exception proving the rule was a contribution combining both of these topics: J. Isenberg reported about his

B. Bertotti et al. (eds.), General Relativity and Gravitation, 83–87.

work with J. Marsden devoted to the proof that the (properly reduced) York map, which naturally arises in the Hamiltonian treatment of the gravitational initial data, is canonical (Isenberg and Marsden, 1983).

There were two other talks closely related to the traditional initial value problem. As is well known, the initial value equations considerably simplify when treated on maximal hypersurfaces. M. Carfora (1983) returned to the old question what asymptotically flat spacetimes admit maximal slicings. He has pointed out that the strong energy condition (which is needed as an input in most existence theorems on maximal slicings) does not need to hold pointwise, but rather in an average. He has further shown that the approach of maximal slices to a null cone is governed not only by the magnitude of the extrinsic curvature, but also by the size of the region from which the main contributions to the extrinsic curvature come. The second talk was given by V. Moncrief (1983), the foremost practitioner of the art of proving results about Einstein's *spacetimes* by looking at the data on *spatial* hypersurfaces. Working along these lines, Moncrief has extended the 2-parameter Taub - NUT family of causality violating spacetimes into an infinite parameter family of spacetimes having similar generic properties. He has also shown (with J. Isenberg) that any (analytic, electro-) vacuum spacetime with a compact null surface ruled by closed null generators necessarily has a Killing symmetry. The importance of this result lies in its tacit suggestion: if only the analyticity and generator closure could be eliminated, one would be able to conclude that non-symmetric (electro-) vacuum spacetimes cannot contain compact Cauchy horizons.

Originally, there was to be a whole group of papers devoted to the initial value problem for generalized theories of gravitation. Only one such result was actually reported at the workshop. P. Teyssandier and Ph. Tourrenc (1983) have classified the Cauchy problem for the two-parameter family of $R + R^2$ metric theories of gravity depending on the value of the parameters adjoining the quadratic invariants R^2 and $R_{\mu\nu}R^{\mu\nu}$ to the scalar curvature R. The higher-order theories of gravitation are, of course, interesting to many people as they naturally arise from quantum corrections to the Einstein theory.

Let me now pass to two papers devoted to the initial value problem associated with other elements than purely spacelike hypersurfaces. In his effort to build in part of the structure of "past null infinity" into the formalism, H. Friedrich (1983) studied the hyperboloidal Cauchy problem for Einstein's field equations written as equations for the conformal four-metric and the conformal factor. The Cauchy data appropriate to this system of equations are prescribed on a smooth spacelike hypersurface which intersects past null infinity at a spacelike 2-surface. Friedrich proved an existence and uniqueness theorem for the development of such data and specified their (Sobolev space) classes. R.A. d'Inverno, G. Holmes and S.T. Swift (1983) discussed their work on the 2 + 2 approach to general relativity. In this approach, the true gravitational degrees of freedom are identified with the conformal 2-structure associated with the decomposition of data with respect to two families of spacelike two-surfaces. Alternatively, one can also study the corresponding double null problem. d'Inverno and his collaborators stressed that, within this approach, the gauge conditions (for Maxwell's theory) are algebraic and that the gravitational 2 + 2 data are unconstrained. A satisfactory Hamiltonian formulation of the gravitational 2 + 2 problem is unfortunately not yet known.

Witten's proof of the positive energy theorem stimulated two contributions to the workshop. G.T. Horowitz and A. Strominger (1983) discussed some aspects of the physical interpretation of Witten's expression for gravitational energy in the presence of sources. In particular, they presented a one-parameter family of decompositions of the Witten expression into a "matter contribution" and a "gravitational contribution" and gave their Newtonian limit. O.M. Moreschi and G.A.J. Sparling formulated a positive energy theorem in the framework of the Kaluza-Klein theory in 5 dimensions. In this way, they were able to derive inequalities relating the mass and charges of an Einstein-Maxwell system with black holes in the interior. In my opinion, one can expect many more consequences for Hamiltonian gravity to be drawn from Witten's expression in the near future.

I come now to papers related to the Lagrangian and Hamiltonian formulation of general relativity rather than to the initial value problem. A.E. Fisher (1983) discussed conservation laws and differential identities for a generally covariant Lagrangian field theory describing the interaction between gravity, gauge fields and matter fields in the language of principal fibre bundles. D.C. Salisbury and K. Sundermeyer (1983) strove to derive possible phase space dynamics of generally covariant (and maybe gauge covariant) systems by paying attention to the maximal subgroup of transformations realizable in a phase space.

Two other talks in this class dealt with surface terms haunting the general theory of relativity. J.M. Charap and J.E. Nelson (1983) studied the most general action free of second derivatives of the gravitational (metric or tetrad) variables which yields Einstein's equations without boundary conditions. C.K. Raju and N. Dadhich (1983), motivated by the problem of gravitational screening, looked at permissible jumps in the electromagnetic and gravitational variables.

In retrospect, it seems quite difficult to perceive an underlying unity hiding behind the diversity of contributions presented at the workshop. One is tempted to blame the situation on the temporary lack of agreement among the people working in the field on what is an outstanding problem which should be solved by a concerted effort. On the other hand, watching the doughnut rather than the hole, one can be glad that the techniques developed by looking at the initial value problem and the Hamiltonian formulation of general relativity found their application in such a wide variety of studies.

NOTE

* On sabbatical leave at Imperial College, London, England.

REFERENCES

Papers presented to the GR10 Conference and published in

Bertotti, B., de Felice, F.,and Pascolini, A. (eds): 1983, *10th International Conference on General Relativity and Gravitation. Contributed Papers*, Vol. 1,CNR, Roma.

Carfora, M.: 1983, "Dynamics of maximal foliations in asymptotically flat spacetimes".

Charap, J.M., and Nelson, J.E.: 1983, "Surface integrals and the gravitational action".

d'Inverno, R.A., Holmes, G., and Swift, S.T.: 1983, "Application of the 2+2 approach to general relativity".

Fisher, A.E.: 1983, "A unified approach to conservation

laws in gauge field theories".

Friedrich, H.: 1983, "The hyperboloidal Cauchy problem in general relativity".

Horowitz, G.T., and Strominger, A.: 1983, "Superderivation of Witten's expression".

Isenberg, I., and Marsden, J.: 1983, "The York map is canonical".

Moncrief, V.: 1983, "Symmetries of generalized Taub-NUT spacetimes".

Raju, C.K., and Dadhich, N.: 1983, "Is gravitational screening possible?".

Salisbury, D.C., and Sundermeyer, K.: 1983, "The eighth route to Einstein's geometrodynamic law".

Teyssandier, P., and Tourrenc, Ph.: 1983, "The Cauchy problem for the $R+R^2$ theories of gravity without torsion".

THE MOTION OF COMPACT BODIES AND GRAVITATIONAL RADIATION

Thibaut Damour

Groupe d'Astrophysique Relativiste
Equipe de Recherche du CNRS n. 176
Observatoire de Paris-Meudon
92190 Meudon (France)

ABSTRACT. We consider the problem of the gravitational interaction of two compact bodies (neutron stars or black holes). We outline a new method where one matches an "external" gravitational field, obtained by iterating a Post Minkowskian Approximation scheme, to the field near each compact body. Equations of motion for each body are derived from the vacuum field equations by means of an Einstein-Infeld-Hoffmann-Kerr-like approach simplified by the use of complex analytic continuation. Because of the "no incoming radiation" condition incorporated in the Post Minkowskian Approximation scheme these equations of motion have a retarded functional form. A slow motion expansion allows one to transform these equations into ordinary differential equations up to the second-and-a-half-Post-Newtonian order. Solving the latter equations, with the help of a second-Post-Newtonian generalized Lagrangian, we find a secular acceleration of the mean orbital longitude of each member of a gravitationally bound binary system. This kinematical behaviour agrees with the phenomena observed in the Hulse-Taylor pulsar: PSR 1913+16.

1. INTRODUCTION

The so-called "problem of motion", i.e. deriving and solving the equations of motion of a system of gravitationally interacting bodies, is, since Newton, the cardinal problem of any theory of gravity. Within

B. Bertotti et al. (eds.), General Relativity and Gravitation, 89–106.

General Relativity this problem has had a long and checkered history since the paper by Einstein, in November 1915, on the motion of the perihelion of Mercury which even preceded the discovery of the full field equations. This problem has been completely transformed and renewed in the seventies by the discovery, in our galaxy, of binary systems containing *compact bodies*. By *compact body* (also called condensed body) I shall mean in the following either a black hole or a body whose physical radius is of the order of its gravitational radius: $2Gm/c^2$ (m being the mass, G Newton's constant and c the velocity of light) like a neutron star. Therefore the presence of compact bodies implies the existence of regions where the gravitational field becomes very strong and where it is therefore necessary to take into account the full non-linear structure of Einstein's equations. Moreover the consideration of the gravitational interaction of several bodies of comparable masses forces one to take into account all the propagation effects of gravity (contained in the hyperbolic structure of Einstein's equations). Therefore the problem of the motion of compact bodies involves the full dynamical role of the gravitational field, as coded in the *non-linear hyperbolic structure* of Einstein's equations.

The discovery of the binary pulsar PSR1913+16 by Hulse and Taylor in 1974 led, thanks to the beautiful observational work of Taylor and collaborators (see Taylor and Weisberg, 1982), to an impressively accurate tracking of the absolute orbital motion of one member of a binary system probably constituted of two compact objects. These results have made it necessary for the theorists to reexamine and to solve approximately the "problem of two compact bodies in General Relativity" with an accuracy at least as good as the observational accuracy which reaches 1 part in 10^{13} for the binary pulsar. The urgency of this problem has been increased by the announcement four years ago (Taylor et al., 1979) of the observation of a tiny secular acceleration of the orbital motion of the binary pulsar that was tentatively attributed to some residual radiative effects. However the attribution of this observed kinematical effect to radiative effects rested mainly on heuristic arguments, and it was not clear whether this effect was really a predicted consequence of General Relativity (especially in the case at hand of a system of two compact bodies). Since then much work has been done, aimed at studying the "radiative effects" which occur both near and far gravitationally interacting systems (for reviews see e.g.

Deruelle and Piran (1983) and the contributions of J. Ehlers and M. Walker to these proceedings). As however most of this work does not apply to the case at hand, I would like instead to outline a new method which has been especially devised for dealing with the motion of compact bodies. It has been possible to work out this method up to an accuracy sufficient for a meaningful comparison with the observational results. The final result (see below) agrees with the phenomena observed in the Hulse-Taylor binary pulsar and leads thereby to a profound confirmation of the non-linear hyperbolic field equations of General Relativity.

2. THE TWO COMPACT BODY PROBLEM IN GENERAL RELATIVITY

In the Solar system the relativistic gravitational potentials (curved "gothic" metric minus flat metric: $g^{\frac{1}{2}}g^{ab}-f^{ab}=:h^{ab}$) are always and everywhere smaller than about 10^{-6}. Therefore it is meaningful to set up an approximation scheme based on this *uniform* smallness of h and to keep only the first few terms of a non-linearity expansion of Einstein's equations. On the other hand, in a system containing weakly coupled compact bodies h^{ab} will still be quite small in any "external region" which stays some distance away from each compact body (and extends to infinity) but it will reach values of order unity near and in each compact body ("internal regions"). Therefore it is unwise to attack the latter problem by means of any approximation scheme such as the ones used in the former problem (Solar system) which were based on the uniform smallness of h. New lines of attack have been investigated by Manasse (1963), Demianski and Grishchuk (1974), D'Eath (1975), Kates (1980) and Damour (1983a). The common feature of all these approaches is to make use of *two* approximation schemes, one appropriate to the weak field domain (external region) and one appropriate to the strong field domain (internal region), and to combine them, by some matching procedure, so as to propagate the information between the two domains of validity of the approximation schemes. Related ideas and techniques have also been useful in understanding the transition between Near-Zone and Wave-Zone gravitational fields (Burke, 1971; Thorne, 1969; Anderson, 1980; Walker and Will, 1980).

In the following we shall describe successively: the setting up of an approximation scheme appropriate to strong field domains (§3 internal problem), another one appropriate to weak field domains with proper consideration of propagation effects (§4 external problem), the construction of the solution of the external problem which matches the internal solution (§5), the derivation of the equations of motion (§6), one way to solve the equations of motions (§7) and the results so obtained (§8).

3. THE INTERNAL PROBLEM

The physical fact which makes it possible to develop an approximation scheme in the internal regions, in spite of the fact that the gravitational field is very strong there, is that in a frame of reference linked with one compact body the influence of its companion will show up only in tiny tidal perturbations which will only slightly deform the body and therefore produce only small corrections to the (supposedly known) strong gravitational field of an *isolated* compact body. This result, which was expected to hold, at least for the first order effects, on the basis of Newtonian arguments, has been proved to be valid in the case of initially spherically symmetric (non rotating) compact bodies and when account is taken of the higher order (non linear) effects (Damour, 1983a). The mathematical tool used is the Regge-Wheeler approach (Regge and Wheeler, 1957; Campolattaro and Thorne, 1970; Zerilli, 1970) generalized to include non linear effects. The useful outcome is that the structure of the gravitational field *outside* each compact body is (with $\underline{g}^{ab} := g^{\frac{1}{2}} g^{ab}$; $a,b=0,1,2,3$; $i=1,2,3$):

$$\underline{g}^{ab}(\overline{x}) = \underline{g}_S^{ab}\left(\frac{c^2\overline{x}^i}{Gm}\right) + \sum_{p\geq 2,\, q\geq 1} \frac{G^{p+q}}{c^{2p+2q}} m^p m'^q \underline{g}_{p,q}^{ab}\left(\frac{c^2\overline{x}^i}{Gm}, \overline{x}^\circ, k\right), \quad (1)$$

where the first term in the right hand side of eq. (1) is the Schwarzschild metric (m being the *Schwarzschild mass* of the isolated compact body and similarly m' the Schwarzschild mass of its companion), and where k denotes a set of dimensionless numbers of order unity (relativistic Love numbers). It can be shown that the lowest order term, in the metric outside the body, in which the internal structure of the body begins to show up is of the order of $(G^6/c^{12})\, km^5 m'/(R^3|\overline{x}|^3)$, R being the distance to the companion. Therefore a system of initially non-rotating compact bodies presents the following remarkable property (which can be called *effacing principle* or *Cheshire cat principle*):

when terms of order G^6/c^{12} are neglected the internal structures of the compact bodies do not show up in their external metric and therefore the external metric depends only on the constant Schwarzschild masses of the isolated compact bodies (the "grins" of the cats).

4. THE EXTERNAL PROBLEM

Because the external metric is weak we can assume that it can be obtained by means of one of the usual approximation methods based on a non-linearity expansion around flat space (f^{ab}):

$$\underline{g}^{ab}-f^{ab}=:h^{ab}=Gh_1^{ab}+G^2h_2^{ab}+G^3h_3^{ab}+\ldots \tag{2}$$

Then, in harmonic coordinates, h must fulfill the following partial differential equations (the first one being obtained by expanding Einstein's equations):

$$\Box_f h^{ab}=N^{ab}(h)=N_2^{ab}(h,h)+N_3^{ab}(h,h,h)+\ldots, \tag{3}$$

$$h^{ab}{}_{,b}=0. \tag{4}$$

In order to solve eq. (3) by successive iterations we need to give ourselves some *boundary conditions* for h at the boundary of the external region, i.e. far from the system and near each compact body. As boundary conditions far from the system we take the Fock (1959) (also called Kirchhoff-Sommerfeld) *no incoming radiation condition* at Minkowski past null infinity. As boundary conditions near each compact body we take, up to order G^3 included, the *Dominant Schwarzschild* conditions, which are deduced from the result (1) of the internal problem by applying the most general allowed coordinate transformation between the internal coordinates $\bar{x}$ and the external coordinates x. These conditions state that the singular behaviour of each $h_n^{ab}(n\leqq 3)$, when x approaches the world-line z(s) which is the external image of the "central" internal line $\bar{x}=0$, is dominated by a boosted Schwarzschild behaviour:

$$\sim G^n m^n/|x-z|^n+0(1/|x-z|^{n-1}).$$

These conditions are very simple because they

contain (as a consequence of the "Cheshire cat principle"), near each body, only the (supposedly given) mass m and a, still unknown, world-line z(s). However this simplicity has to be paid for by the mere asymptotic nature of these conditions. Nevertheless it is possible to show that the preceding boundary conditions are sufficient, given two worldlines z(s), z'(s'), to characterize, up to order G^3 included, at most *one* (in the geometric sense) metric in the class of metrics that admit a formal expansion in series of powers of G and that, moreover, the worldlines must fulfill some equations of motion for a solution to exist. Because of this uniqueness result one can use any convenient auxiliary mathematical technique if it allows one to construct a particular solution of (3) and (4) which satisfies the preceding boundary conditions.

The resulting metric will necessarily be, up to order G^3 at least, the metric outside two compact objects, if the latter metric admits a power series expansion in G. The next section presents such a well defined and consistent construction technique, using complex analytic continuation, where the compact bodies appear as schematized by a mass and a world--line.

5. CONSTRUCTION OF THE THIRD ORDER RELAXED FIELD

It has been found convenient to make use, as an auxiliary mathematical technique, of a process of complex analytic continuation, based on some functions $Z_A(x)$ (where $x \in R^4$ and $A \in \mathbb{C}$) introduced by Riesz (1949):

$$Z_A(x) = (H_n(A))^{-1} (-f_{ab} x^a x^b)^{\frac{A-n}{2}}, \tag{5a}$$

when x is future directed in a n-dimensional Minkowski space (signature -++...), and when x is not so:

$$Z_A(x) = 0, \tag{5b}$$

with the coefficient being (Γ denoting the usual Eulerian function):

$$H_n(A) = \pi^{\frac{n-2}{2}} \cdot 2^{A-1} \cdot \Gamma\left(\frac{A}{2}\right) \cdot \Gamma\left(\frac{A+2-n}{2}\right). \tag{5c}$$

It had been suggested by Havas and Goldberg (1962)

that it might be useful to apply Riesz's functions Z_A to the problem of motion in General Relativity. One way of applying this technique in a manner which is both mathematically well defined and physically meaningful (because of its link to the "compact body problem") has recently been set up. It consists in solving by successive iteration the following integral equation:

$$h^{ab}=-Z_2 * (GT_A^{ab}(f+h)+N^{ab}(h)) \tag{6}$$

where the non-linear terms N(h) have been defined in eq. (3) and where $T_A(\underline{g})$ is defined as:

$$T_A^{ab}=\sum_{m,m'} 16\pi m\times \\ \times\int_{-\infty}^{+\infty} ds\; Z_A(x-z(s))\frac{dz^a}{ds}\frac{dz^b}{ds}(\underline{g}(z))^{1/4}(-\underline{g}_{cd}(z)\frac{dz^c}{ds}\frac{dz^d}{ds})^{-\frac{1}{2}}. \tag{7}$$

In eq. (6) the integral operator $-Z_2*$ (the star denoting the convolution) is nothing but the usual retarded integral (inverse of the flat d'Alembertian $\Box_f$). The role of this operator is to solve eq. (3) while incorporating the Fock no incoming radiation condition at Minkowski past null infinity. We shall assume here that this condition is then indeed satisfied (for a discussion of this point and of other delicate issues see Ehlers, 1983). In eq. (6) the term T_A which adds up to the non linear terms can be thought of as a fictitious stress-energy tensor density, the role of which is to incorporate the Dominant Schwarzschild conditions (§4). Eq. (6) is solved by a *Post--Minkowskian* iteration scheme which starts with:

$$h_{A1}^{ab}=-Z_2 * T_A^{ab}(f) \tag{8}$$

and then gives the successive powers of G in the formal expansion, eq. (2), of h as retarded integrals of an effective source which is constructed from the coefficients of the lower powers of G. This approximation scheme is called *Post-Minkowskian*, by opposition to the often used *Post-Newtonian* approximation schemes, because, at each step, it uses retarded integrals (which embody the causality properties of Minkowski space) instead of instantaneous integrals (which embody the causality properties of Newtonian physics). (These schemes have also been called respectively: Fast Motion and Slow Motion).

It has been shown (Damour, 1983a) that all the integrations appearing in the first three iterations are well defined, by analytic continuation in A, and that they lead to a finite third order iterated metric when A is continued to zero:

$$h_{0(3)} = (Gh_{A1} + G^2 h_{A2} + G^3 h_{A3})_{A=0} . \tag{9}$$

When A=0, the function T_A vanishes off the world-lines and therefore $h_{0(3)}$ is an approximate solution of the harmonically relaxed Einstein vacuum equations which is a *functional* of two arbitrary world-lines. The problem to give a useful expression for the latter functional has been first solved (at second order) in a seminal paper by Bertotti and Plebanski (1960) (see also Kovacs and Thorne (1977) and Westpfahl and Hoyler (1980)). An explicit expression has been obtained by Bel et al. (1981) (second order). A useful expression for the third order metric has been derived recently (Damour, 1983b).

6. THE HARMONICITY CONDITION AND THE THIRD ORDER EQUATIONS OF MOTION

In the preceding section we have sketched the construction of an approximate solution of the harmonically relaxed Einstein vacuum equations which is a functional of two arbitrary world-lines: $h^{ab}(x;z(s);z'(s'))$. Therefore the harmonicity condition that we must still satisfy:

$$H^a(x;z(s);z'(s')) := \partial_b h^{ab} = 0 , \tag{10}$$

yields constraints on the two world-lines. This was in fact the original approach of Einstein, Infeld and Hoffmann (1938), as generalized to the Post-Minkowskian approach case by Kerr (1959), for deriving the equations of (global) motion of two interacting bodies from the field in *vacuum* and from the *vacuum* field equations alone. These last two properties make this approach very useful in the compact body case where, because of the strong self gravitational fields, it would be very troublesome to derive the global equations of motion by integrating the local ones

($T^{ab}{}_{;b}=0$). On the other hand the original Einstein-Infeld-Hoffmann-Kerr approach necessitates the computation of exceedingly intricate surface integrals. However the analytic continuation approach sketched above, when consistently used for deriving the harmonicity condition (10), leads to a drastically simplified way of computing the equations of motion. It is then possible to derive, from the third order relaxed metric, semi-explicit expressions for the third order equations of motion:

$$\frac{d^2z^a}{ds^2} = \sum_{n=1}^{3} G^n W^a_n(z(s);z'(s')), \tag{11}$$

where the W's are retarded functionals of the two world-lines. When eq. (11) is satisfied, the harmonicity condition (10) will be fulfilled and thereby $h(x;z;z')$ will then be an approximate solution of the full Einstein vacuum equations incorporating the Dominant Schwarzschild conditions. Therefore, if $h_{0(3)}$ satisfies also the Fock conditions, it will necessarily be, by the uniqueness result quoted in section 4, the third order gravitational field outside two compact bodies (provided the latter field admits a power series expansion in G). (The satisfaction of the Fock conditions is hopefully implied by the construction procedure together with the past behaviour of the solutions of eq. (11), but this has still to be proven).

Because of its retarded functional structure it is a priori very difficult to extract useful information from eq. (11). At second order it has been possible, using the framework of Predictive Relativistic Mechanichs (for a review see Bel and Fustero, 1976), to associate to eq. (11) some manifestly Poincaré invariant ordinary differential equations (Bel et al., 1981). In the case of small relative velocities the explicit quasi-Newtonian form of the third order equations of motion has been obtained up to order c^{-5} by making a near-zone expansion of $h_{0(3)}$ (Damour and Deruelle, 1981a; Damour, 1982).

It is to be emphasized that the latter near-zone expansion does not imply at all a contradiction with the Post Minkowskian approach but is only a simple way of computing an approximate expression of eq. (11). On the other hand if one wished to iterate one step further the field equations it would be a priori illicit to use such a near zone expanded lower order field in the non-linear termsof eq. (6). The final equations of motion (order G^3 and c^{-5}; usually called second and a half Post Newtonian order) are expressed only in terms of *instantaneous* (same harmonic time x^0) positions and velocities (for the spin orbit terms see Damour, 1982):

$$\frac{d^2z^i}{dt^2} = A_0^i(z-z') + c^{-2}A_2^i(z-z',v,v') + c^{-4}A_4^i(z-z',v,v') + + c^{-5}A_5^i(z-z',v-v') + 0(c^{-6}). \tag{12}$$

Let us denote the instantaneous coordinate distance between the two worldlines by:

$$R := |z^i(t) - z'^i(t)| := \left((z-z')^2\right)^{\frac{1}{2}},$$

the unit radial vector from the second to the first world-line by:

$$N^i := R^{-1}(z^i(t) - z'^i(t)),$$

and the instantaneous relative velocity by:

$$V^i := v^i(t) - v'^i(t).$$

Then we can write down the following explicit expressions for the different terms in eq. (12) (m and m' being respectively the two masses):

$$A_0^i = -Gm'R^{-2}N^i,$$

$$A_2^i = Gm'R^{-2}\Big[N^i\big(-v^2 - 2v'^2 + 4(vv') + \tfrac{3}{2}(Nv')^2 + 5(Gm/R) + 4(Gm'/R)\big) + (v^i - v'^i)(4(Nv) - 3(Nv'))\Big],$$

$$A_4^i = B_4^i + C_4^i,$$

with:

$$B_4^i = Gm'R^{-2}\Big[N^i(-2v'^4 + 4v'^2(vv') - 2(vv')^2 + \tfrac{3}{2}v^2(Nv')^2 + + \tfrac{9}{2}v'^2(Nv')^2 - 6(vv')(Nv')^2 - \tfrac{15}{8}(Nv')^4 +$$

$$+(Gm/R)\left(-\frac{15}{4}v^2+\frac{5}{4}v'^2-\frac{5}{2}(vv')+\frac{39}{2}(Nv)^2-39(Nv)(Nv')+\frac{17}{2}(Nv')^2\right)+$$
$$+(Gm'/R)(4v'^2-8(vv')+2(Nv)^2-4(Nv)(Nv')-6(Nv')^2))+$$
$$+(v^i-v'^i)(v^2(Nv')+4v'^2(Nv)-5v'^2(Nv')-4(vv')(Nv)+$$
$$+4(vv')(Nv')-6(Nv)(Nv')^2+\frac{9}{2}(Nv')^3+$$
$$+(Gm/R)\left(-\frac{63}{4}(Nv)+\frac{55}{4}(Nv')\right)+(Gm'/R)(-2(Nv)-2(Nv')))\Big],$$

$$C_4^i=G^3m'R^{-4}N^i\left(-\frac{57}{4}m^2-9m'^2-\frac{69}{2}mm'\right),$$

$$A_5^i=\frac{4}{5}G^2mm'R^{-3}(V^i(-V^2+2(Gm/R)-8(Gm'/R))+$$
$$+N^i(NV)\left(3V^2-6(Gm/R)+\frac{52}{3}(Gm'/R)\right)).$$

7. SOLVING THE EQUATIONS OF MOTION

In order to compare the theory with the observations we must solve, or at least control sufficiently well the solution of the preceding equations of motion (12). This can be done in four steps (Damour, 1983c):

1st step: one first proves a generalization of the center of mass theorem which is valid at the order c^{-5} included. This result allows then to eliminate the position and velocity of the second body (the "companion") and to restrict the attention to the motion of the first body in the center of mass frame. One obtains autonomous equations for the first body:

$$\frac{d^2z^i}{dt^2}=B_0^i(z)+c^{-2}B_2^i(z,v)+c^{-4}B_4^i(z,v)+c^{-5}B_5^i(z,v). \tag{13}$$

2nd step: it has been shown (Damour and Deruelle, 1981b; Damour, 1982) that the equations of motion of the *binary* system, considered at order c^{-4} only, could be derived from a *generalized* Lagrangian $L_{(4)}(z,z',dz/dt,dz'/dt,d^2z/dt^2,d^2z'/dt^2)$, function not only of the simultaneous positions and velocities but also of the simultaneous accelerations of the two bodies. (What is meant here by "derived from $L_{(4)}$" is the following: the Euler-Lagrange equations of $L_{(4)}$ contain d^3z/dt^3 and d^3z'/dt^3 but only in the terms of order c^{-4}, then the use of Newton's equations of motion for these terms leads to eq. (12) truncated at order c^{-4}). The group symmetries of this generalized Lagrangian allow one to construct Noetherian conserved quantities for the binary system at order c^{-4}. The elimination by the first step, of the variables of the companion leads then to prove the existence of four first integrals of the c^{-4} truncated motion of the first body described by the following equations:

$$\frac{d^2 z^i}{dt^2} = B_0^i(z) + c^{-2} B_2^i(z,v) + c^{-4} B_4^i(z,v). \tag{14}$$

3rd step: using the latter 4 first integrals one can integrate eq. (14) by means of two quadratures (this leads to introducing some hyperelliptic integrals and functions).

4th step: it is then convenient to consider the last term of the right hand side of eq. (13): $c^{-5}B_5^i$ as a perturbing term added to eq. (14). A suitably modified method of variation of arbitrary constants leads then to a c^{-5} approximate solution of eq. (13).

8. SECULAR EFFECTS IN THE MOTION OF TWO COMPACT BODIES

The preceding solution to the motion of the first body leads to the following formula for the motion of the coordinate radius $r := |z|$:

$$r(t) = S\left(2\pi \int_0^t \frac{du}{P(c_1(u), c_2(u))} + c_0(t),\ c_1(t),\ c_2(t)\right), \tag{15}$$

where $S(a,b,c)$ is a *periodic* hyperelliptic function of its first variable a (with period 2π), $P(c_1, c_2)$ is a complete hyperelliptic integral, and the $c_i(t)$ $(i=0,1,2)$ are slowly varying "constants" which contain secular terms but only linearly in time. These results alone suffice for showing that the date of the Nth passage of the first object at its periastron (the slowly varying minimum of $r(t)$) has a parabolic dependence on N:

$$t_N = t_0 + P_0\, N + \frac{1}{2} P_0 \dot{P}_0 N^2. \tag{16}$$

The explicit computation of the "secular acceleration" $\dot{P}_0$ leads to evaluating several complete hyperelliptic integrals. However the latter integrals can be well approximated by simpler circular integrals which yield the final result:

$$\dot{P}_0 = -\frac{192\pi}{5c^5}\left(\frac{2\pi G}{P_0}\right)^{5/3}\frac{mm'}{(m+m')^{1/3}}\left(1+\frac{73}{24}e_0^2+\frac{37}{96}e_0^4\right)(1-e_0^2)^{-7/2}, \qquad (17)$$

where e_0 denotes $(1+2(m+m')c_1^0(c_2^0)^2/(G^2(mm')^3))^{\frac{1}{2}}$.

Because the c^{-5} center of mass of the system is unaccelerated, and, because in the Post Minkowskian approach the coordinate time t is a proper time far away from the system we can conclude that the quantity $\dot{P}_0$, eq. (17), which measures the secular decrease of the time of return to the periastron, must be identified (modulo a constant Doppler factor close to unity due to the relative motion of the center of mass of the system and of the center of mass of the Solar system) with the obervational quantity denoted by $\dot{P}_b$ in Taylor et al. (1979).Similarly it is seen that e_0 is, within the accuracy now available, to be identified with the observational parameter e. The same is true for the two masses: $m=m_p$, $m'=m_c$.

The result (17) agrees both with the standard, heuristically predicted (Peters and Mathews, 1963; Wagoner, 1975) "quadrupole formula for the secular acceleration", and with the observations of the Hulse--Taylor pulsar (Taylor and Weisberg, 1982). Several recent works have been aimed at justifying the same formula (see references in §1). However most of these do not meet, at the same time, the following two requirements that the method presented here was aimed at satisfying:(i) the derivation should apply to compact bodies, and, (ii) one should compute the *direct* effect of the non-linear retarded gravitational interaction on the absolute orbital motion of a member of a binary system (without relying on heuristic balance equations, or on badly defined "local energy loss" formulae).

9. CONCLUSIONS AND OPEN ISSUES

In contradistinction with the "Solar system tests" of General Relativity which give very good *quantitative* confirmations of Einstein's theory (accuracy of 10^{-3}, see Will (1981) and the contribution of R.Hellings to these proceedings) but which test mainly the linearized structure of gravitation theories (so that they cannot disprove many alternative theories of gravity), the "Radiative two-compact-body problem" in General Relativity and its predicted secular acceleration $\dot{P}_0$, yields an excellent *qualitative* test of the *non-linear hyperbolic structure* of Einstein's equations. This test includes:

1) a direct test of the cubically non linear structure of the weak field expansion of Einstein's equations,

2) an indirect test of the full non-linear structure of Einstein's equation ("Cheshire cat principle"),

3) the first test of the existence of irreversible phenomena linked with the finite velocity of propagation of gravity and with time-asymmetric boundary conditions,

4) a very selective test of alternative theories of gravity (Will-Eardley, 1977; Weisberg-Taylor, 1981; Will, 1981).

In view of all this richness enwrapped in the study of the radiative two compact body problem in General Relativity it is important to stress some of the open issues that should be clarified before one can safely rely on the agreement between the observed $\dot{P}_b$ and the predicted $\dot{P}_0$.

From the observational point of view, one should try to check all the plausible assumptions that enter into the model: assumptions concerning the nature of the unseen companion, the cleanliness of the system, the motions of the binary system and of the Solar system in the Galaxy and the distance between the two systems. One way of checking some of these assumptions would be to look more closely at the "periodic" effects in the time arrival analysis. In the long term, one would very much like to be able to observe other clean relativistic binary systems so as to strongly confirm the validity of the agreement between $\dot{P}_b$ and the right hand side of eq. (17).

From the theoretical point of view there remain several issues that are left open at the present state of the art of approximation methods in General Relativity. Among them: estimating the errors entailed by the approximation method (especially those coming from the use of a flat retarded propagator), extending the validity of the formal part of the method to all orders, checking that one can fit an explicit model of a compact object in the previously constructed external gravitational field, understanding the relevance of the flat space Fock conditions in the final curved space-time, studying the connection with other approximation methods, and with rigourous results or plausible "Ansätze" (especially concerning the asymptotic behaviour of the space-time, see Ehlers (1983) and the contribution of A.Ashtekar to these proceedings).

ACKNOWLEDGEMENTS. It is a pleasure to acknowledge useful discussions with J.Anderson, A.Ashtekar, J. Ehlers, B.Schmidt and M.Walker. I thank the Max Planck Institute for Astrophysics, Garching, for hospitality during the preparation of this paper.

REFERENCES

Anderson, J.L.: 1980, "New derivations of the quadrupole formulas and balance equations for gravitationally bound systems", *Phys. Rev. Lett.*, 45, pp. 1745-1748.

Bel, L., Damour, T., Deruelle N., Ibañez J., and Martin J.:1981, "Poincaré invariant gravitational field and equations of motion of two point like objects: the post-linear approximation of General Relativity", *Gen. Rel. Grav.*, 13, pp. 963-1004.

Bel, L. and Fustero, X.: 1976, "Mécanique relativiste prédictive des systèmes de N particules", *Ann. I.H.P.*, A25, pp. 411-436.

Bertotti, B., and Plebanski, J.: 1960, "Theory of gravitational perturbations in the fast motion approximation", *Ann. Phys.*, 11, pp. 169-200.

Burke, W.L.: 1971, "Gravitational radiation damping of slowly moving systems calculated using matched asymptotic expansions", *J.Math. Phys.*, 12, pp. 401-418.

Campolattaro, A., and Thorne, K.S.: 1970, "Non radial pulsation of general relativistic stellar models. V. analytic analysis for $\ell=1$", *Ap.J.*, 159, pp. 847-858.

Damour, T.: 1982, "Problème des deux corps et freinage de rayonnement en relativité générale", *C.R.Acad. Sci. Paris*, 294, série II, pp. 1355-1357.

Damour, T.: 1983a, "Gravitational radiation and the

motion of compact bodies", in N.Deruelle and T.Piran (eds), *Gravitational radiation*, North Holland, Amsterdam, pp. 59-144.

Damour, T.: 1983b, "Radiation damping in General Relativity", in Hu Ning (ed.), *Proceedings of the Third Marcel Grossmann Meeting*, Science Press, Peking, (in press).

Damour, T.: 1983c, "Gravitational radiation reaction in the binary pulsar and the quadrupole formula controversy", *Phys. Rev. Lett.*, 51, pp.1019-1021.

Damour, T. and Deruelle, N.: 1981a, "Radiation reaction and angular momentum loss in small angle gravitational scattering", *Phys. Lett.*, 87A, pp. 81-84.

Damour, T. and Deruelle, N.: 1981b, "Lagrangien généralisé de deux masses ponctuelles, à l'approximation post-post-newtonienne de la relativité générale", *C.R.Acad. Sci. Paris*, 293, série II, pp. 537-540.

D'Eath, P.D.: 1975, "The interaction of two black holes in the slow motion limit", *Phys. Rev.*, D12, pp. 2183-2199.

Demianski, M. and Grishchuk, L.P.: 1974, "Note on the motion of black holes", *Gen. Rel. Grav.*, 5, pp. 673-679.

Deruelle, N., and Piran, T. (eds): 1983, *Gravitational Radiation*, (Les Houches, 1982) North Holland, Amsterdam.

Ehlers, J.: 1983, "Weak field approximations and equations of motion in General Relativity", updated version of the Max Planck Institute Report: MPI-PAE/Astro 138, 1977.

Einstein, A.: 1915, "Erklärung der Perihelbewegung des Merkur aus der Allgemeinen Relativitäts theorie", *Sitzungsber. d.K. Preuss. Akad. d. Wiss.*, II, pp. 831-839. (18 November).

Einstein, A., Infeld L., and Hoffmann, B.: 1938, "The gravitational equations and the problem of motion", *Ann. Math.*, 39, pp. 65-100.

Fock, V.A.: 1959, *Theory of Space, Time and Gravitation*, Pergamon, London, section 92.

Havas, P. and Goldberg, J.N.: 1962, "Lorentz-invariant equations of motion of point masses in the General Theory of Relativity", *Phys. Rev.*, 128, pp. 398-414.

Kates, R.E.: 1980, "Gravitational radiation damping of a binary system containing compact objects calculated using matched asymptotic expansions", *Phys. Rev.*, 22, pp. 1871-1878.

Kerr, R.P.: 1959, "The Lorentz-covariant approximation method in General Relativity", *Nuovo Cim.*, 13, pp. 469-502.

Kovács, S.J. and Thorne, K.S.: 1977, "The generation of gravitational waves. III. Derivation of bremsstrahlung formulae"., *Ap.J.*, 217,pp.252-280.

Manasse, F.K.: 1963, "Distortion in the metric of a small center of gravitational attraction due to its proximity to a very large mass", *J. Math. Phys.*, 4, pp. 746-761.

Peters, P.C. and Mathews, J.: 1963, "Gravitational radiation from point masses in a Keplerian orbit", *Phys. Rev.*, 131, pp. 435-440.

Regge, T. and Wheeler, J.A.: 1957, "Stability of a Schwarzschild singularity", *Phys. Rev.*, 108, pp. 1063-1069.

Riesz, M.: 1949, "L'intégrale de Riemann-Liouville et le problème de Cauchy", *Acta Math.*, 81, pp.1-223.

Taylor, J.H., Fowler, L.A. and Mc Culloch, P.M.: 1979, "Measurements of general relativistic effects in the binary pulsar PSR 1913+16", *Nature*, 277, pp. 437-440.

Taylor, J.H. and Weisberg, J.M.: 1982, "A new test of General Relativity: gravitational radiation and the binary pulsar PSR 1913+16", *Ap.J.*, 253, pp. 908-920.

Thorne, K.S.: 1969, "Nonradial pulsations of general relativistic stellar models. IV. The weak field limit", *Ap.J.*, 158, pp. 997-1019.

Wagoner, R.V.: 1975, "Test for the existence of gravitational radiation", *Ap.J.*, 196, pp. L63-L65.

Walker, M. and Will, C.M.: 1980, "Gravitational radiation quadrupole formula is valid for gravitationally interacting systems", *Phys. Rev. Lett.*, 45, pp. 1741-1744.

Weisberg, J.M. and Taylor, J.H.: 1981, "Gravitational radiation from an orbiting pulsar", *Gen. Rel. Grav.*, 13, pp. 1-6.

Westpfahl, K. and Hoyler, H.: 1980, "Gravitational bremsstrahlung in post-linear fast-motion approximation", *Lett. Nuovo Cim.*, 27, pp.581-585.

Will, C.M.: 1981, *Theory and experiment in gravitational physics*, Cambridge University Press, Cambridge.

Will, C.M. and Eardley, D.M.: 1977, "Dipole gravitational radiation in Rosen's theory of gravity: observable effects in the binary system PSR 1913+16", *Ap.J.*, 212, pp. L91-L94.

Zerilli, F.J.: 1970, "Gravitational field of a particle falling in a Schwarzschild geometry analyzed in tensor harmonics"; *Phys. Rev.*, D2, pp. 2141-2160.

THE QUADRUPOLE APPROXIMATION TO GRAVITATIONAL RADIATION

Martin Walker

Max-Planck-Institut für Physik und Astrophysik
Institut für Astrophysik
Karl-Schwarzschild-Str. 1, 8046 Garching, FRG

The purpose of this essay is to comment on the status of the quadrupole formulae for gravitational radiation and radiation reaction.

1. INTRODUCTION

The quadrupole formulae (there are two quite different ones!) arise in the description of radiative effects in Einstein's relativistic theory of gravity. The importance of classical gravitational radiation is twofold. Firstly, gravitational waves convey information to a distant observer about the motion and distribution of matter in regions which may be invisible to the observer by means of electromagnetic radiation. Therefore, direct measurement of the amplitude, polarization and spectrum of gravitational radiation far away will enable us to infer properties of matter as yet unseen. Secondly, since gravitational waves carry energy, momentum, and higher moments away from their sources, radiation emission influences the evolution of these sources (see, eg., Taylor et al., 1979; Ritter, 1980; Paczynski and Sienkiewicz, 1981).

The gravitational radiation theorist therefore has two tasks. These are to derive from first principles, firstly, a representation of the dependence of that part of the space-time metric describing the gravitational waves far from the sources on variables describing the matter distribution within the sources, so that inferences about the latter can be drawn from

B. Bertotti et al. (eds.), General Relativity and Gravitation, 107–123.

measurements of the former; secondly, to derive the rate of change of matter variables, characterizing the sources, due to emission of radiation (eg. the period of binary system). These tasks are very difficult. Einstein's gravitational field equations are sufficiently complicated that it is not known, for example, whether either a functional describing the dependence of the radiative part of the metric on matter variables, or equations of motion (as opposed to laws of motion; ch. Havas, 1979, p. 76; Ehlers, 1980) for matter variables, exist in the exact theory. Explicit representation of the metric as a functional of matter variables has been achieved only in the simplest of cases, eg. spherically symmetrical vacuum, (Schwarzschild, 1916), in which the sole matter variable is the mass. No metrics at all are known which describe the field either outside or inside bounded radiating sources (but cf. Bičák, 1968; Kinnersley and Walker, 1970; Schmidt, 1981). We are therefore obliged to resort to approximation.

A great variety of different approximation schemes have been used to make two different attacks on the two tasks. I shall first describe attempts to treat a restricted but fairly general class of radiating systems. Then I shall describe a few attempts to work specific, special problems. The literature on these topics is vast, and my references to it will be incomplete. For recent reviews see eg. Ehlers (1977, 1980), Walker and Will (1980a), and Damour (1983).

The tasks I have described would appear to require rather different techniques, because for the first we want the metric far away from the sources, while for the second we want it inside the sources. This can mean that different approximation procedures may have to be employed in different regions of space-time, the two then being matched together in some intermediate domain. Nevertheless, most procedures thus far employed share certain basic features, which are the following.

2. APPROXIMATION PROCEDURES

One first chooses a model of the material sources, ie. selects matter variables m_A, A=1, 2,..., characterizing these sources. The most common choices are the masses and world lines (which are to be determined afterwards) of point particles (so that the stress-energy tensor is distributional, with support on the world lines), or the density, equation of state and spatial flow vector of an isentropic perfect fluid. An important difference between these two is that the former is a mathematical fiction (point particles do not exist in relativistic gravity, because of horizons), whereas the second is well defined mathematically. Unfortunately, existence and uniqueness theorems for solutions of the Cauchy problem for the Einstein-Euler system, which would guarantee the existence of a relativistic celestial mechanics of fluid balls, are not available. The difficulty lies in the hydrodynamics of fluid balls; local existence theorems for Euler's equations for compressible fluids hold only inside the domain of dependence of an initial surface contained in the support of the density (B. Schmidt and D. Christodoulou, private communication). This is of no use for celestial mechanics. It may well be that the best way to set up a relativistic celestial mechanics is with black holes, but I shall not discuss this possibility further here.

Having chosen matter variables, one then has the freedom of choice of a coordinate chart for the domain where one is working. This choice is usually referred to in the context of approximation procedures as a gauge condition (G). A common idea is to choose coordinates which put the field equation (E) into a form which can be recast into an integral equation (IE) by means of a known integral identity (an example will be given presently). This integral equation alone is not equivalent to the field equation until the gauge condition has been satisfied: (E)$\Leftrightarrow$(IE) $\cup$ (G). One hopes now that the integral equation defines the metric as a functional of the matter variables, that the gauge condition determines the space-time dependence of the matter variables, and that the resulting metric satisfies Einstein's gravitational field equation. None of these hopes is nourished by theorems, but they are nevertheless pursued, by applying a method of successive approximations to the integral equation. It is assumed, but not established, that this iterative procedure generates a sequence of approximations which are asymptotic in an appropriate sense to a solution of the integral equation. That is to say, for the first few steps, each further step improves the accuracy of approximation.

There is a single theorem on iterative approximation procedure of the above kind in the literature (Christodoulou and Schmidt, 1979). It applies to the Cauchy problem for the harmonically reduced gravitational field equation, to which the Kirchhoff integral formula for the flat space scalar wave operator is applied. The theorem establishes that iteration on the resulting integral equation does indeed generate an asymptotic approximation to the solution if the stress-energy tensor is specified arbitrarily ("painted onto space-time"), and the Cauchy data for the gravitational field vanishes. The former condition is catastrophic since the true motion is not known a priori; it is intended to be determined later. The latter condition (which can be relaxed) means that the initial value constraints are not satisfied, and the approximate metric, to the future of the Cauchy surface, but outside the future of the support of the stress-energy in the Cauchy surface, remains flat at all orders of approximation. It follows that for the Cauchy problem, if the initial value constraints are not satisfied, the reduced field equation, together with the local equation of motion (vanishing covariant divergence of the stress-energy tensor), are not equivalent to the full field equation. This theorem as it stands does not apply to any existing approximation scheme.

Having chosen matter variables, and a gauge condition to cast the field equations into a tractable form, the next thing to ask is, how many iteration steps must be carried out in order to achieve the accuracy required to include the radiative effects one wants to describe? At this stage it is convenient to restrict the class of physical systems one aims to deal with. The most common restrictions, and those which I shall make here, are firstly, that the system evolve on a time scale long compared to a light crossing time, and secondly, that the system be embeddable in a sequence having a Newtonian limit, and not be far from this limit. Under these circumstances, one expects it to make sense to expand the radiative part of the metric in multipoles, and that successively higher multipoles make smaller and smaller contributions to the radiation field. As is well known, the lowest order multipole that can be expected to contribute to gravitational radiation is quadrupole. There are cases in which this vanishes, so that octupole radiation dominates (cf. Wagoner, 1973, 1977), but I shall consider only the quadrupole case. The question above about the accuracy of approximation required then becomes: how many times must I iterate the integral equation in order to be sure that further iterations make no contribution to the quadrupole part of the radiation field which are as large as those I already have? Fortunately, this is one of the few points on which almost everyone in the subject agrees. It does, however, depend on the choice of gauge.

I shall continue the discussion in harmonic gauge, in which the coordinate functions satisfy the scalar wave equation, not becuase I think that this is in any sense the best possible choice, but because it is one for which theorems on the global existence of harmonic charts can be established (as consequences of theorems in Choquet-Bruhat et al., 1979), and is the gauge in which most work on existence and uniqueness of solutions has been done (see, eg. Choquet--Bruhat and York, 1980). Therefore one has some hope of making contact with rigorous results. A heuristic argument (Walker and Will, 1980a), based on the fact that it is non-spherical motions of matter bending outgoing light rays differently in different directions which introduces shear into outgoing null hypersurfaces, together with the assumption of a global energy balance, leads to the conclusion that, for a bounded system of mass M and size b, one must work to order $(GM/bc^2)^2$ for the far field calculation, and to order $(GM/bc^2)^3$ in order to compute radiative effects on the motion of sources. In harmonic gauge, these accuracies are achieved after two iterations of the reduced field equation for the former, and after three iterations for the latter. This heuristic argument is not a derivation of either quadrupole formula. It serves only to establish the accuracy of approximation which must be achieved.

In an harmonic chart, the principal part of the Einstein operator is the scalar wave operator. Splitting off the flat space scalar wave operator from this, and putting everything else on the right hand side, one can obtain an integral equation in two different ways. Assuming that the metric components satisfy Sommerfeld-type conditions along past harmonic light cones, one can use the retarded Green function for the flat space scalar wave operator (see, eg. Ehlers, 1977). Alternatively, one can give intial data on a Cauchy surface and use the Kirchhoff integral identity. A conceptual difficulty with the former way is that experience with Schwarzschild's spacetime suggests that the past harmonic light cones in a spacetime with nonvanishing mass go out to spatial infinity, and not to past null infinity. But asymptotic flatness at spatial infinity implies sufficiently rapid fall-off of the metric that Sommerfeld conditions along harmonic cones would be satisfied independently of whether there were any incoming radiation or not. It is therefore unclear what these conditions mean in the exact theory. The initial value approach does not have this particular difficulty.

The first practical difficulty with obtaining an explicit representation of the approximate metric as a functional of matter variables arrives with the very first iteration step, in which the harmonic flat space metric is inserted into the integral. The retarded integral has to be evaluated somehow. To this end, two different approaches have been pursued.

One approach is to take the fictitious distributional stress-energy tensor model of matter, so that the retarded integration simply evaluates the integrand on the intersection of the past harmonic cone of the field point with the world lines of the source. In addition to the lack of an exact formulation which is being approximated with point particle sources, the catch with this approach is that the nonlinearity of the integrand generates meaningless divergent expressions at the second and all subsequent iterations. Therefore, some kind of auxiliary procedure, which is not itself part of the relativistic theory of gravity, must be introduced in order to extract finite results. The most far-reaching results along these lines have been achieved by Damour and collaborators (see Bel et al., 1981; Damour, 1983; and Damour's contribution to these proceedings). Damour's analytic continuation procedure is unambiguous and consistent with the field equations up to the fourth iteration. Unlike the approach using extended bodies, his applies also to compact bodies such as neutron stars or black holes (see also Kates, 1980).

The second approach to obtaining an explicit representation of the retarded integral is not to treat the retardation exactly, but to approximate it somehow. One can then hope to represent the metric in terms of successions of Newtonian-like potentials. The catch this time, even using extended bodies, is that after the first iteration step, the support of the integrand extends out to infinity, again due to the theory's nonlinearities. Expansion of the retardation in a Taylor series generates positive powers of the integration variable in the integrand, which sooner or later lead to divergences (as first pointed out by Ehlers et al., 1976). In early attempts to compute radiative effects on the source motion in this way, divergences appeared already at the second iteration step, before the desired accuracy had been reached (Chandrasekhar et al., 1965, 1969, 1970; Epstein and Wagoner, 1975; Anderson and DeCanio, 1975; see also Papapetrou and Linet, 1981). Subsequent work has banished divergences to at least the fourth iteration (Ehlers, 1977; Kerlick, 1980; Breuer and Rudolph, 1981, 1982; Anderson et al., 1982). It would appear that the approach from initial data on a Cauchy surface advocated by Schutz (Schutz, 1980; Futamase and Schutz, 1983; Futamase, 1983) generates only finite integrals. The procedures which approximate the retardation for extended bodies so far require that the Newtonian potential of the density distribution be small everywhere, hence apply only to non-compact bodies.

3. THE QUADRUPOLE FORMULAE

Since the first task requires only two iterations of the harmonic integral equation in order to obtain the leading contribution to the quadrupole part of the far radiation field, it would appear to be the easier one. A problem here, however, is that the harmonic light cones deviate arbitrarily greatly from the true space-time light noces as one recedes far from the source. An argument has been sketched by Walker and Will (1980b) which uses a harmonic retarded approximation inside a radius at which the harmonic and true cones have not yet deviated appreciably, then matches this at a large but finite radius to a Bondi-Sachs (Bondi et al., 1962; Sachs, 1962) radial expansion outside (see also Walker, 1983a). The result is that the asymptotic news tensor (time derivative of the transverse, traceless, r^{-1}-part of the asymptotic metric along an outgoing null hypersurface) is minus twice the third time derivative of the traceless mass quadrupole moment of the matter. Time derivatives of the latter are computed using the (osculating) Newtonian equations of motion holding at that (retarded) time, these being the equations of motion consistent with the second iteration of the harmonically reduced field equations (see Ehlers, 1977; Walker and Will, 1980a,b). I propose to call this the far-field quadrupole formula. The formally estimated relative error in this formula is the magnitude of the largest velocity in the source. An alternative argument for the validity of the far-field quadrupole formula using the Landau-Lifschitz complex has been proposed by Anderson (1980). The best way to think of the far-field quadrupole formula, however, is as a consequence of the constraint equations for the characteristic initial value problem on a complete future light cone in the context of the Newtonian limit. This last approach was motivated, after the GR10 conference, by a preprint of Winicour (1983), and is being pursued by the author and James Vickers. It appears likely that one will be able to do better than make just formal error estimates (cf. also Kuusk and Unt, 1983).

As for the second task, a number of independent calculations using quite disparate techniques (point--particles, extended bodies, different gauges - eg. Thorne, 1969; Chandrasekhar et al., 1965, 1969, 1970; Schäfer, 1982, 1983; ...) lead to the conclusion that there exists a gauge (not harmonic; see Miller, 1974; Schäfer, 1982) in which the secular effects of radiation emission on the sources can indeed be described by use of Burke's (1971) radiation-reaction potential. This potential is obtained from the fifth time derivative of the mass quadrupole moment, the time derivatives again being computed using the Newtonian equations of motion corresponding to the configuration of the system at the time considered. This may be called the radiation-reaction quadrupole formula.

Neither the far-field nor the radiation-reaction quadrupole formula requires, for its derivation, a global energy-momentum balance between sources and fields. It may well be that an exact energy-momentum balance holds only for scattering (for the electromagnetic case, see Walker, 1983b). Integration of, for example, the (vanishing) coordinate divergence of the Landau-Lifschitz complex, in some chart(s), over a space-time volume bounded to the future and past by asymptotically null spacelike hypersurfaces yields an exact "balance", the flux integral through future null infinity being precisely the Bondi-Sachs (Bondi et al., 1962; Sachs, 1962) energy-momentum flux (Streubel, 1978). The integrals over the spacelike hypersurfaces are not so easy to interpret, however, except perhaps in the Newtonian limit (cf. Anderson, 1980). Similar remarks apply to a suggestion of Horowitz and Strominger (1983) to use Witten-like spinorial arguments (see, eg., Horowitz and Perry, 1982).

4. SPECIAL CASES

The second attack on the problem of the description of gravitational radiation has been to work specific problems. Three of these will be briefly mentioned here.

The only problem which has received a definitive treatment is gravitational Bremsstrahlung radiation (Thorn and Kovács, 1975; Crowley and Thorne, 1977; Kovács and Thorne, 1977, 1978). This is one of very few calculations which takes account both of the fact that characteristics are not flat, and that propagation is not sharp. An inadequate summary of the work is that the far-field quadrupole formula is valid in that domain where one would have expected it to hold.

Rosenblum (1980, 1981) was the first to attempt the computation of radiation-reaction effects for small-angle scattering. It appeared at first sight that one could avoid the third iteration by considering only very weak scattering. This, however, was subsequently discovered to be incorrect (Will's workshop contribution described by Cohen, 1981; Westpfahl and Hoyler, 1980; Bel et al., 1981; Kates and Madonna, 1982; Damour, 1983). The third-iterated equations of motion, specialized to this case and to the second iteration, turn out to be conservative. There is no loss of mechanical energy in the second approximation to small-angle scattering, contrary to an initial claim to the contrary (Rosenblum, 1980). With three iterations, the calculation agrees with the radiation--reaction quadrupole formula. The discrepancy with further calculation by Rosenblum (1983) remains unexplained.

Thirdly, there is the far-field computation of the radiation emitted during the very early stage of the free fall from rest of two fluid balls, initiated by Cooperstock and colleagues (Cooperstock and Lim, 1977; Cooperstock and Hobill, 1979). Since the initially tidally stressed fluid balls fall from rest, there are hydrodynamical difficulties here in addition to those associated with computing perturbative solutions of the gravitational field equations. The excess pressure required to hold the balls in their initially static configuration has to be turned off "by hand" (ie. with a time-dependent equation of state) before the balls can fall freely. If this be done throughout the fluid in a time short compared with the hydrodynamical time scale, then the fluid will not have had time to evolve at the start of free fall, and all hydrodynamical variables except the pressure will retain their static values for a short time. This assumption makes the hydrodynamical equations of motion trivial to solve, and the result of doing so is the solution written down by Cooperstock and Lim (1977, equation (3.4)), and employed in the subsequent papers.

The result of the perturbation calculation is consistent with an application of the far-field quadrupole formula to the initial tidally induced oscillations of the balls (Walker and Will, 1982; Will, 1983). The analysis from an altered set of perturbed field equations by Cooperstock (1982) leaves this conclusion unchanged (equations (2.4), (2.8), and (2.12) of this paper are equivalent to equation (1) of Walker and Will, 1982; the explicit time dependence in equation (2.14) comes from the solution of the hydrodynamical equations of motion mentioned above).

5. CONCLUSION

We have therefore the following situation. A number of independent derivations, employing quite different techniques which, although not rigorously justified, do not seem to contain internal contradictions, lead to both quadrupole formulae. At the same time, there are no counterexamples.

How can this situation be improved? There are situations to which the quadrupole formulae do not apply (eg. relativistic scattering: Kovács and Thorne, 1978; systems moving on a time scale comparable with their light crossing times: Press, 1977; systems whose dominant radiation pattern is not quadrupole: Wagoner, 1969, 1977). It would be useful to stake out their domain of validity more precisely. This will involve getting a better grip on the errors made in the various approximation schemes, together with a relation of approximate concepts and formulae to rigorous parts of the theory (Bondi et al., 1962; Sachs, 1962; Dixon, 1979; Ehlers and Rudolph, 1977; Schattner, 1979). An analogous treatment of higher multipole radiation would also be welcome, especially in view of potential astrophysical applications (eg. Bekenstein, 1973). Most desirable of all would be surveyable, exact solutions of the gravitational field equations describing radiating isolated radiating systems. Unfortunately, this would appear to lie in the very distant future.

ACKNOWLEDGEMENTS. This article could not have been written without the help of numerous colleagues. I am particularly grateful to James Anderson, Demetrios Christodoulou, Thibaut Damour, Nathalie Deruelle, Jürgen Ehlers, Ron Kates, John Madore, Bernd Schmidt, Bernard Schutz, James Vickers and Clifford Will for discussions, comments and suggestions. Errors or omissions are, of course, mine alone.

REFERENCES

Anderson, J.L.: 1980, "New Derivations of the Quadrupole Formulas and Balance Equations for Gravitationally Bound Systems", *Phys. Rev. Lett.*, 45, 1745-48.

Anderson, J.L., and De Canio, T.A.: 1975, "Equations of Hydrodynamics in General Relativity in the Slow Motion Approximation", *Gen. Rel. Grav.*, 6, 197-238.

Anderson, J.L., Kates, R.E., Kegeles, L.S., and Madonna, R.G.: 1982, "Divergent Integral of Post-Newtonian Gravity: Nonanalytic Terms in the Near-Zone Expansion of a Gravitationally Radiating System Found by Matching", *Phys. Rev.*, D25, 2038-48.

Bekenstein, J.D.: 1973, "Gravitational-Radiation Recoil and Runaway Black Holes", *Astrophys.J.*, 183, 657-664.

Bel, L., Damour, T., Deruelle, N., Ibanez, J., and Martin, J.: 1981, "Poincaré-Invariant Gravitational Fields and Equations of Motion of Two Pointlike Objects: The Postlinear Approximation of General Relativity", *Gen. Rel. Grav.*, 13, 963-1004.

Bičák, J.: 1968, "Gravitational Radiation From Uniformly Accelerated Particles", *Proc. Roy. Soc.*, A302, 201-24.

Bondi, H., van der Burg, M.G.J., and Metzner, A.W.K.: 1962, "Gravitational Waves in General Relativity, VIII. Waves from Axi-symmetric Isolated Systems", *Proc. Roy. Soc.*, A269, 21-52.

Breuer, R.A., and Rudolph, E.: 1981, "Radiation Reaction and Energy Loss in the Post-Newtonian Approximation of General Relativity", *Gen. Rel. Grav.*, 13, 777-93.

Breuer, R.A., and Rudolph, E.: 1982, "The Force Law for the Dynamic Two-Body Problem in the Second Post-Newtonian Approximation of General Relativity", *Gen. Rel. Grav.*, 14, 181-211.

Burke, W.L.: 1971, "Gravitational Radiation Damping of Slowly Moving Systems Calculated Using Matched

Asymptotic Expansions", *J. Math. Phys.*, 12, 401-418.

Chandrasekhar, S.: 1965, "The Post-Newtonian Equations of Hydrodynamics in General Relativity", *Astrophys. J.*, 142, 1488-1512.

Chandrasekhar, S., and Nutku, Y.: "The Second Post-Newtonian Equations of Hydrodynamics in General Relativity", *Astrophys. J.*, 158, 55-80.

Chandrasekhar, S., and Esposito, F.P.: 1970, "The 2 1/2-Post-Newtonian Equations of Hydrodynamics and Radiation Reaction in General Relativity", *Astrophys. J.*, 160, 153-180.

Choquet-Bruhat, Y., Christodoulou, D., and Francaviglia, M.: 1979, "On the Wave Equation in Curved Spacetime", *Ann. Ist. H. Poincaré*, 31,399-414.

Choquet-Bruhat, Y., and York, J.W. Jr.: 1980, "The Cauchy Problem" in A. Held (ed.), *General Relativity and Gravitation* Vol. 1, Plenum, New York, pp. 99-172.

Christodoulou, D., and Schmidt, B.G.: 1979, "Convergent and Asympotic Iteration Methods in General Relativity", *Commun. Math. Phys.*, 68, 275-89.

Christodoulou, D., and O'Murchadha, N.: 1981, "The Boost Problem in General Relativity", *Commun. Math. Phys.*, 80, 271-300.

Cohen, J.M.: 1981, "Summary of the Workshop on Radiation Processes in Strong Gravitational Fields", in R. Ramaty and F.C. Jones (eds), *Proc. 10th Texas Symp. on Relativistic Astrophys. Baltimore*, *Ann. N.Y. Acad. Sci.* 375, 459-64.

Cooperstock, F.I., and Lim, P.H.: 1977, "Axially Symmetric Two-Body Problem in General Relativity, II. Free Fall", *Phys. Rev.*, D15, 2105-22.

Cooperstock, F.I., and Hobill, D.W.: 1979, "Axially Symmetric Two-Body Problem in General Relativity, III. Bondi Mass Loss and the Failure of the Quadrupole Formula", *Phys. Rev.*, D20, 2995-3003.

Cooperstock, F.I.: 1982, "Axially Symmetric Two-Body Problem in General Relativity, IV. Boundary

Conditions, Time Scales, and the Quadrupole Formula", *Phys. Rev.*, D25, 3126-36.

Crowley, R.J., and Thorne, K.S.: 1977, "The Generation of Gravitational Waves, II. The Postlinear Formalism Revisited", *Astrophys. J.*, 215, 624-635.

Damour, T.: 1983, "Gravitational Radiation and the Motion of Compact Bodies", in N.Deruelle and T. Piran (eds), *Gravitational Radiation*, North-Holland, Amsterdam, pp. 59-144.

Dixon, W.G.: 1979, "Extended Bodies in General Relativity; their Description and Motion", in J.Ehlers (ed.), *Isolated Gravitating Systems in General Relativity, Proc. Int. School of Phys. "Enrico Fermi", Course LXVII, Varenna*, North-Holland, Amsterdam, 1979, pp. 156-219.

Ehlers, J.: 1977, "Weak Field Approximations and Equations of Motion in General Relativity", in R.Ruffini, J.Ehlers, and C.W.F.Everitt (eds), *Proc. Int. School od General Relativistic Effects in Physics and Astrophysics: Experiments and Theory (3rd Course)*, Institute Report MPI-PAE/Astro 138, M.P.I. f.Astrophysik, D-8046 Garching, F.R.G.

Ehlers, J.: 1980, "Isolated Systems in General Relativity", in J. Ehlers, J.J. Perry, and M.Walker (eds), *Ninth Texas Symposium on Relativistic Astrophysick*, *Ann. N.Y.Acad. Sci.*, 336, 279-94.

Ehlers, J., Rosenblum, A., Goldberg, J.N., and Havas, P.: 1976, "Comments on Gravitational Radiation Damping and Energy Loss in Binary Systems", *Astrophys. J. Lett.*, 208, L77-L81.

Ehlers, J., and Rudolph, E.: 1977, "Dynamics of Extended Bodies in General Relativity: Center of Mass Description and Quasirigidity", *Gen. Rel. Grav.*, 8, 197-217.

Epstein, R., and Wagoner, R.V.: 1975, "Post-Newtonian Generation of Gravitational Waves", *Astrophys. J.*, 197, 717-23.

Futamase, T.: 1983, "Gravitational Radiation Reaction in the Newtonian Limit", to appear in *Phys. Rev.D.*

Futamase, T., and Schutz, B.F.: 1983, "The Newtonian

and Post-Newtonian Approximations are Asymptotic to General Relativity", to appear in *Phys. Rev.D.*

Havas, P.: 1979, "Equations of Motion and Radiation Reaction in the Special and General Theory of Relativity", in J. Ehlers (ed.), *"Isolated Gravitating Systems in General Relativity", Proc. Int. School of Physics "Enrico Fermi", Course LXVII, Varenna*, North-Holland, Amsterdam, pp. 74-155.

Horowitz, G.T., and Perry, M.: 1982, "Gravitational Energy Cannot Become Negative", *Phys. Rev. Lett.*, 48, 371-74.

Horowitz, G.T., and Strominger, A.: 1983, work referred to by Horowitz in N.Deruelle and T.Piran (eds), *Gravitational Radiation*, North Holland, Amsterdam, p. 431.

Kates, R.E.: 1980, "Gravitational Radiation Damping of a Binary System Containing Compact Objects Calculated Using Matched Asymptotic Expansions", *Phys. Rev.*, D22, 1871-78.

Kates, R.E., and Madonna, R.G.: 1982, "Gravitational Radiation Reaction from Small Angle Scattering of Slowly-Moving Compact Bodies Calculated by Matched Asymptotic Expansions", *Phys. Rev.*, D25, 2499-2508.

Kerlick, G.D.: 1980, "Finite Reduced Hydrodynamic Equations in the Slow-Motion Approximation to General Relativity. Part I. First Post-Newtonian Equations", *Gen. Rel. Grav.*, 12, 467-482, "Part II. Radiation Reaction and Higher Order Divergent Terms", *ibid.*, 12, 521-543.

Kinnersley, W., and Walker, M.: 1970, "Unformly Accelerated Charged Mass in General Relativity", *Phys. Rev.*, D2, 1359-70.

Kovács, S.J., and Thorne, K.S.: 1977, "The Generation of Gravitational Waves. III. Derivation of Bremsstrahlung Formulae, *Astrophys.J.*, 217, 252-280.

Kovács, S.J., and Thorne, K.S.: 1978, "The Generation of Gravitational Waves. IV. Bremsstrahlung", *Astrophys. J.*, 224, 62-85.

Kuusk, P., and Unt, V.: 1983, "Energy Loss by Isolated Self-gravitating Systems", *Estonian SSR Acad. Sci. preprint.*

Miller, B.D.: 1974, "The Effect of Gravitational Radiation-Reaction on the Evolution of the Riemann S-type Ellipsoids", *Astrophys. J.*, 187,609-20.

Paczynski, B., and Sienkiewicz, R.: 1981, "Gravitational Radiation and the Evolution of Cataclysmic Binaries", *Astrophys. J. Lett.*, 248, L31-34.

Papapetrou, A., and Linet, B.: 1981, "Equations of Motion Including the Reaction of Gravitational Radiation", *Gen. Rel. Grav.*, 13, 335-359.

Press, W.H.: 1977, "Gravitational Radiation from Sources Which Extend into Their Own Wave Zone", *Phys. Rev.*, D15, 965-8.

Ritter, H.: 1980, "Z Cha - New Evidence for Gravitational Waves?", *Astron. Astrophys.*, 86,204-11.

Rosenblum, A.: 1978, "Gravitational Radiation Energy Loss in Scattering Problems and the Einstein Quadrupole Formula", *Phys. Rev. Lett.*, 41,1003-5.

Rosenblum, A.: 1981, "Gravitational Radiation Energy Loss in Scattering Problems and the Einstein Quadrupole Formula", *Phys. Lett.*, 81A, 1-4.

Rosenblum, A.: 1983, "Energy Loss Due to Small-Angle Scattering in General Relativity", *J.Phys. A: Math. Gen.*, 16, 2751-56.

Sachs, R.K.: 1962, "Gravitational Waves in General Relativity. VIII. Waves in Asymptotically Flat Space-Time", *Proc. Roy. Soc.*, A270, 103-126.

Schäfer, G.: 1982, "The Equations of Motion for an Astrophysical Binary with Accuracy $1/C^5$", *Progr. Theor. Phys.*, 68, 2191-93.

Schäfer, G.: 1983, "Strahlungsrückwirkung der Gravitation in der Quadrupolnäherung", *Mitteilungen der Astr. Ges.*, 58, 135-37.

Schattner, R.: 1978, "The Center of Mass in General Relativity", *Gen. Rel. Grav.*, 10, 377-393.

Schattner, R.: 1978, "The Uniqueness of the Center of Mass in General Relativity", *Gen. Rel. Grav.*, 10, 395-399.

Schmidt, B.G.: 1981, "The Decay of the Gravitational Field", *Commun. Math. Phys.*, 78, 447-54.

Schutz, B.F.: 1980, "Statistical Formulation of Gravitational Radiation Reaction", *Phys. Rev.*, D22, 249-59.

Schwarzschild,K.: 1916, "Über das Gravitationsfeld eines Massenpunktes nach der Einsteinschen Theorie", *Sitzber. Deut. Akad. Wiss. Berlin, Kl. Math-Phys. Tech.*, 189-96.

Streubel, M.: 1978, *"Conserved Quantities Related to Asymptotic Symmetries for Isolated Systems in General Relativity"*, Dissertation, Universität Hamburg (available as Institute Report MPI-PAE/ Astro 165, M.P.I. f. Astrophysik, D-8046 Garching, FRG).

Taylor, J.H., Fowler, L.A., and McCulloch, P.M.: 1979, "Measurements of General Relativistic Effects in the Binary Pulsar PSR 1913+16", *Nature*, 277, 437-40.

Thorne, K.S.: 1969, "Nonradial Pulsations of General--Relativistic Stellar Models. IV. The Weak-Field Limit", *Astrophys. J.*, 158, 997-1019.

Thorne, K.S., and Kovács, S.J.: 1975, "The Generation of Gravitational Waves. I. Weak-Field Sources", *Astrophys. J.*, 200, 245-262.

Wagoner, R.V.: 1969, "Low Frequency Gravitational Radiation from Collapsing Systems", *Phys. Rev.*, D19, 2897-901.

Wagoner, R.V.: 1977, "Gravitational Radiation From Slowly-Rotating Collapse: Post-Nowtonian Results", in *Simposio Internazionale sulla Gravitazione Sperimentale (Experimental Gravitation), Pavia, Atti dei Convegni Lincei* 34, Accademia dei Lincei, Roma, 1977, pp. 117-135.

Walker, M.: 1983a, "Isolated Systems in Relativistic Gravity", X.Fustero and E.Verdaguer (eds), to appear in *Proc. XIVth GIFT International Seminar on Theoretical Physics, San Feliu de Guixols,* World Scientific Publishers, Singapore.

Walker, M.: 1983b, "Global Energy-Momentum Conservation for Scattering of Point Charges", *Phys. Rev. Lett.*, 50, 636-39.

Walker, M., and Will, C.M.: 1980a, "The Approximation of Radiative Effects in Relativistic Gravity: Gravitational Radiation Reaction and Energy Loss in Nearly Newtonian Systems", *Astrophys. J. Lett.*, 242, L129-33.

Walker, M., and Will, C.M.: 1980b, "Gravitational Radiation Quadrupole Formula is Valid for Gravitationally Interacting Systems", *Phys. Rev. Lett.*, 45, 1741-44.

Walker, M., and Will, C.M.: 1982, "Axially Symmetric Gravitational Two-Body Problem of Cooperstock, Lim, and Hobill", *Phys. Rev.*, D25, 3433-37.

Westpfahl, K., and Hoyler, H.: 1980, "Gravitational Bremsstrahlung in Post-Linear Fast Motion Approximation", *Lett. Nuovo Cimento*, 27, ser. 2, 581-5.

Will, C.M.: 1983, "Tidal Gravitational Radiation from Homogeneous Stars", to appear in *Astrophys. J.*

Winicour, J.: 1983, "Newtonian Gravity on the Null Cone", *University of Pittsburgh, preprint.*

GRAVITATIONAL RADIATION AND THE 'QUADRUPOLE' FORMULA REPORT OF WORKSHOP A1

Jürgen Ehlers* and Martin Walker

Max-Planck-Institut für Physik und Astrophysik, Institut für Astrophysik
Karl-Schwarzschild-Str. 1
8046 Garching, FRG

1. INTRODUCTION: THE QUESTIONS

In connection with the quadrupole formula, *three* related but different *questions* can be asked: 1) How does the gravitational radiation field (in some wave zone or at future null infinity) of a nearly isolated system depend on the motion and structure of its sources? 2) How is the motion and the structure of a source emitting gravitational waves affected by this emission, i.e. what are the radiative corrections to the source's motion? 3) Is there a conservation law linking the energy-momentum carried "to infinity" by gravitational radiation to the loss of energy-momentum by the source?

A tentative partial answer to the first question is given by the *far field quadrupole law*: For a source which moves smoothly in a nearly Newtonian manner and whose configuration changes little during a light-crossing time, the gravitational radiation power at a time t in the wave zone is approximately equal to

$$(G/5c^5)\dddot{Q}_{jk}\dddot{Q}_{jk}, \tag{1}$$

where $\dddot{Q}_{jk}$ denotes the third time-derivative of the trace-free part

$$Q_{jk}=\int(x_j x_k-\tfrac{1}{3}\delta_{jk}x_\ell x_\ell)\,dm$$

B. Bertotti et al. (eds.), General Relativity and Gravitation, 125–137.

of the mass quadrupole moment, computed by means of an osculating Newtonian approximation to the source's motion at the retarded time corresponding to t. (The motion must be sufficiently smooth and quasistationary for the differential retardation- and multipole expansions to be valid. It is understood that the quadrupole term is to be replaced by the lowest order non-vanishing multipole term in those cases where the quadrupole term happens to be zero).

A tentative partial answer to the second question is given by the *radiation reaction quadrupole law:* For a source obeying the conditions stated above, the secular rate of decrease of the Newtonian total energy due to gravitational radiation reaction is approximately equal to a suitable time average of the power given in (1), again computed in a Newtonian manner, provided the system is nearly periodic. (Without the last condition, the energy decrease rate differs from (1) by a total time derivative of a function depending on Q_{jk} and its derivatives).

These answers are *tentative* since so far they have been established by non-rigorous, formal approximations only. They are *partial* insofar as they concern the energy only; similar statements have been made about momentum and, perhaps even more tentatively, angular momentum, and about other dynamical variables.

Each of the two quadrupole laws by itself is meaningful independently of an *overall energy conservation law*. Taken together, they imply such a law and thus provide an answer to the third question. Conversely, an energy conservation law implies a partial answer to the second question in terms of one to the first question, and vice versa.

The fundamental problem which underlies all three questions is that of the *relation between source and field* in Einstein's theory which, because of the simultaneous presence of nonlinearity and retardation, so far has not only not been "solved", but has hardly been brought in contact with "rigorous general relativity". However, since the 9th Conference on General Relativity and Gravitation (GR9) some progress has been made in the development of *approximation methods* designed to answer the above (and related) questions under various assumptions, and in the analysis of *pertinent examples*. Some of these improvements have been reported at GR10 in the invited lectures by Thibaut Damour and by Martin Walker (published in this volume), and some of them have been discussed and criticized in the special session and the related workshop A1 "Equations of motion, gravitational radiation and asymptotic structure of spacetime", as will be reported below.

It appears that an *exact* answer to either of the first two questions requires such an answer to the other one, too. However, approximation methods indicate (see Walker's contribution) that the computation of the radiation field to some accuracy requires a less precise knowledge of the metric and the source variables in the near zone than a comparably accurate computation of the radiation reaction. It must be admitted, though, that errors have so far been judged on the basis of crude plausibility arguments only (e.g. Synge, 1970; Thorne and Kovács, 1975; Ehlers, 1977), if at all; rigorous errors estimates are neither available nor in sight.

2. THE TALKS

At the special session, a brief introduction by the chairman was followed by talks of Fred Cooperstock, Clifford Will, Jim Anderson, Ronald Kates, and Bernard Schutz. Closely related to the special session were talks given during the corresponding workshop by Martin Walker and by Gerhard Schäfer. In the following report, comments by J. Ehlers which were not made during the session but which may elucidate the issues will be put in brackets and marked by a star:(*...).

The first two contributions concerned the question whether the far field quadrupole law does or does not apply to the free-fall problem posed in 1974 by Fred Cooperstock (Cooperstock, 1974; Cooperstock and Lim, 1977; Cooperstock and Hobill, 1979): Two fluid balls are assumed to be in static equilibrium, held apart by a strut, with membranes to contain the fluid, until time t=0. Between t=0 and t=t*, the strut and membranes are dissolved via an (unspecified) time-dependent equation of state. Afterwards the balls fall freely. Because of the static initial state there is no question in this case concerning the absence of incident radiation. However, the behaviour of the fluid balls in the free-fall phase - for which the quadrupole formula is to be tested primarily - depends on how the transition from equilibrium to free fall takes place, and it is with respect to the hydrodynamically possible transitions that Walker and Will (Walker and Will, 1982) disagree with Cooperstock (Cooperstock, 1982a, b).

In his talk, Cooperstock agreed that Walker and Will (1982) and Will (1983) obtain a fluid motion and far field consistent with the far field quadrupole law during the stress-breaking phase and the very first stage of free fall, provided the turn-off time t* is chosen to be short compared to the hydrodynamic time--scale t_H of the fluid balls. He emphasizes, however, that the case he and his collaborators are interested in is the opposite one, in which t* is long compared to t_H, the supporting stresses are removed "gently", and the fluid balls fall freely towards each other while undergoing quasistatic tidal deformations only. In this case the equations of Newtonian gravity would imply that the quadrupole formula yields an energy flux of order (size of ball/separation of balls)^2 times that which follows directly from the far field, obtained by solving the field equation iteratively[(1)]. Cooperstock thus arrives at a contradiction with the far field quadrupole law. He maintains that his larger flux is due to nonlinear terms of higher than 1 post--Newtonian order which have not been taken into account in derivatives of the quadrupole formula, and that these terms lead to a compactness-dependent emission rate (Cooperstock, 1982a, b).

Clifford Will devoted part of his contribution to a discussion of how the time-dependence of the matter variables in the free-fall problem of Cooperstock immediately after t* can be found from the Newtonian hydrodynamic equations. He showed that for rapid dissolution, $t^* \ll t_H$, the tidally induced oscillations of the balls give rise, via the quadrupole formula, to an emission rate which has the same parameter-dependence as the "large" rate obtained by Cooperstock. This holds for the initial stages of free fall; later the quadrupole contribution from the overall motion of the balls dominates (Will, 1983). Will also asserted that under the opposite assumption, $t^* \gg t_H$, a detailed investigation of the hydrodynamics during the dissolution phase is unavoidable to determine the time-dependence of the flow after t*. Will also discussed gravitational radiation emitted by tidally induced quadrupole distortions of homogeneous fluid balls of density ρ_0 falling into black holes. He showed that only incompressible "Kelvin" normal modes are excited by a tidal potential. The quadrupole mode has frequency

$$\omega = (16 G \rho_0 / 15)^{\frac{1}{2}},$$

and this makes possible a fairly simple analytic treatment, both of this problem and of that of Cooperstock et al.

(* On tha basis of this work, it appears still unclear whether Cooperstock's claim (Cooperstock, 1982a, b) is true that during later stages of free fall, for suitable dissolution mechanism, the emission rate is larger than the quadrupolar one because of a compactness-dependence having nothing to do with tidal distortions. The published work of Cooperstock et al. does, in my opinion, not justify this claim, for at least three reasons: 1) For times long enough in order that the balls have fallen through a distance much larger than their diameters, Cooperstock's approximation method as it stands is presumably not valid since his zeroth-order (and thus formally dominant) metric is the initial, static one which does not "follow the bodies". One would have to include at least a parametric time-dependence in the Weyl-metric functions ν, γ, μ as done, e.g., by Kates (1982) and that is bound to affect the evolution of the metric at higher orders. 2) Since according to Cooperstock, a formally higher-order term of the metric gives rise to a larger flux contribution than lower-order ones, it appears to be necessary at least to make plausible why the next higher-order term will not produce a still larger contribution (error estimate!). In other methods of approximation it appears now to be possible to judge errors at least by estimating the size of the next-order term (Walker's contribution to the present volume, and Anderson et al., 1982; Futamase and Schutz, 1983; Futamase,1983)[(2)], although that has not yet been done, as far as I am aware. 3) There appears to be a contradiction between the assumption that the bodies undergo quasistatic tidal deformations only during free fall, and the assumption that the pressure gradient balances the original, static, internal self-gravity within each body when free fall commences. The first assumption requires that each body has adjusted its shape and pressure distribution to the tidal field of the other body when free fall begins, and that is excluded by the second assumption. Now, Cooperstock's higher-order far-field energy flux is due to the component of the fluid momentum density orthogonal to the axis of symmetry. Neither the first nor the second assumption, however, leads to that linearly time-dependent momentum density component which Cooperstock uses to obtain his "super-quadrupolar" flux. - If Cooperstock were right, the compactness of the bodies would enter the emitted flux, and thus presumably also the radiation reaction, even in systems like PSR 1913+16, where the coupling between the bodies is weak and the motion nearly Newtonian, but the structures of the bodies are relativistic. This claim appears to be in conflict with the results of Damour (published in this volume) and Kates (1982). A comparison of Cooperstock's computations with other schemes is not easy because of different gauges, different iterations and different order--assignments).

After Will's contribution Peter Havas said that no derivation of the farfield quadrupole law has been presented yet which is "consistent" in the sense that the motion of the sources is non-conservative in the correct manner, corresponding to the fact that radiation carries energy-momentum to infinity. He also argued that the close agreement between the measured rate of decrease of the orbit-period of the binary pulsar and the predictions based on the radiation-reaction quadrupole law applied to a pair of Newtonian orbiting point masses can only be regarded as a coincidence until all other possible (non-radiative) causes of a period-decrease have been eliminated. He and colleagues are re-analyzing the pulse arrival time data with a view to looking for alternative explanations.

The third speaker, Jim Anderson, began his talk by listing some of the difficulties one is faced with in trying to derive answers to the three questions formulated in the introduction to this report: i) One has to determine the propagation of the gravitational field out to "infinity" according to a wave-operator whose coefficients contain this very field; use of a flat, background, harmonic wave operator is wrong since the "harmonic null cones" are space-like and tend to spatial infinity. ii) The so-called "radiation condition" of Fock (1959) requires some fall-off of the metric on past, "harmonic" null cones; thus it is not obvious and probably not true that this condition excludes incoming gravitational radiation.(* Consider e.g., an initially static star which is excited by an incident gravitational wave pulse and then reradiates and settles down again (Price and Thorne, 1969). There is hardly any doubt that in this case Fock's condition holds, although it is solely the incident radiation which "causes" the motion of the "source"). One therefore has to exclude incident radiation in some other way. iii) A useful approximation method would be such that terms which are formally of higher order are, in fact, smaller than lower-order terms throughout the domain of validity of the approximation. This condition is clearly violated by iteration schemes using the flat space, harmonic wave operator because of (lnr)/r-terms. iv) The assumption frequently made that the metric depends analytically on a slowness parameter ("c^{-1}"), is presumably not true and leads to divergences. (See, e.g., the work of Chandrasekhar et al., Anderson and Decanio, Kerlick). Anderson then indicated, with reference to his recent published (Anderson, 1979, 1980; Anderson et al., 1982) and submitted (Anderson, 1984a, b; Anderson and Madonna, 1984; Anderson and Heyl, 1984) papers, how all of these difficulties may be overcome. He uses an energy-momentum balance equation based on the conservation law for the Landau-Lifshitz complex, integrated over a four-dimensional spacetime region bounded in the wave-zone by two future-directed (physical) null cones and in the near zone by two spacelike hypersurfaces. The balance equation is to be evaluated by means of a nearly Newtonian inner expansion and a post-Minkowskian outer expansion, linked to each other by matching. In the outer expansion, a modification of the harmonic gauge condition and a stepwise determination of the "true" retarded time (method of strained coordinates) are employed to avoid logarithmic terms, and in the inner expansion two time scales are used to separate the Newtonian time-dependence from secular changes of orbit elements due to radiation reaction. Anderson requires the field to depend causally on the source in the sense that if the latter is stationary prior to some time, then its field is also stationary prior to an appropriately corresponding (approximate) retarded time. Such a field is expected to contain no incoming radiation provided the source variables do not grow with time in the infinite past. Model calculations have been carried out to demonstrate that and how the method works. Parts of this overall plan have been applied to general relativity; the work is being continued. So far, this work supports both quadrupole laws.

The fourth speaker, Ron Kates, described his extension (Kates, 1982), by a method of matched asymptotic expansions, of work of Morgan and Bondi (1970) on quasistationary axially symmetric spacetimes to describe both outgoing radiation and radiation damping. His analysis requires neither a treatment of the internal structures of the sources nor a weak-field assumption. In the case of weakly self-gravitating systems, however, his field quantities can be expressed as integrals over material sources, and the usual far-field and radiation reaction quadrupole laws result. This work covers, in particular, the free-fall problem of Cooperstock et al. According to Kates the far-field quadrupole law *does* hold for this problem irrespective of the dissolution mechanism, and even for times of the order of the free-fall time of the system.

The last speaker, Bernard Schutz, outlined the approach by himself and T. Futamase (Futamase and Schutz, 1983; Futamase, 1983) to the computation of radiation-reaction effects in weakly self-gravitating, quasi-stationary systems by means of a statistical initial value formulation, and its intended application by Clifford Will and himself to the radiation reaction of a particle freely falling in an external field. First, an asymptotic expansion of a one-parameter family of initial data for an Einstein-Euler-system (gravitational field + perfect fluid) obeying the contraints is to be constructed from a family of Newton-Euler initial data, so that the two-families agree asymptotically if a slowness parameter tends to zero. Then, the metric off the initial hypersurface is to be computed via a retarded integral with respect to a flat background metric, truncated at the intersection of the past null cone with the initial hypersurface. Instead of imposing a radiation condition "at infinity", an averaging over free gravitational radiation initial data is to be carried out, and the expected metric and matter variables are to be represented as a power series in the slowness parameter. Schutz emphasizes that his approach is a *local* one which concerns a relatively compact domain of spacetime containing the history of the source in a finite time-interval. No "infinity" ($\mathscr{I}$, e.g.) enters, and thus the integrals are finite. The computations confirm the radiation--reaction quadrupole formula in the case of bound, nearly Newtonian systems. The authors expect to be able to extend their approach to cover the free-fall problem and the far-field computation. (*This interesting approach still faces serious difficulties: In order for a family of Einsteinian spacetimes to have a Newtonian limit, the radiative degrees of freedom have to disappear in a specific manner. Some of the data which are free in the relativistic case and which have been assumed by Schutz et al. to vanish on average, do not vanish for a nearly Newtonian system, but are determined by the source at each instant. Also, the evaluation of the field off the initial hypersurface requires delicate assumptions concerning the dependence of metric- and matter variables on the spacetime coordinates and the slowness parameter. Not only is it unknown whether such assumptions are true; so far, according to my understanding, such assumptions as are required to carry out the computations reported by the authors have not been spelled out in sufficient detail. Thus, I am unable to verify that, by this approach, post-Newtonian-approximations have been justified as valid asymptotic approximations. It remains to be seen whether these gaps can be closed).

It should be mentioned here that Martin Walker outlined, in the related workshop, a new method developed by Metin Gürses and himself (Gürses and Walker, 1984), to evaluate the flat space, retarded integrals which appear in several post-Minkowskian approximation schemes. The method consists of breaking up the past null cone, in a manner that depends on the position of its vertex, into parts stretching into the near zone and parts stretching out to infinity, and to approximate the integrals over the former parts by integrals over suitably chosen domains of constant time, and those over the latter parts by integrals over parts of past-directed null cones. This method has been applied to a non-linear scalar, special-relativistic model theory for which the new iteration scheme gives finite results at all orders and reproduces the leading radiation-reaction formulae of the older scheme. Work to extend the method to general relativity is under way.

Also, Gerhard Schäfer reported briefly on calculations in which he obtained the metric and the equations of motion of point particles up to the $2\frac{1}{2}$ post-Newtonian-level by means of a formal iteration and regularization method (Schäfer, 1982). He follows essentially the work of Chandrasekhar and Esposito (1970), with the modifications that the sources are taken as point particles (this necessitates the use of regularizations whose mathematical justification is not attempted), and the Chandrasekhar-gauge is kept at all orders, which implies a simplification of the computations of radiative corrections. Both quadrupole laws are formally confirmed; the conceptual and mathematical objections to previous, similar derivations have not been considered, however.

3. CONCLUDING REMARKS[(3)]

It appears that although we still do not have rigorous results concerning the quadrupole laws, approximation methods have been developed to meet separately all the conceptual and mathematical objections that have been raised against earlier "derivations": 1) The difficulties concerning the propagation of radiation out to infinity (wrong null cones, logarithmic terms) may be overcome by the techniques developed by Thorne et al. (stepwise correction of Green functions), or those of Anderson (Anderson, 1979, 1980, 1984a, b; Anderson and Madonna, 1984; Anderson and Heyl, 1984), or by the use of characteristic initial value problems (as indicated by Winicour, 1983). 2) A mathematically transparent, systematic method to treat "point-like" sources beyond the linear approximation (and which can be extended to arbitrarily high orders[2]) has been developed by Damour (contribution to the present volume); this approach has further clarified, in the spirit of Einstein-Infeld-Hoffmann, the relation between equations of motion and field equations. 3) Techniques to match wave-zone ("outer"), and near-zone ("inner") expansions first used in general relativity by Burke, have been extended by Kates, Anderson and others; they have clarified the origin of infinities in slow motion methods and made plausible that the results of these earlier methods are valid as long as they give finite results. 4) Ways to express absence of incoming radiation without recourse to the Fock-Sommerfeld condition on harmonic null cones have been proposed ("causal" dependence of field on sources combined with asymptotic initial stationary; averaging over radiation initial data). 5) The concept of Newtonian limit of a family of general relativity-spacetimes has been made precise and related to approximation methods (Künzle and Nester, 1983; Futamase and Schutz, 1983; Ehlers, 1981; Winicour, 1983).

What is still missing, in my opinion, is one coherent treatment which combines the strengths of the various approaches and avoids their weaknesses.

Contact between approximation methods and rigorous parts of the theory (Bondi-Sachs-Penrose asymptotics/ /outer approximations; Dixon theory of motion/point- -like particles; initial value problems/Newtonian limit; positivity of energy-momentum/balance equation), though still very weak, may perhaps be established by means of the new approaches.

Several formal approximation methods which are internally consistent support the quadrupole laws, under conditions specified in the first section. On the other hand, the derivations underlying proposed counter-examples have been shown to suffer from internal inconsistencies.

NOTES

* J.Ehlers was the chairman of workshop A1 "Equations of motion, gravitational radiation and

asymptotic structure of spacetime" and of the special session on the "quadrupole formula".

(1) This result was arrived at by a correction, originally due to Walker and Will (1982) and incorporated in Cooperstock (1982a, b), of an error contained in Cooperstock and Lim (1977).

(2) The approximations scheme of T.Damour can be continued to higher orders without divergences, too. (Personal communication from T.Damour).

(3) By J.Ehlers.

REFERENCES

Anderson, J.L.: 1979, "Approximation methods in general relativity", in J.Ehlers (ed.), *Isolated Gravitating Systems in General Relativity*, Proc. Int. School of Physics "Enrico Fermi", Course 67, North-Holland, Amsterdam, pp. 289-306.

Anderson, J.L.: 1980, "New derivations of the quadrupole formulas and balance equations for gravitationally bound systems", *Phys. Rev. Lett.*, 45, 1745.

Anderson, J.L.: 1984a, "A model calculation of radiation damping by the energy balance method", to appear in *Gen. Rel. Grav. J.*

Anderson, J.L.: 1984b, "Approximate causal solutions for a class of wave equations with backscatter", to appear in *J.Math. Phys.*

Anderson, J.L., and Heyl, A.: 1984, to appear.

Anderson, J.L., Kates, E., Kegeles, S., and Madonna, R.G.: 1982, "Divergent integrals of post-Newtonian gravity: nonanalytic terms in the near-zone expansion of a gravitationally radiating system found by matching", *Phys. Rev.*, D25, 2038.

Anderson, J.L., and Madonna, R.G.: 1984, to appear.

Chandrasekhar, S., and Esposito, F.P.: 1970, "The 2½ post-Newtonian equations of hydrodynamics and radiation reaction in general relativity", *Astrophys. J.*, 160, 153.

Cooperstock, F.I.: 1974, "Axially symmetric two-body problem in general relativity", *Phys. Rev.*, D10, 3171.

Cooperstock, F.I.: 1982a, "Axially symmetric two-body problem in general relativity. IV. Boundary conditions time scales, and the quadrupole formula", *Phys. Rev.*, D25, 3126.

Cooperstock, F.I.: 1982b, "Replay to the comments of Walker and Will regarding the axially symmetric two-body problem", *Phys. Rev.*, D25, 3438.

Cooperstock, F.I., and Hobill, D.W.: 1979, "Mass loss and the failure of the quadrupole formula", *Phys. Rev.*, D20, 2995.

Cooperstock, F.I., and Hobill, D.W.: 1979, "Axially symmetric two-body problem in general relativity. III. Bondi mass loss and the failure of the quadrupole formula", *Phys. Rev.*, D20, 2995.

Ehlers, J.: 1977, "Weak-field approximations and equations of motion in general relativity", in R. Ruffini, J.Ehlers, and C.W.F.Everitt (eds), *Proc. Int. School of General Relativistic Effects in Physics and Astrophysics: Experiments and Theory"*, Institute Report MPI-PAE/Astro 138, MPI für Astrophysik, D-8046 Garching, FRG.

Ehlers, J.: 1981, "Über den Newtonschen Grenzwert der Einsteinschen Gravitationstheorie", in J.Nitsch, J.Pfarr, and E.-W.Stochow (eds), *Grundlagenprobleme der Modernen Physik*, B.I.-Wissenschaftsverlag, Mannheim.

Fock, V.: 1959, *The theory of space, time and gravitation*, Pergamon Press, London, p. 342.

Futamase, T.: 1983, "Gravitational radiation reaction in the Newtonian limit", *Phys. Rev.*, D28, 2373.

Futamase, T., and Schutz, B.F.: 1983, "Newtonian and post-Newtonian approximations are asymptotic to general relativity", *Phys. Rev.*, D28, 2363.

Gürses, M., and Walker, M.: 1984, "A finite slow-motion approximation method for self-gravitating systems", *Phys. Lett.*, A101, 15.

Kates, R.E.: 1982, "Gravitational radiation reaction

in quasi-static axially symmetric systems with possibly strong fields", Institute Report, Institüt für Astrophysik, MPA20, to appear in *Phys. Rev.*, D.

Künzle, H.P., and Nester, J.M.: 1983, "Hamiltonian formulation of gravitating perfect fluids and the Newtonian limit", Institute Report MPA58, to appear in *J. Math. Phys.*

Morgan, T., and Bondi, H.: 1970, "Transfer of energy in general relativity", *Proc. Roy. Soc. London*, A320, 277.

Price, R., and Thorne, K.S.: 1969, "Non-radial pulsation of general relativistic stellar models, II, *Astrophys.J.*, 155, 163.

Schäfer, G.: 1982, "Radiation reaction and energy loss for gravitationally bound systems", *Prog. Theor. Phys.*, 68, 2191.

Synge, J.L.: 1970, "Equations of motion in general relativity", *Proc. Res. Inst. Atmos.*, 69, sect. A, II-38.

Thorne, K.S., and Kovács, J.: 1975, "The generation of gravitational fields, I. Weak-field sources", *Astrophys. J.*, 200, 245.

Walker, M., and Will, C.M.: 1982, "Axially symmetric two-body problem of Cooperstock, Lim and Hobill", *Phys. Rev.*, D25, 3433.

Will, C.M.: 1983, "Tidal gravitational radiation from homogeneous stars", *Astrophys. J.*, 274, 858.

Winicour, J.: 1983, "Newtonian gravity on the null cone", University of Pittsburgh preprint.

ALGEBRAIC COMPUTING IN GENERAL RELATIVITY

Ian Cohen, Inge Frick and Jan E. Åman *

Institute of Theoretical Physics
University of Stockholm
S-113 46 Stockholm (Sweden)

1. INTRODUCTION

Algebraic Computing i.e. the use of computers to manipulate formulae, has been used in general relativity since about 1965. A very early example was the program GRAD ASSISTANT (Fletcher, 1965) that could calculate the Ricci tensor from a given not too complicated metric.

Since then several reviews of the field have appeared (Barton and Fitch 1972 ; Cohen et al., 1976; d'Inverno 1975, 1980, 1983). In this paper we give a review of the situation as of Summer 1983 but restrict ourselves, as far as references are concerned, mainly to new material. For a more complete list of references, we refer to d'Inverno (1980, 1983) that give a very good description of the situation as of 1980 and contain an extensive bibliography.

In the second section we give a short presentation of algebraic computing in general. Section 3 contains an excerpt from a typical computer session. After this follows, in Section 4, a discussion of the various standard uses of algebraic computing in general relativity and in Section 5 we review recent work where algebraic computing has been used. This is followed in Section 6 by a presentation of the classification project as an example of the interaction of computing facilities and mathematical development in advanced applications of algebraic computing. Finally Section 7 contains an up-to-date comparison of algebraic computer systems.

B. Bertotti et al. (eds.), General Relativity and Gravitation, 139–162.

2. ALGEBRAIC COMPUTING

Before we discuss the use of algebraic computing in general relativity, we first very briefly discuss algebraic computing in general. For more thorough introductions we recommend Pavelle et al., (1981), Yun and Stoutemyer (1980), Buchberger et al., (1982). The last reference is rather technical but serves as a good introduction to the literature in the field.

What do we mean by the term algebraic computing, or symbolic manipulation or computer algebra as it is also called? A simple definition of algebraic computing is: The manipulation of formulae, in which *not* every quantity is replaced by a number. Here are some (very simple) examples:

1. $\partial\{\exp[y \sin(x)]\}/\partial x = \exp[y \sin(x)]\, y \cos(x)$

2. $\int x^2 \sin(x)\,dx = -x^2\cos(x) + 2x \sin(x) + 2\cos(x)$

3. $(Adx + Bdy)\wedge(Cdx - Ddy) = -(AD + BC)\,dx\wedge dy$

The field of algebraic computing as such is a dynamic field of research. It may be divided up into three main areas:

1. Mathematical theory, development of algorithms. This area is often called "Computer Algebra" in a more restricted sense.

2. System development. Design and programming.

3. Applications. This is probably what most people understand by "Algebraic Computing".

As an example of 1, we mention the algorithm which has been developed for analytical integration. This algorithm can decide if a function in the class of rational functions of the elementary functions can be integrated within this class and if so gives the primitive function. Other examples are algorithms that (efficiently) factorize multivariate polynomials.

As regards 2, we will give a review of those algebraic computer systems, which we think might be of interest in general relativity, later in this paper.

As regards 3, the main application areas of algebraic computing have been in theoretical physics and above all in the areas of quantum electrodynamics, celestial mechanics and, in fact, general relativity.

This is because especially these areas have needed analytical results, and one has sometimes been forced to do enormous hand calculations.

An obvious example in general relativity is when one wants to know if a give metric is vacuum. A numerical calculation which gives an approximate answer is usually not satisfactory even if it is correct to 8 digits; one would rather have an exact answer.

3. AN EXAMPLE

Some algebra systems have facilities which allow standard calculations to be performed even by users with very little knowledge of computers. As an example we investigate a simple metric.

In the following excerpt of the conversation with the computer our explanations, which were added afterwards, are placed on the right. What we typed into the computer, we have underlined for clarity. Everything else is what the computer answered.

The metric (found by Petrov in 1962) is here given in the form of a null-tetrad i.e. the line element is given as:

$$ds^2 = 2W^0W^1 - 2W^2W^3$$

where the one-forms W for this particular metric have been previously typed into the computer.

SHP> <u>(WMAKE FORMSU)</u> Here we ask the computer to type them out.

$$W^0 = du$$

$$W^1 = (-1/2vx+1/2e^{x}x)du + dv$$

$$W^2 = 1/2(2)^{1/2}e^{1/2x}dx + 1/2(2)^{1/2}e^{-u+1/2x}idz$$

$$W^3 = 1/2(2)^{1/2}e^{1/2x}dx - 1/2(2)^{1/2}e^{-u+1/2x}idz$$

SHP> <u>(WMAKE DS2)</u> Calculate and type out the line element

$$ds^2 = (-vx+e^{x}x)du^2 + 2dudv - e^{x}dx^2 - e^{-2u+x}dz^2$$

SHP> <u>(WMAKE RIC)</u> Calculate the Ricci tensor.

Vacuum solution

SHP> (WMAKE PSI) Calculate and type out the Weyl spinor, PSI.

$PSI_0 = 3/4 + 1/4x + 1/4e^{-x}v$

$PSI_1 = -1/4(2)^{1/2}e^{-1/2x}$

We told the computer earlier not to print the zero components.

SHP> (PETROV) We end the example with a Petrov classification.

Petrov type is III

4. APPLICATIONS IN GENERAL RELATIVITY

In this section we discuss the most frequent applications of algebraic computing in general relativity.

The most basic and common algebraic calculation in general relativity is to compute the Ricci or Einstein tensor from a given line element. This is used for example, to check if a given metric is vacuum. Another use is the classical method of obtaining interesting new solutions by investigating the field equations for a line element with certain properties, e.g. some symmetry. For example one may obtain a set of equations describing vacuum or perfect fluids. Since such calculations tend to become quite laborious it is an obvious case where computer algebra should be used.

For most applications it has turned out to be more efficient to calculate components of tensors with respect to a moving frame rather than a coordinate frame (Campbell and Wainwright, 1977). The most important difference between calculations of coordinate components and frame components with algebraic computing is that while even fairly simple metrics in coordinate form create problems (due to needing large primary computer memory) more complicated metrics can successfully be handled with moving frames. Normally the frame is chosen to be either orthonormal (Lorentz tetrad) or to be a (complexified) null tetrad. The use of the Newman-Penrose formalism (Campbell and Wainwright, 1977; Åman, 1982; Dautcourt and Jann, 1980; Mc Crea, 1982b) or of spinors can be regarded as extensions of calculations with the use of a null tetrad.

Another common application is the Petrov classification. Here the Newmann-Penrose formalism is normally preferred. (Åman, 1982; Campbell and Wainwright, 1977;

d'Inverno and Russel-Clark, 1971; Dautcourt and Jann, 1980). The same is true for the Segre classification.

Most modern techniques for finding new, physically interesting metrics are based on moving frames, the Newman-Penrose formalism or similar formalisms. A frequently used method involves specifying the properties of the spin coefficients (e.g. assume some of them to be zero and require that the remainder have derivatives only with respect to certain tetrad direction; then match this to certain properties of the Weyl spinor, PSI, the (tracefree) Ricci spinor, PHI, and the curvature scalar LAMBDA). However, for most of the known generating techniques (Kramer et al., 1980, chapter 30) no algebraic computing program has been written.

A slightly more complicated application is where power series expansions are involved as in the case where one makes a power series expansion about a known exact solution (Levy, 1968; Cohen, 1976). Many attempts have been made to find the metric expressed as a power series starting from a given energy momentum tensor in general relativity. One of these methods has been programmed (Åman, 1977).

Not all calculations with tensors involve explicit tensor components; one is often interested in regarding tensors as formal objects with certain properties as in the following (simple) example:

prove that $g^{ij}R_{ijkl}=0$ if R_{ijkl} is the Riemannian tensor and g^{ij} is symmetric.

This type of automatic tensor algebra has not been performed much so far for the simple reason that algebraic computing programs that do tensor algebra are quite difficult to write. There are now two systems available that are able to handle tensors as abstract objects, rather than just doing calculations with explicit tensor components. MACSYMA has a simple package by Pavelle (1) and SHEEP has an advanced but partly experimental package by Hornfeldt (1983). Examples of the use of automatic tensor algebra can be found in Pavelle and Hornfeld, 1980.

5. SOME RECENT USES OF ALGEBRAIC COMPUTING IN GENERAL RELATIVITY

Through the years, algebraic computing has been used fairly regularly to manipulate axisymmetric solutions. Examples of this are the work of Hoenselaers

(Hoenselaers, 1979, 1981, 1982a; Hoenselaers and Ernst, 1983), Ernst (Hoenselaers and Ernst, 1983) and Mc Crea (Jordan and Mc Crea, 1982; Mc Crea, 1981, 1982a). Hoenselaers has developed a program POLYNOM (Hoenselaers, 1982b) that, although limited in what it can do, is very efficient in calculating the enormous polynomials that appear in, for example, the Tomimatsu-Sato metrics of higher order. He has used this to discuss, alone and together with Ernst, the properties of the Tomimatsu-Sato metrics. In other works he discusses some metrics derivable with the help of transformations of the Ernst equation. Ernst has also developed a program (Ernst, 1983). His program runs on a single user micro-computer and handles differential forms. Mc Crea discusses some cylindrical solutions and possible Petrov types near a singularity. For his calculations he has developed a REDUCE program for the Newman-Penrose formalism (Mc Crea, 1982b).

Another area of exact solutions, where algebraic computing has been used, is in the discussion of (mainly cosmological) solutions of specific Bianchi classes. Examples of this are the work of Wainwright (Wainwright 1979, 1981; Collins and Wainwright, 1982), Krasinski (1981, 1983a), Sviestins and Bradley (Bradley and Sviestins, 1983) and Rosquist (1983a,1983b). Wainwright has developed a package for CAMAL that does general relativity calculations (Wainwright, 1978), Krasinski has developed his own system ORTOCARTAN (Krasinski and Perkowski, 1981a, 1981b) and Sviestins, Bradley and Rosquist use SHEEP.

Hehl, Baekler and Lenzen (Hehl and Lenzen, 1982; Baekler and Hehl, 1982) have studied lagrangians for theories with torsion and Poincaré gauge fields using ORTHOCARTAN (Krasinski and Perkowski,1981a, 1981b) and Mc Crea's programs (Mc Crea, 1982b) in REDUCE. Mc Crea has studied some solutions in connection with these theories (Mc Crea, 1983).

We should also mention that Krasinski has studied Riemannian tensors with higher symmetry than the corresponding metric (Krasinski, 1983b) and that the ADM formalism has been programmed by Mossiaux et al. both in REDUCE and SHEEP (Moussiaux et al., 1982).

Calculations in quantum gravity are normally similar to calculations in other forms of quantum field theory. For this reason, the programs used are often programs developed especially for elementary particle physics like SCHOONSHIP (Strubbe, 1974) and REDUCE (Hearn, 1971; Grimm and Kuhnelt, 1980) (REDUCE is now a general purpose system but it started as a program for Feynman integrals). An example of this is the work of Capper and coworkers who have used SCHOONSCHIP in a series of articles, see e.g. Capper and Dulwich, 1982.

We conclude this section by giving a list of active users of algebraic computing in general relativity. We do not claim that this list is complete, but we have tried to give a representative selection. Hovever we know that this list is incomplete as regards MACSYMA users in USA, as it has proved impossible to obtain more information.

TABLE I. List of active users.

Address	Persons	Problems	Language
Munich FRG	C. Hoenselaers	Axisymmetric solutions	POLYNOM (own system)
Chicago USA	F.J. Ernst	Ernst equation	own micro-computer system
Waterloo Canada	J. Wainwright	Cosmological models Exact solutions	CAMAL (own GR program)
Konstanz FRG	A. Krasinski	Cosmological models	ORTHOCARTAN (own system)
Stockholm Sweden	J. Frick A. Åman L. Hornfeld	Classification Supergravity Perturbations	SHEEP (own system) (REDUCE)
QMC, London England	M. MacCallum H. Joly	Classification Exact solutions	SHEEP (CAMAL)
Dublin Ireland	D. McCrea	Exact solutions	REDUCE (own GR programs)
Southampton England	R. d'Inverno	Exact solutions	SHEEP
Namur Belgium	A. Moussiaux	ADM formalism	REDUCE, SHEEP
Cologne FRG	F.W. Hehl	Gauge fields	REDUCE, ORTHOCARTAN
MIT USA	R. Pavelle	Alternate theories	MACSYMA

QMC, London England	D. Capper	Quantum Gravity	SCHOONSCHIP
Potsdam GDR	G. Dautcourt K.P. Jann		REDUCE (own GR programs)
Leningrad USSR	S. Tarasevich		GRATOS (own FORTRAN based system)
Bern Switzerland	A. Held	Vacuum Petrov type N with twist	ADC (FORMAC)

6. THE EQUIVALENCE PROBLEM, A CASE HISTORY

One of the most important observations we have made, is that in order to apply algebraic computing to large problems, one ought to re-evaluate the mathematical method or theory used, from an algorithmic point of view. As an example of this we would like to briefly review the history of the development of the equivalence problem in general relativity, which illustrates the interaction of available computing facilities and theoretical development. By the equivalence problem we mean the problem of deciding whether two metrics locally describe the same geometry or not, i.e. if there exists a coordinate transformation transforming one into the other.

The equivalence problem has a long history. The first paper dealing with the problem was Christoffel's who in a paper published in 1879, described an algorithm that (for the nondegenerate case, but it was later generalized by others) could convert the set of differential equations describing the coordinate transformation to an equivalent set of algebraic equations. There is a built-in classification process in the algorithm: it finds the (local) invariants for each metric. By equating invariants for the two metrics one then obtains the algebraic equations that have to be satisfied for the two metrics to be equivalent. Whether this set of algebraic equations have a solution or not cannot be determined algorithmically but the problem is considerably simpler than the original one.

Christoffel's algorithm can in the worst case involve computing the $n(n+1)$-th covariant derivative of the Riemann tensor of a n-dimensional space. By reformulating the algorithm using differential forms, Cartan (1946) simplified it so that the worst case "only" involves taking $n(n+1)/2$ covariant derivatives.

Even in the best case the first covariant derivative must be computed, something that often requires long calculations. This algorithm was a nice piece of mathematics, but it was not of any practical use because of the enormous calculations needed.

Now enter the algebraic computing programs. In the early 70's our group in Stockholm investigated the possibility of using the then available algebraic computing programs to implement the Cartan algorithm. No implementation was made at this time as the calculations would have been too lengthy for the computer. Instead we proceeded in two ways: a more efficient computer language, SHEEP (d'Inverno and Frick, 1982), was developed by Frick and at the same time a more efficient form of Cartan's algorithm was searched for.

It turned out that by about 1965 Brans (1965) had written some papers on the subject, where he makes an important improvement: the n variables and the n(n-1)/2 tetrad rotation parameters are treated separately.Using this, Karlhede (1980) recast the algorithm for the case of four dimensions,in the language of spinors. He viewed the classification process as a continuation of the Petrov classification. By using Bran's idea, Karlhede proved that at most 7 covariant derivatives would be necessary in the worst case (except possibly for conformally flat solutions). In practice it has never been necessary to compute more than the second covariant derivative. When Åman (Åman and Karlhede, 1981) then implemented the classification, he had an efficient algorithm and at the same time a suitable computer language in which to implement it. During the implementation, the algorithm was further improved (in cooperation with Karlhede) to calculate only the symmetrized parts of spinors whenever satisfactory (Åman and Karlhede, 1981). The original implementation could only handle vacuum solutions but a later addition by Joly and MacCallum allows non-vacuum solutions to be handled as well.

As a result of all this, we now have a set of computer programs that can classify metrics by giving all their local invariants (Åman,1982; MacCallum,1983a). These programs are efficient enough to be able to handle quite complicated metrics with available computer resources. It has already been used to classify a number of metrics (Karlhede and Åman, 1981). These computer programs will now be used to build a database classifying the known metrics. This project will be undertaken jointly by d'Inverno at Southampton and MacCallum at QMC, London,in close collaboration with the present authors. For a proposed format for this database see MacCallum (1983b).One use of this database will be to check whether a new metric really is new.

In conclusion: to be able to implement a useful classification method it would not have been enough to simply have developed SHEEP or just improved the algorithm, both were necessary.

7. COMPARISON OF SYSTEMS

At this point one might ask what are the available computer systems for performing algebraic calculations? We will in this section try to answer this question from the point of view of a worker in general relativity.

Firstly, we must point out that quite a few such systems have been developed over the past twenty years. There have been several reasons for this. Up to about five years ago, many a potential user, after investigating the field, found that none of the existing systems were available on local computers. Alternatively, it was often soon discovered that even if an algebraic system was available, it was not able to do the desired calculations. In both cases this has sometimes led to an algebraic system being developed within the department of the intended user. Most such projects have had a tendency to blossom for a few years and then die out,the final algebraic system seldom having been used outside its department of origin. However a few notable exceptions have actually broken this barrier and are still being used by various groups throughout the world.

Another category of system is the one which has been developed within a computer science department of a university or company. The driving force has here been more that of any research and development project, i.e. the belief that one can construct a product that in some way or ways is superior to all existing products. However, even here, algebraic computing projects have blossomed and died away.

When one is reviewing a system, it is important to keep this in mind. Basically this means establishing to what degree the said system is supported, i.e. being improved. To recommend a system, one would at last desire the situation whereby bugs (faults) discovered by users are corrected. However in the best case, one would like to use a system which has maximum support, that is, it is under constant improvement and development. Thus rather than tabulate all known algebraic systems, we have made a final selection of four systems which have active support groups.

In the following guide we have designated some algebraic systems as "large". By this we mean that polynomial expressions are represented in the computer in a generalized form. This in turn means that polynomial division is possible, which is the basic requirement for most advanced computer algebra like analytical integration. However, representing polynomials in this way, means a considerable increase in the time taken to perform a calculation. It also means that expressions take up much more computer memory. This coupled with the fact that such systems will invariably contain implementations of advanced computer algebra facilities, which themselves often require as much memory as the original systems, will result in a final system needing large amounts of memory. In contrast, "small" systems represent polynomials in an efficient way, and calculations will be much faster. Also as polynomial division is not available, they do not contain any extra facilities, which require much space. In fact, large systems can be around ten times slower than small systems and require over three times as much memory, but have many advantages as regards user facilities.

By implementation language we mean the language in which the algebraic system is written, STANDARD LISP, PSL and MACLISP are different dialects of LISP.

By interactive computing we mean the situation whereby the user types in just one command at a time which is then processed by the computer. For algebraic computing this is far more satisfactory than the usual alternative, whereby the whole program must be processed each time (called batch processing).

It is almost impossible to construct a program which works satisfactorily first time round. The process of correcting programs is called debugging, and it is important for the system to have routines which facilitate this work.

As regards the availability of systems, it should be pointed out that each algebraic system is only available on a limited number of computers and only works under certain operative systems. However as regards the four selected systems below, their availability is continually being increased. In our table we have given versions supported by the designers of the systems. Nonetheless, there usually exist versions for many other computers, so potential users who do not have access to precisely the computers shown in the table, can obtain more information from the respective designers.

In the following discussion we refer to descriptions of the systems, but note also the addresses in the appendix.

7.1. REDUCE

REDUCE (Hearn, 1971) is a large general purpose algebra system which has been under development for around 20 years. The philosophy of its creator is most attractive. He wishes REDUCE to be a multipurpose, multinational, generally available system containing implementation of the best algorithms at any given time. However, anyone who has come in contact with REDUCE will be aware of two major weaknesses. Firstly, a poor manual. Secondly, no applications library. The first fault has been rectified in the new version out this summer, and plans are under way to arrange a centrally distributed library. However as yet the system is not distributed with a collection of programs to perform calculations in general relativity. Many packages exist however, see Dautcourt and Jann, (1980) and addresses in the appendix.

7.2. MACSYMA

MACSYMA (1) is also a large general purpose system. It has been developed at MIT, which has meant that a formidable amount of intellectual expertise and effort has gone into its creation. It is very pleasant to use on-line, and has an excellent users' library including some routines for performing calculations in general relativity.

So far MACSYMA has only been available for some users in north America, but very recently its maintainance and development have been taken over by a private company, which should mean far greater availability but, alas, at a cost. It has been used very little up to now to perform calculations in general relativity, and thus it is difficult to judge its potential use in the field.

7.3. SHEEP

SHEEP (d'Inverno and Frick, 1982) is undoubtedly the most successful special purpose system designed to perform calculations in general relativity, and has the advantages of by far the most extensive set of programs for the field, see also Hornfeld,(1983) and Åman, (1982). It is under continual development and there are plans to make it available on the same range of computers for which REDUCE is available. Its main disadvantages are poor documentation and that (at present) it is only available on the computers of one manufacturer.

7.4. CAMAL

CAMAL (Fitch and Cohen, 1979) was developed over 10 years ago at Cambridge University, largely to perform calculations in celestial mechanics and general relativity. It is an efficient system using little computer memory and time even to do very large calculations. It is supported at the level that system bugs are corrected, but actual enhancement of the system is only taking place slowly. A package of programs to do calculations in general relativity has also been developed (Wainwright, 1978). The main disadvantage is that CAMAL is not interactive, i.e. programs must be run in batch.

In Table II we compare these four systems.

TABLE II. Comparison of some systems

	REDUCE	MACSYMA	SHEEP	CAMAL
Interactive	yes	yes	yes	no
Readability of output	fair	good	good	fair
GR packages	fairly good	good	very good	good
Users Documentation	good	very good	fair	good
Speed	average	average	fast	very fast
Memory Requirements	large	very large	average	small
Debugging Facilities	fairly good	very good	fair	good
Support	very good	very good	good	fair

Origin	theor. phys and comp. science dept.	comp. science dept.	theor. phys. dept.	comp. science dept.
Cost	USD 200	USD 1000	USD 100	USD 100
Implementation Language	STANDARD LISP and PSL	MACLISP	STANDARD LISP	BCPL
Machines Available	IBM, DIGITAL (VAX, PDP-20) Many others	HONEYWELL -MULTICS Symbolics (3600) DIGITAL (VAX, PDP-20)	DIGITAL (VAX, PDP-10/20)	IBM DIGITAL (VAX) Others

Besides these four algebraic systems there are a number of other systems which we ought at least to mention for the sake of completeness.

SCHOONSCHIP (Strubbe, 1974) is an oddity among algebraic systems. It is a small system which was developed at CERN to perform gigantic calculations in quantum electrodynamics. For calculations of this type, e.g. graviton scattering (Copper and Dulwich, 1982), it is far superior to other systems, not least in its ability to handle hundreds of thousands of terms. However there are many disadvantages. SCHOONSCHIP has not been maintained since the mid-seventies and is only available on CDC computers. It is not easy to learn how to use it, at least in comparison with other systems, and is completely unsuitable for "ordinary" algebraic calculations.

FORMAC (Bahr, 1977) is a small algebraic system for IBM computers. It was developed between 1963 and 1973. For those with some programming knowledge of FORTRAN, the time required to learn to use FORMAC is negligible and this has unfortunately proved to be a strong incentive for prospective users. However FORMAC can only perform very limited calculations as the basic design of the system is too primitive. SYMBAL (Engeli, 1975) is also a small algebraic system available on a number of leading computers and is extremely efficient. However, no development has taken place since 1975, and our impression is that SYMBAL has no apparent advantage if one compares it with CAMAL, whereas the latter is, as mentioned earlier, a special purpose system for general relativity. As regards the system SAC (Collins) and ALTRAN (Brown, 1977), they do not appear to be able to manage large calculations in physics.

We turn now to large general purpose systems, which naturally are far fewer. SCRATCHPAD (Jenks, 1974) was for many years regarded as the main rival to MACSYMA, but it would appear that this very large algebraic system has now been written off by IBM and will not be released for general use. Also the recent CALTECH project to create a large modern system with the name SMP (Cole and Wolfram, 1981) has run into difficulties, and it is not clear if and when and in what condition this system will be released.

We ought also to mention three recent small special purpose systems designed by relativists. They are, ORTHOCARTAN (Krasinski and Perkowski, 1981a, 1981b) which is similar to SHEEP, POLYNOM (Hoenselaers, 1982b), a very limited (but efficient) polynomial system and GRATOS (Tarasevich, 1981), a FORTRAN-based system. All are, at present, being enthusiastically developed by their respective creators, but it is too early to say if they will stay the course, and none at present is near the level of development of SHEEP or CAMAL.

Finally, we would mention a very important development, namely the implementation of algebraic system on micro-computers. The first such attempt was in 1979 (Stoutemyer, 1980), but cannot be considered of interest for general relativity due to memory limitations. However micro-computers with memories of the same order of magnitude as large computers are appearing, and work is under way to implement the leading algebraic systems on them, as well as construct new algebraic systems taking into account recent developments. In fact REDUCE is now available on a number of these large-address, 16-bit micro-computers, and one such personal computer is selling together with REDUCE for under USD 15.000 (Fitch), a price which should decrease considerably in the near future. Even relativists have begun to develop programs for their own personal computers, see Ernst, (1983) for a very interesting ongoing project. In fact we would expect the above discussion to be the last comparison of algebraic computer systems to worry about memory requirements, speed (just let your personal computer run all night) or availability. These properties will be less significant when, for little more than a month's salary, one can buy a computer capable of doing relativity calculations that would take months to do by hand. In a few years time you will all have such a machine on your desk!

8. CONCLUSIONS

Algebraic computing has continued to be used more and more in general relativity. Projects that would once have been considered intractable are now tackled. These projects are usually undertaken by people with some specialist knowledge also in algebraic computing. In fact, an important conclusion to be drawn from these projects is that one must be prepared to re-evaluate the mathematical method or theory used from an ALGORITHMIC point of view.

However the bulk of usage is to perform more routine calculations, although they sometimes may be rather large. These calculations require only a minimum knowledge of algebraic computing and computers able to do them will be available in a few years time on everyone's desk.

ACKNOWLEDGEMENTS. We wish to thank Bertel Laurent (Stockholm), Malcom MacCallum (QMC, London), and John Fitch (Bath), for helpful discussions. We wish also to thank Ray d'Inverno (Southampton) for much help in initiating this work.

APPENDIX

In this appendix we give addresses of persons to contact in order to obtain programs and further information:

REDUCE and more information is available from:
Anthony C. Hearn
The Rand Corporation
1700 Main Street
Santa Monica, CA 90406
USA
tel: +1 213 3930411

REDUCE Programs for general relativity
G. Dautcourt
Zentralinstitut für Astrophysik
der Akademie der Wissenschafte
1502 Potsdam-Babelsberg
DDR

REDUCE Programs for Newman-Penrose formalism:
J.D. McCrea
Dept. of Mathematical Physics

University College
Belfield
Dublin 4
Ireland

REDUCE Programs to do calculations with differential forms:
E. Schrufer
University of Bonn
Bonn
BRD

MACSYMA and more information is available from:
Richard Pavelle
Symbolics Inc.
257 Vassar Street
Cambridge, Mass 02139
USA
Tel: +1 617 8644660

SHEEP and more information is available from:
Inge Frick
Institute of Theoretical Physics
University of Stockholm
Vanadisv. 9
S-113 46 Stockholm
Sweden
Tel: +46 8 228160

CAMAL and more information is available from:
John Fitch
School of Mathematics
University of Bath
Claverton Down
Bath BA2 7AY
UK
Tel: +44 225 61244

CAMAL programs for general relativity:
John Wainwright
Department of Applied Mathematics
University of Waterloo
Waterloo
Ontario N2L 3G1
Canada
Tel: +1 519 8851211

Information about SAC is available from:
G.E. Collins
Computer Science Department

University of Wisconsin
1210 W. Dayton Street
Madison, Wisconsin 53706
USA

Information about ALTRAN is available from:
Computing Information Library
Bell Telephone Laboratory
600 Mountain Ave.
Murray Hill, New Jersey 07974
USA

Information about ORTHOCARTAN is available from:
Andrzej Krasinski
N. Copernicus Astronomical Centre
Polish Academy of Science
Bartycka 18
00716 Warzawa
Poland

Information about POLYNOM is available from:
Cornelius Hoenselaers
Max-Planck-Institut für Astrophysik
Karl-Swarzschild-Strasse 1
8046 Garching bei München
BRD

Information about GRATOS is available from:
S.V. Tarasevich
Inst. for Theoretical Astronomy
10 Kutuzov Quay
Leningrad 191187
USSR

Information about FORMAC is available from:
Kurt Bahr
Gesellschaft für Mathematik
und Datenverarbeitung G/IFV
Rheinstrasse 75
D-6100 Darmstadt
BRD

Information about SYMBAL is available from:
Max Engeli
FIDES Trust Company
P.O. Box 656
CH-8027 Zurich
Switzerland

NOTES

* The lecture was delivered at the Conference by I. Cohen.

(1) This was adversited in Physics Today, May 1983.

REFERENCES

Åman, J.E.: 1977, "A computer program for Synge's approximation method", *Abstracts of GR8 Contributed Papers*, University of Waterloo, Waterloo, p. 359.

Åman, J.E.: 1982, "Manual for CLASSI - Classification programs for geometries in General Relativity", Preprint (Stockholm).

Åman, J.E., and Karlhede, A.: 1980, "A computer-aided complete classification of geometries in General Relativity. First results", *Phys. Lett.*, 80 A, 229

Åman, J.E., and Karlhede, A.: 1981, "An algorithmic classification of geometries in General Relativity", *Proceedings of the 1981 ACM Symposium on Symbolic and Algebraic Computation*

Baekler, P., and Hehl, F.W.: 1982, "A micro-deSitter spacetime with constant torsion: a new vacuum solution of the Poincaré gauge field theory", *International Symposium on Gauge Theory and Gravitation*, Nara, Japan, August 1982.

Bahr, K.A.: 1977, "New release of FORMAC73 available", SIGSAM Bull., 11, No. 4, 6.

Barton, D., and Fitch, J.P.: 1972, "Applications of algebraic manipulation programs in physics", *Rep. Prog. Phys.*, 35, 235

Bertotti, B., de Felice, F., and Pascolini, A.(eds): 1983, *10th International Conference on General Relativity and Gravitation. Contributed Papers*, CNR, Rome, Vols. 1-2

Bradley, J.M., and Sviestins, E.: 1983, "Some rotating, time-dependent Bianchi type VIII cosmologies with heat flow", *Gen. Rel. Grav.* to be published.

Brans, C.H.: 1965, "Invariant approach to the geometry of spaces in general relativity", *J. Math. Phys.*, 6, 94.

Brown, W.S.: 1977, "ALTRAN user's manual", Bell Laboratories Report.

Buchberger, B., Collins, G.E., and Loos, R. (eds.): 1982, *Computer Algebra*, Computing Supp., 4, Springer, Berlin.

Campbell, S.J., and Wainwright, J.: 1977, "Algebraic computing and the Newman-Penrose formalism in general relativity", *Gen. Rel. Grav.*, 8, 987.

Capper, D.M., and Dulwich, J.J.: 1982, "On the one-loop finiteness of quantum gravity off mass-shell", Preprint, Queen Mary College, London QMC-82-14.

Cartan, E.: 1946, *Leçons sur la Geometrie des Espaces de Riemann*, Gautier-Villars, Paris.

Cohen, H.I., Lehringe, O. and Sundblad, Y.: 1976, "The use of algebraic computing in general relativity", *Gen. Rel. Grav.*, 7, 269.

Cohen, I.: 1976, "Slowly pulsating axisymmetric systems in general relativity", *J. Comput. Phys.*, 23, 397.

Cole, C.A., and Wolfram, S.: 1981, "SMP Handbook", Caltech Report.

Collins, C.B., and Wainwright, J.: 1982, "On the role of shear in general relativistic cosmological models", Preprint University of Waterloo.

Collins, G.E.: Unpublished programs.

Dautcourt, G., and Jann, K.P.: 1980 "User's guide to REDUCE subroutines for algebraic computations in general relativity". Preprint (Potsdam and Leipzig".

d'Inverno, R.A.: 1975, "Algebraic computing in general relativity", *Gen. Rel. Grav.* 6, 567.

d'Inverno, R.A.: 1980, "A review of algebraic computing in general relativity", in A. Held (ed.) *General Relativity and Gravitation*, Vols. 1-2, Plenum, New York.

d'Inverno, R.A.: 1983, "Computer methods in general relativity" in E. Schmutzer (ed.) *Proceedings of the 9th International Conference on General Relativity and Gravitation*, VEB Deutscher Verlag der Wissenschaften, Berlin (DDR), p. 93.

d'Inverno, R.A. and Frick, I.: 1982, "Interacting with SHEEP", *Gen. Rel. Grav.*, 14, 935.

d'Inverno, R.A., and Russel-Clark R.A.: 1971,"Classification of the Harrison metrics", *J. Math. Phys.*, 12, 1258.

Engeli, M.E.: 1975, "An enhanced SYMBAL system", *SIGSAM Bull.*, 9, No. 4, 21.

Ernst, F.J.: 1983, Illinois Institute of Technology preprint.

Fitch, J.P.: 1983, Private communication.

Fitch, J.P., and Cohen, H.I.: 1979, "Using CAMAL for algebraic computing in general relativity", *Gen. Rel. Grav.*, 11, 411.

Fletcher, J.G.: 1965, "GRAD ASSISTANT", report UCRL-14624-T.

Grimm, R., and Kuhnelt, H.: 1980, "Using REDUCE in problems of supersymmetry and supergravity", *Comput. Phys. Commun.*, 20, 77.

Hearn, A.C.: 1971, "Applications of symbol manipulation in theoretical physics", *Commun. ACM*, 9, 511.

Hehl, F.W., and Lenzen, H.J.: 1982, "Vacuum solutions with double duality properties of the Poincaré gauge field theory. II". in *Third Marcel Grossmann Meeting on the Recent Developments in General Relativity*, Shanghai, August 1982.

Hoenselaers, C.: 1979, "Weyl conform tensor of the Tomimatsu-Sato $\delta=3$ metric", *Gen. Rel. Grav.*, 11, 325.

Hoenselaers, C.: 1982a, "Brute force method for solving Ernst's equations and limits of the Kinnersly-Chitre solution", *J. Phys.*, A15, 3531.

Hoenselaers, C.: 1982b, "POLYNOM", Preprint (Munchen).

Hoenselaers, C., and Ernst, F.J.: 1983, "Remarks on the Tomimatsu-Sato metrics", Preprint.

Hornfeldt, L.: 1983, "Manual for STENSOR", Preprint (Stockholm).

Jenks, R.D.: 1974, "The SCRATCHPAD language", *SIGSAM Bull.*, 8, No. 2, 20.

Jordan, S.R., and McCrea, J.D.: 1982, "The gravitational field of a rotating infinite cylindrical shell", *J. Phys.*, A 15, 1807.

Karlhede, A.: 1980, "A review of the geometrical equivalence of metrics in general relativity", *Gen. Rel. Grav.* 12, 693.

Karlhede, A., and Åman, J.E.: 1981, "A classification of the vacuum D metrics", Preprint (Stockholm).

Kramer, D., Stephani, H., MacCallum, M.A.H., and Hertl, E.: 1980, *Exact Solutions of Einstein's Field Equations*, Deutscher Verlag der Wissenschaften and Cambridge University Press.

Krasinski, A.: 1981, "Spacetimes with spherically symmetric hypersurfaces", *Gen. Rel. Grav.*, 13 1021.

Krasinski, A.: 1983a, "The universe with varying topology of spacial slices", in Bertotti et al., 1983, p. 841.

Krasinski, A.: 1983, "Symmetries of the Riemann tensor", in Bertotti et al., 1983, p. 290.

Krasinski, A., and Perkowski, M.: 1981a, "ORTOCARTAN - a new computer program for analytical calculations in general relativity", *Gen. Rel. Grav.*, 13, 67.

Krasinski, A., and Perkowski, M.: 1981b, "ORTOCARTAN - a new computer program for analytical calculations in general relativity", *Comput. Phys. Commun.*, 22, 269.

Levy, H.: 1968, "Gravitational induction", *Proc. Cambridge Philos. Soc.*, 64, 1081.

MacCallum, M.A.H.: 1983a, "Classifying metrics in theory and practice", in V. De Sabbata and E.

Schmutzer (eds.), *Unified Field Theories of more than 4 Dimension Including Exact Solutions*, Proceedings of the 8th school of cosmology and gravitation at Erice, World Scientific Publishing, Singapore, p. 352.

MacCallum, M.A.H.: 1983b,"Proposed format for recording the invariant characterisation of exact solutions", in Bertotti et al., 1983, p. 301

McCrea, J.D.: 1981, "The Petrov type of a static vacuum space-time near a normal-dominated singularity", *J. Phys.*, A 14, 1351.

McCrea, J.D.: 1982, "A stationary cylindrically-symmetric electrovac space-time", *J. Phys.*, A 15 1587.

McCrea, J.D.: 1982, Unpublished programs.

McCrea, J.D.: 1983, "Static vacuum, cylindrical and plane symmetric solutions of the quadratic Poincarre field equation", *J. Phys.*, to appear.

Moussiaux, A., Tombal, P. and Demaret, J.: 1982, "Algebraic computation of Hamiltonian formalism in general relativity" *Gen. Rel. Grav.*, to be published.

Pavelle, R., and Hornfeld, L.: 1980, "Gauss-Bonnet invariants and indicial tensor manipulation on MACSYMA" in E. Schmutzer (ed.) *Proceedings of the 9th International Conference on General Relativity and Gravitation*, VEB Deutscher Verlag der Wissenschaften, Berlin (DDR), p. 111.

Pavelle, R., Rothstein, M., and Fitch, J.: 1981, "Computer algebra", *Scientific American*, 245, No. 6, 136.

Rosquist, K.: 1983a,"Exact rotating and expanding radiation-filled universe", *Phys. Lett.*, A 97, 145.

Rosquist, K.: 1983b, "Regularized field equations for Bianchi type II spatially homogeneous cosmology", *Class. and Quant. Grav.*, to be published.

Stoutemyer, D.R.: 1980, "PICOMATH-80", *SIGSAM Bull.*, 14, No. 3, 5.

Strubbe, H.: 1974, "Manual for SCHOONSCHIP, A CDC 6000/7000 program for symbolic evaluation of algebraic expressions", *Comput. Phys. Commun.*, 8, 1.

Tarasevich, S.V.: 1981, "GRATOS - The general relativity analytical tensor operations system", Manual (Leningrad)

Wainwright, J.: 1978, "CAMAL programs for GRT: a user's guide, Report, University of Waterloo, Waterloo.

Wainwright, J.: 1979, "Gravitational wave pulse in a spationally homogeneous universe", *Phys. Rev.*, D 20, 3031.

Wainwright, J.: 1981, "Exact spatially inhomogeneous cosmologies", *J. Phys.*, A 14, 1131.

Wainwright, J., and Goode, S.W.: 1980, "Some exact inhomogeneous cosmologies with equation of state $\rho=\sigma$", *Phys. Rev.*, D 22, 1906.

Yun, D.Y.Y., and Stoutemyer, D.R.: 1980, "Symbolic mathematical computation", in Belzer et al. (eds.) *Encyclopedia of Computer Science and Technology* Vol. 15, p. 235.

COMPUTATIONAL RELATIVITY: NUMERICAL AND ALGEBRAIC APPROACHES
REPORT OF WORKSHOP A4

Larry Smarr*

Departments of Astronomy and Physics
University of Illinois
Urbana, IL 61801, USA

The current use of computers in general relativity was reviewed at GR10. Today several groups have working codes which can calculate the coupled general relativity/hydrodynamics equations in two spatial dimensions plus time. These codes are compared and a critical review is made of the areas needing improvement. The potential uses of numerical relativity are discussed. In some situations the gravitational field can be chosen to be an analytically known solution, and hydrodynamics can be done in that background field. Regge calculus is still in its infancy numerically, but interesting new work has been done. Algebraic computing is becoming a mature technology, with the emphasis moving toward making these powerful software tools available to those relativists without previous computer training.

1. INTRODUCTION

There were two sessions of workshop A4 on numerical relativity at the Tenth International Conference on General Relativity and Gravitation . As chairman of that workshop, I am writing a summary of the state of the use of computers in general relativity. Rather than summarize the papers given (abstracts are published in Bertotti et al., 1983), I will attempt to outline the major efforts in this area and give a prospectus for the future. Because of the unending exponentiation in the power of computers, it is clear that relativity

B. Bertotti et al. (eds.), General Relativity and Gravitation, 163–183.

will be transformed just as so many other fields have been. However, relativity has been more resistant than other fields because of its proud heritage of analytic methods. What our field must recognize is that computational tools are not enemies of analytic tools. Indeed, each needs the other if continued progress is to be made. I believe that this marriage is occurring in each of the subfields I will survey: numerical relativity, relativistic hydrodynamics, Regge calculus, and algebraic computing. My review will emphasize the first topic, because there is a very good review talk on algebraic computing in this volume by Ian Cohen.

2. NUMERICAL RELATIVITY

Numerical relativity discretizes the spacetime continuum manifold into a spacetime lattice, the events of which are organized into a spacetime coordinate system. The differential Einstein equations have their derivatives replaced by a finite difference approximation. This transforms the differential field equations into a nonlinear coupled system of algebraic equations. The unknown variables are the components of the 4-metric tensor at the lattice points. As the lattice is refined to smaller scales, the solutions converge to the continuum solution of Einstein's differential equations. Note that, in contrast to the Regge calculus discussed in Section 4, the algebraic system of equations in numerical relativity are not the *exact* Einstein field equations on the chosen spacetime lattice.

In practice, this lattice of events is built up from an initial data hypersurface by constructing a family of hypersurfaces to the future of that data. There are two fundamentally different approaches to this construction: the spacelike and the null. In the first, the spacetime is built up from spacelike hypersurfaces starting on a Cauchy initial value surface (for a detailed definition see Eardley and Smarr, 1979). In the second, characteristic initial data is evolved using null hypersurfaces. Since there were no results on the null approach reported at GR10, I shall not discuss this important technique here, but refer the interested reader to papers by Stewart and Friedrich (1982) and Isaacson et al. (1983).

2.1. One Spatial Dimension Codes

There are well documented codes which can solve the Einstein/hydrodynamics equations in one spatial dimension plus time. The time slicing is maximal for noncosmological spacetimes or spatially constant mean extrinsic curvature for cosmologies. The details of the equations and numerical techniques depend on the topology of the space slices. For examples see: SPHERICAL (Shapiro and Teukolsky, 1980), PLANAR (Centrella and Wilson, 1983a,b), and CYLINDRICAL (Piran, 1980). If the reader is not familiar with the 3+1 split of spacetime terminology which is used in this section, consult the review article of York (1979).

The use of these codes has been manyfold: 1) to test out differential (i.e. maximal slicing) or numerical (i.e. finite differencing schemes) techniques for use in multidimensional codes, 2) to elucidate the formation process of neutron stars and black holes, 3) to model gravitational radiation generation in the extreme of highly nonspherical collapse (e.g. cylindrical), 4) to calculate relativistic shock wave formation and interactions, and 5) to study highly anisotropic or inhomogeneous cosmology. Although these codes are physically highly idealized, they can yield extremely accurate solutions to fully relativistic problems. They can be used on small computers by researchers with a minimum of training. There is a great future in such codes which I will come back to in Section 2.4.

2.2. Two Spatial Dimension Codes

My emphasis will be on the multidimensional codes in which at least two spatial dimensions can be represented. These allow one to determine the spacetime representing nonspherical isolated gravitating systems with finite energy, which can radiate gravitational waves. The stress tensor source may vanish, as in black holes represented by Einstein-Rosen bridges or in pure gravitational wave spacetimes, or the stress tensor may be given by some matter field. In the latter case, the evolution equations for the matter must also be solved. Because the vacuum problem has been extensively reviewed elsewhere (Eppley, 1979; Smarr, 1979), I will concentrate only on the matter source codes.

There are two major groups (for shorthand the US and the Japanese groups) which have built Einstein/hydro codes and whose work was reviewed at GR10. These two groups have taken nearly opposite approaches which illustrate the wide variety of choice one has in the construction of numerical relativity codes. References to two groups work can be found in Evans (1984), Evans et al., (1984), Nakamura (1981, 1983).

Both codes assume a perfect fluid, typically with

a gamma law equation of state. Although both groups codes are axisymmetric, the US group (Evans, Smarr, and Wilson) does not allow for rotation, while the Japanese group (Nakamura, Sato, Maeda, et al.) does. Both groups use maximal slicing, with the Japanese group experimenting some with a hypergeometric slicing. The slicing condition sets the trace of the extrinsic curvature tensor equal to zero and forces the lapse function to satisfy an elliptic differential equation.

The selection of shift vectors leads to a dramatic difference between the form of the equations used by the two groups. The Japanese group sets the shift equal to zero, which, in their rotating case, means that all six of the 3-metric components are nonzero. This leads to a very large number of terms in any quantity, such as the 3-Ricci tensor, which appears in the Einstein equations. In contrast, the US group uses the "Wilson gauge" to set the shift vector. This gauge, in the non-rotating case, requires only two nonzero 3-metric components. These can be naturally chosen to be the overall 3-metric conformal factor and the Brill (1959) wave function q. Thus, in a spherical spacetime the radial coordinate is just the conformal Schwarzschild coordinate (with $q=0$).In a pure gravitational wave spacetime, the 3-metric is exactly that of the Brill wave on each spatial slice. This approach greatly aids the physical interpretation of the numerical results. As a consequence of their gauge choice, the Japanese do not have to solve for the shift vector, while the US group must solve a vector elliptic equation for the shift vector on each time slice.

Since the Einstein evolution equations are constrained, there is great flexibility in which set of equations one uses to solve for the components of the 3-metric (see review by Piran, 1980). The Japanese use a totally unconstrained evolution. This means that they solve six coupled pairs of hyperbolic equations for each of the 3-metric components and their corresponding extrinsic curvature components. The constraint equations are then evaluated on each time slice as a measure of deviation of the evolved data from the constrained path. The particular form of the evolution equations that they use is called "(2+1)+1" and is based on Geroch's (1971) three dimensional formalism for spacetimes with one (axial) Killing vector. The US group, in contrast, uses the full York (1979) formalism to turn the constraint equations into one scalar and one vector elliptic equation. These are solved on each time slice, leaving only one hyperbolic pair of evolution equations to be solved for. Thus, they perform a fully constrained evolution.

In summary, the current Japanese code follows directly from approach of a zero shift, totally unconstrained evolution, used by Eppley (1979) and Smarr (1979) in the two black hole collision problem. There is no attempt in this method to isolate the gravitational wave degrees of freedom in the form of the 3-metric. The US group follows instead from the opposite approach, pioneered by Wilson (1979), in which one tries to simplify the form of the 3-metric as much as possible from the outset. In particular, the method then naturally provides metric variables for the constraint equations to constrain and wave variables for the evolution equations to evolve. Which method is superior remains to be shown, I believe. There still is a large amount of work to do in which the two methods are tried on the same problem with identical sized grids. Hopefully, cooperation between these two groups in the future will allow these comparisons.

In addition to the differences between the gauge and the mix of evolution/constraint equations used by the two groups, there is also a major difference between the numerical aspects of the two programs. The Japanese group has used a cylindrical coordinate system with a coarse spatial grid of 28×28 zones in the (R,Z) direction. This does not allow adequate zoning in the radial direction to determine the gravitational field in any detail, particularly gravitational radiation in black hole formation problems. Also, their finite differencing is first order (donor cell hydro, Lax-Friedrich for Einstein equations) which is very diffusive on a coarse grid. Realizing the limitations of their current numerical scheme, the Japanese group does not attempt to determine gravitational radiation information, but instead uses the code in a survey mode to determine the final states of gravitational collapse for a wide range of initial data. One should note that it is very easy for their group to greatly improve the fineness of their grid zoning and the accuracy of their numerical algorithm for solving the evolution equations, while not having to change the form of their differential equations at all. After these improvements are made, we will have a much better idea of the utility of their method.

The US group uses spherical coordinates and a much finer radial zoning (100 radial grid zones). The hydro differencing is closer to second order (see Wilson, 1979) with careful attention given to expansions of all functions near the origin (Evans et al., 1984). This allows the code to be used in a higher accuracy mode to study particular problems in great detail, including gravitational radiation generation and propagation. Unfortunately, because of the large number of elliptic equations which must be solved for on each time slice, the code takes hours on a supercomputer for each spacetime evolved. Thus, it currently has not been used in survey mode to determine which problems are of most interest. Furthermore, it can not be used to calculate rotating problems. Clearly, what is needed is a combination of coarse survey runs to sketch out parameter space followed by the high accuracy runs on those spacetime which promise to be the most interesting.

2.3. The Crucial Role of Resolution

The issue of accuracy and resolution is a central and crucial one. The computation of gravitational radiation emission can be in error by a large factor, if one uses a coarse grid and low order hydro finite differencing. To illustrate the concerns see Figure 1. This shows a typical moment in a nonspherical gravitational collapse to a neutron star. Note that the core of the star has bounced along the z-axis and is rebounding outward, while most of the star in the equatorial plane is still collapsing inward. Consider also that the star is very inhomogeneous in density. This indicates that resolving the shape of the density contours and therefore the quadrupole moment of the star requires high spatial resolution. But since the luminosity in gravitational waves goes as $(\Delta Q)^2/(\Delta t)^6$, the time resolution must also be quite accurate. For instance, if the true time scale over which the quadrupole moment is reversed from oblate to prolate is a factor of two shorter than is resolved by the code, then the gravitational luminosity can be off by a factor of 64!

Even worse, Haugan et al. (1982) have shown that the relative phase of motion of each piece of matter in the star is crucial to the determination of the correct gravitational wave form. In their analysis, the coherent wave form has been built up from a lot of local oscillators. Destructive phase cancellation of the generated gravitational radiation can occur because of the light travel time differences across the star. Again, a poorly resolved matter flow will not have a hope of giving accurate enough relative phase information to ensure a correct final wave form.

Finally, almost every nonspherical gravitational collapse produces a curved shock front. As this shock

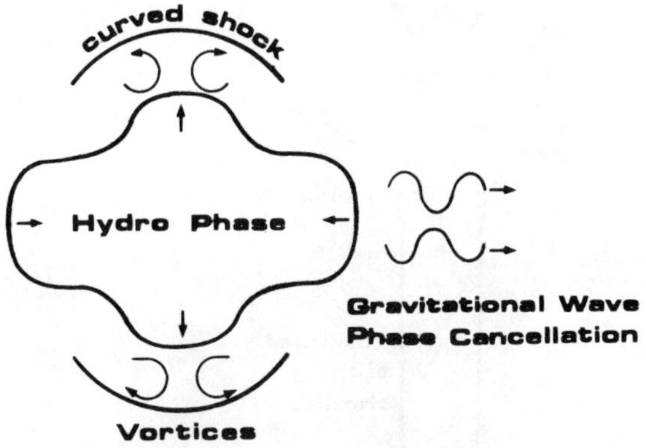

Fig. 1. A schematic diagram of a typical moment in the gravitational collapse of an oblate star to form a neutron star. The z-axis of symmetry is vertical. The core has just bounced and is rebounding along the z-axis. However, the collapse continues over most of the core. Outgoing curved shock fronts are generating vortices behind them. The outgoing gravitational waves are experiencing destructive cancellation because of light travel time differences across the core.

front propagates outward it reverses the direction of flow of the infalling matter. This means shocks can cause a very rapid change in the quadrupole moment and therefore a high gravitational radiation luminosity. Furthermore, curved shocks generate large scale vortices in the region behind the shock fronts. These rotational motions may have a critical effect on the emissivity in higher multipole moments. The correct resolution of the shape and strength of these shock fronts requires many grid zones. As an example of how a fine grid resolves shock fronts which are missed by a coarse grid, see Smarr and Hawley (1983).

In summary, while a coarse grid will allow one to calculate the evolution of initial data, the resulting approximation to the true spacetime may be very far off in both the energy radiated and the shape of the wave form produced.

There are even more difficulties when a black hole forms during the evolution. Figure 2 illustrates these new problems. The maximal slicing allows the time slices to penetrate inside the hole, while

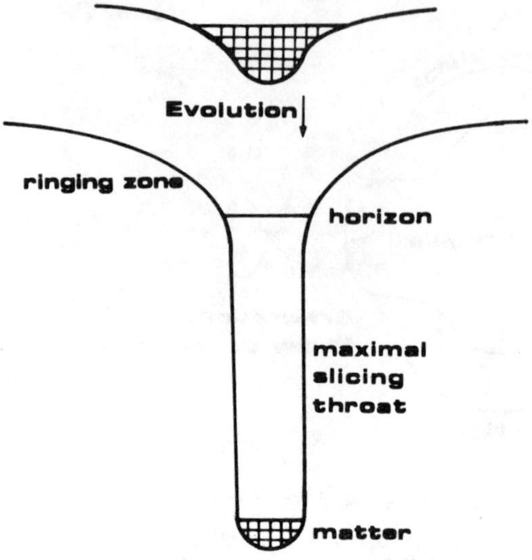

Fig. 2. Two phases of the evolution of the gravitational collapse of a star to form a black hole. The embedding diagram of the equatorial plane is shown. Maximal time slicing produces a long cylindrical throat with the matter at the bottom. The entire space from outside of the event horizon to about 15M is the wave generation or "ringing" zone.

avoiding the singularity to the future (see; e.g. Eardley and Smarr, 1979). The price one pays for this convenience is that the matter ends up at the bottom of a very long cylinder. The proper radius of this cylinder is 3M/2, while the length of the cylinder steadily increases with time (Eastabrook et al.,1973). The event horizon must be located somewhere above the cylinder. The location of either the true event horizon or the apparent horizon can only be determined numerically. From the event horizon out to about 15M, the gravitational radiation is generated in the "ringing zone" of the black hole. From perturbation theory (Press, 1971), we know that the wavelength of the radiation emerging from the hole will be 15-20M. This means we must locate our outer numerical boundary at least at 100M, if we are to follow the generation and propagation of a full wavetrain.

The numerical difficulties which are associated

with these physical and mathematical properties of black holes are extreme. The enormous stretching, during the evolution, of the proper distance from the center of the star to the outer grid point at 100M, means a very large number of radial grid points are going to be needed. These zones are "wasted" in the early phases of the calculation, but must be "held-in--reserve" for accurately resolving the final stretched configuration. Since the waves are redshifted as they emerge from the hole, there must be a higher density of grid points in the near hole region to resolve the shorter wavelength waves as they are generated. Similar remarks apply if one is going to accurately resolve the location, shape, area, and dynamics of the event horizon. It may be that sophisticated rezoning techniques will be needed if efficient and accurate black hole formation codes are to be created.

2.4. Future Numerical Relativity Codes

In many cases, Nature will not restrict herself to axisymmetry. As Miller (1983) emphasized in his presentation at GR10, many rotating initial data will result in collapse to a configuration with a Kerr parameter a/M>1. Almost surely, he argued, such a situation will lead to a nonaxisymmetric rotating bar which could be a copious producer of gravitational radiation. To model this four dimensional dynamic strong field spacetime is beyond the capability of any existing code. It is also beyond the capability of a Cray-1 supercomputer. Because all of the arguments on grid resolution made above apply to the third spatial dimension as well, it it clear that a fully 3-d code would require 50-100 times the storage and calculation time of today's 2-d codes. Fortunately, there is hope for attaining large increases in memory and speed in the next generation of computers, such as the Cray X-MP. These hardware increases coupled with continued algorithm improvements make accurate 3-d codes conceivable within the next five years.

In the meantime, calibration and higher resolution will be watchwords of 2-d codes. Now that there are several working codes, they must be turned into reliable tools for research in general relativity. All possible analytic or perturbation solutions, whether of flat spacetime or of relativistic objects, must be computed with the codes to get the finite difference approximations calibrated (see, e.g. Eppley, 1979). One should expect agreement to less than a percent everywhere. Horizon finders and gravitational wave form calculators must be put through rigorous documented tests. The same spacetime should be calculated at finer and finer resolution to determine when all the physics has been reliably determined. Ideally, several groups using different codes should calculate the same problem. As an example of this calibration procedure applied to general relativistic hydrodynamics, see Hawley et al., (1983b,c).

The 1-d codes can be made essentially analytic in their accuracy. They should be rewritten as time implicit codes with adaptive grids (see e.g. Winkler and Norman, 1984). This not only allows for extremely high spacetime resolution, but it also allows for several time scales or length scales to be present in the problem. That will be a critical feature as more microphysics (detailed equations of state, electromagnetic radiation coupled to hydro, viscosity, magnetic fields, etc.) is added to the 1-d codes. Because of their smaller size, one can add enough physics to 1-d codes to be able to study realistic astrophysical situations. The study of realistic gravitational collapse or early universe phenomenology is in its infancy. These 1-d codes will be powerful tools in the development of our knowledge in the next few years.

2.5. The Uses of Numerical Relativity

With the advent of high resolution calibrated numerical relativity codes, many issues of principle in general relativity can be attacked. The tool is the only one which can solve for the general globally hyperbolic solution of Einstein equations (Eardley and Smarr, 1979). Our knowledge of such general dynamic strong field spacetimes is primitive at present. Detailed gravitational waveforms can be calculated from a wide variety of situations. For the first time we will have a glimpse of what gravitational wave astronomy may uncover (see Drever's contribution). Already straight line motion of strong field bodies has been calculated (Smarr, 1979; Wilson, 1979). These spacetimes need to be recalculated with accurate codes and compared in detail with the work on equations of motion reported here by Damour. The quadrupole moment formalism controversy (see Walker's contribution) may find useful input when gravitational radiation is calculated from a numerical spacetime by the several contending methods. Since strong field spacetimes can also be calculated by numerical relativity, we will find which of the weak field formalisms break down first as the gravity becomes strong. Finally, the issues of singularity structure and cosmic censorship (see Christodoulou's contribution) will get many new examples from numerical relativity. By relaxing the maximal slicing condition, one can "sneak up" on spacelike singularities and perhaps determine if Cauchy horizons are being formed (Eardley and Smarr, 1979; Eardley, 1979).

Therefore, I conclude that numerical relativity is coming into its own. There is an enormous amount of work to do yet, but there seems to be no road blocks in the way. The field is currently manpower and computer resources limited, but the latter is changing for the better very rapidly. It seems likely that GR11 will see dramatic advances in numerical relativity.

3. RELATIVISTIC HYDRODYNAMICS

Another use of computers is to solve laws of physics in a given gravitational field. That is, the dynamics of the matter does not feed back through the Einstein equations to modify the spacetime curvature. This approximation needs to be justified by some physical reasoning in each separate instance. Fortunately, there are a large class of astrophysical situations where such an approximation is quite accurate. I will briefly mention a number of problems of this type which were reported on at GR10.

There is strong evidence that a supermassive (one hundred million solar mass) black hole exists at the center of many active galactic nuclei. Accretion onto this hole powers the enormous luminosity of quasars and the radio jets emerging from radio galaxies. Only a few solar masses a year need to accrete in order to reach quasar power levels. Therefore, it seems to be a good approximation to investigate the physics of matter in a fixed Kerr background.

Two competing methods were described for using computers to calculate the three dimensional problem of a star passing close to a black hole and being tidally distorted. Luminet and Marck (1983) reported a general relativistic improvement of a Newtonian model of Carter and Luminet. Here the star is assumed to maintain an ellipsoidal (3 unequal axes) shape and move on a geodesic around the hole. The relativistic equations for the tidal deviations and the bulk compression are computed along the trajectory. An alternate approach, reported by Bicknell and Gingold (1983) in session B1, uses a 3-d Newtonian smoothed particle hydro code to compute the full hydrodynamic behavior of the star. They find that the shape does not remain ellipsoidal and that shocks play a crucial role in preventing high central compressions. Unfortunately, their 3-d resolution is not adequate to get a definitive answer and they neglect relativistic effects in the hydrodynamics. In spite of the defects of the two approaches, it is very impressive to see the beginnings of 3-d dynamic Newtonian hydro in black hole spacetimes being attacked today.

John Hawley, in collaboration with Smarr and Wilson (1983a), reported in session B1 on a fully general relativistic hydrodynamics code which can calculate rotating axisymmetric fluid flow around a Kerr black hole. Although the code cannot calculate 3-d hydro, it does not suffer from the drawbacks of leaving out relativistic effects or having low resolution. The code is used to study gas accreting in bulk onto a black hole. When pressure forces are operative, it is found that strong intersecting shocks can form well inside r=6M. These shocks can reverse the flow of gas into the hole and form a rotating fat disk orbiting the hole. Future uses of the code will be to study the general accretion physics in the near hole region. Note that such a code can become a subroutine in a numerical relativity code to solve the axisymmetric rotating Einstein equations.

Multidimensional hydro codes are typically Eulerian, that is they fix the grid in space and let the matter flow past. Unfortunately, it turns out that in such an approach it is difficult to model special relativistic flows accurately. This is because the special relativistic equations are horribly nonlinear in the velocity. Therefore, little is known about the dynamics of matter moving very near the spead of light. Lanza et al. (1983) reported on a 1-d Lagrangian code (based on the approach of May and White, 1967), which can accurately model these flows. It assumes no self gravity (Minkowski spacetime background) and a simple radiation or Synge equation of state. Very interesting results on the formation and damping of relativistic shock waves were obtained. This approach should lead to a high resolution 1-d numerical relativity code which can study how gravity alters the picture.

Finally, fully 3-d cosmological numerical simulations are being carried out by Centrella and Melott and Davis et al. Here the gravitational background is assumed to be Friedmann in the large, but the 3-d dynamics of fragmentation of the matter field is calculated using Newtonian physics. This problem is of crucial importance because of the growing evidence, reported here by Sato, that the universe's dynamics is governed by dark matter. The nature of this dark matter (heavy neutrinos or some more exotic elementary particles) can be an input to the codes so that different fragmentation scenarios can be calculated and compared to observations. The numerical approach is not to model the dark matter by hydro, but rather as collisionless matter. The two groups have very different numerical approaches, which must be compared to determine which is more appropriate to model the physics of the dark matter.

4. REGGE CALCULUS

There were two very good papers delivered in this session on the Regge calculus. A review of the use of Regge calculus to obtain solutions of Einstein's equations was reported by Brewin (1983) and new techniques for solving the geodesic equations in a Regge spacetime were reported by Williams and Ellis (1983). Both projects require computers to numerically solve the resulting equations.

Regge calculus is often thought to be an alternate way to do numerical relativity. Actually, the two formalisms are very different. In Regge calculus the spacetime is divided up into 4-dimensional blocks *before* the Einstein equations are written down. The metric is assumed to be flat inside of these blocks. The unknown is not the metric tensor, as in numerical relativity, but instead a set of scalars that represent the proper lengths of the legs of the blocks. The gravitational action can be written down in terms of the properties of these legs and extremized *on the discrete spacetime*. Therefore, the resulting nonlinear algebraic system of equations are the *exact* Einstein field equations on that discrete spacetime. There is no need for a coordinate system to label events in the spacetime.

Only the simplest analytically known spacetimes have been calculated using this formalism: Friedmann, Tolman, and Kasner cosmologies and the static Schwarzschild black hole. In this regard, Regge calculus is where numerical relativity was over a decade ago. Part of the difficulty seems to be in choosing an appropriate set of 4-blocks. Brewin reported some advances on this issue in the context of the closed Friedmann model. If one demands that the blocks obey the same symmetry as the continuum space, then there are only three possible tetrahedra models which span the 3-sphere, N=5,16, and 600 (Collins and Williams,

1973). Brewin showed that the evolution of any such model has a finite endpoint some time before the final Big Crunch of the continuum solution. Furthermore, the radius versus time paths of the models are quite far from the analytic solution. Brewin proved that the first problem is generic to the Regge calculus, but the second problem can be relieved by softening the symmetry requirements. This procedure allowed him to subdivide each of the tetrahedra into twelve smaller tetrahedra. The N=7200 model was then quite close to the analytic model.

Williams and Ellis explained how one could use their method for tracing geodesics in a Regge spacetime. Within a flat block, the geodesic goes in a straight line. At the boundaries the geodesic suffers a "refraction" of a certain form. They studied light deflection, perihelion shifts, Thomas precession, and tidal forces in a Schwarzschild Regge geometry. Results were only close to correct if thousands of blocks were used, a remarkably similar finding to Brewin's. Furthermore, things could be very far off if they were not using the a priori knowledge of the Regge geometry. Again, it it not clear whether these difficulties are fundamental or only the result of poor resolution.

The major problem with Regge calculus is that no one seems to know what it is the answer to. It currently does not seem to be competitive with numerical relativity in constructing new solutions of Einstein equations. The fact that thousands of blocks must be used to get convergence to the true solutions, even in homogeneous or static spacetimes does not bode well for calculations of more complicated solutions. There has been no research on how matter shock waves will be treated, how gravitational waves can be resolved, what time slicings will work in black hole interiors, etc. As a result, one has no way to say today whether Regge calculus can reproduce the successes already achieved by numerical relativity. On the other hand, finite element methods, which Regge calculus seems related to, are enormously successful as a competitor to finite difference methods in hydrodynamics and other engineering simulations. Therefore, perhaps it is just a matter of more researchers working on the subject.

Certainly a formalism as elegant as the Regge calculus was meant as an answer to something important. Both Brewin and Ellis speculated that it might lie in the quantum gravity arena, where the spacetime continuum may need to be replaced by a discrete structure. If so then the computer may have a major contribution to make to that specialty of general relativity as well.

5. ALGEBRAIC COMPUTING

The ability to manipulate long strings of symbolic expressions is the heart of general relativity theory. Although many of us pride ourselves for our capacity to fill reams of paper without making mistakes, I do not believe that such a capability should be confused with creative science. Therefore, it was encouraging to see how much progress has been made in the development of easy-to-use algebraic computing software. Since a detailed invited talk by Ian Cohen on the subject is included in the proceedings, and since I am not an expert on the merits of the competing systems, I will make only a few remarks on how our workshop appeared to a relativist eager to use such packages.

What an algebraic computing program does is use a rulebook for algebra, differentiation, integration, tensor calculations, etc. to take input from you and convert it to some other form you desire. It does all the same intermediate steps you would do, but very much faster and without error.It does not substitute in numerical values for variables, but rather deals in strings of those variables. An example would be for you to type in the algebraic form of some metric and it would give back the expression for the Riemann tensor components. It can also check relationships between functions. For instance, it can compare the different components of the Riemann tensor to decide the Petrov class of the spacetime.

One of the most exciting new uses of such packages was described by MacCallum (1983a) at GR10. It is a procedure for classifying all exact solutions of the Einstein equations by their invariants. The project will catalog all known solutions, enabling the user to type in a metric he is interested in and have the program tell him what solutions are isometric to his. The current version of this project is using the algebraic package SHEEP.

Given these general properties of any algebraic package, what makes one package better than another? The first question is whether you can run it on the type of computer that you have access to. Most of these packages are written in LISP, a common general purpose list manipulation language. Unfortunately, there are many dialects of LISP and which version your machine will run may not be the dialect a relativity package was written in. This is referred to as the portability problem. Secondly, the different packages require varying amounts of memory to run and their speed can vary widely. It used to be that most discussions on the merits of different packages were centered on the memory requirements and speed. This does not seem to be as big an issue today because of great improvements in these properties by all the major packages.

The major issue today seems to be usability. We are seeing a shift from a small user base of experienced programmers to a much larger user base of relativists who are inexperienced in using computers. For this to happen, the authors of the package have to spend a lot of their time programming in "user friendliness". Examples of these features are: 1) ability to input and output in "textbook" notation, with familiar symbols, sub and superscripts, etc.; 2) interactive features where a user can decide at the keyboard what he wants next; 3) intelligent prompts in plain language to help the user through the calculation; and 4) online "help" packages so the user can teach himself how to use the package.

Equally important are clearly written user manuals with lots of worked examples. The support that the authors can give the user later is going to become more and more a deciding factor. If we are going to have hundreds of relativists using these packages, a one person support group is not going to be adequate. Thus, packages developed by a research group may become preferable.

The generality of the package has to be considered. Several general purpose packages which cover all types of physics and mathematics exist, such as MACSYMA, REDUCE, and SMP. Many relativists may never use large portions of these packages. However, they may pay for the extra generality in reduced speed and increased memory requirements. A more useful arrangement is to have greater modularity of the packages. A system would have a central "kernel" that contains the "symbol crunching" core of the package. Then in an overlying "shell" are modules that perform specific tasks when linked to the kernel. In this architecture, the user only has to load those modules that he needs for the calculation at hand. A nice step in this direction was the STENSOR package reported on by Hornfeldt of the Stockholm group. It can sit on top of either MACSYMA or SHEEP and manipulate tensor and spinor formulae in textbook conventions.

Given all the above considerations, it seems that the major contenders at this stage are SHEEP, CAMAL, REDUCE, and MACSYMA. A detailed comparison of these and the other packages is given by Cohen in his invited talk. Of these SHEEP seems to be very compact and fast, well supported, easy to obtain, and interactive. It is written just for relativists and can run STENSOR on top of it. For examples of its use see MacCallum (1983b).

The final general trend I will comment on is the revolution in hardware which is occurring today. Until a few years ago, all these packages had to be run on mainframe computers (costing millions of dollars), of which there might be one per university. Often these had to be run in batch mode, so the turn around was very slow. Then the superminis (costing hundreds of thousands of dollars), such as the DEC VAX series, became widely available. Today many departments or active timesharing operating systems. The user can ractive timesharing operating systems. The user can work at a terminal with fairly fast turn around if there are only a few users on the machine. What is appearing this year are the first of the next generation machines called scientific work stations (costing tens of thousands of dollars). These condense a supermini to a desk top machine owned by a single user. Examples are APOLLO, SUN, and the just announced microVAX. So what we are seeing is an evolution in which many research scientists will do most of their research, writing papers, communicating with colleagues, etc. on one machine sitting on their desk. This is happening today and will be commonplace in five years.

It is clear that the relativity algebraic computing packages will have to follow this hardware trend. Indeed, a number of these packages are being ported to the intelligent work stations. However, the integration of these tools into the scientists working environment on the work station must also be accomplished. Since these work stations can also act as post processing graphics work stations for the numerical relativity codes described in Section 2, the relativist of five years from now may not make a major distinction between algebraic and numerical relativity. He or she will simply call it computational relativity.

ACKNOWLEDGEMENTS. I would like to thank all the participants in the workshop on numerical relativity at GR10 for their lively discussions. My review of numerical relativity was greatly helped by Charles Evans and James Wilson. Special thanks to Peter Barnes for lettering the Figures. Financial support was partially provided by the NSF under grants PHY 80-01496 and PHY 83-08826

NOTE

* Alfred P.Sloan Fellow.

REFERENCES

Papers presented to the GR10 Conference and published in:

Bertotti, B., de Felice, F., and Pascolini, A. (eds): 1983, *10th International Conference on General Relativity and Gravitation. Contributed Papers*, Vols 1-2, CNR, Roma.

Brewin, L.: 1983, "Computer generated solutions using Regge calculus", p. 427.

Bicknell, G.V., and Gingold, R.A.: 1983, "Tidal detonation of stars by black holes", p. 690.

Hawley, J., Smarr, L., and Wilson, J.R.: 1983, "Numerical simulation of flat disks around black holes", p. 709.

Lanza, A., Miller, J.C., and Motta, S.: 1983, "Formation and damping of relativistic strong shocks", p. 436.

Luminet, J.P., and Marck, J.A.: 1983, "Tidal effects in Kerr geometry", p. 438.

MacCallum, M.A.H.: 1983a, "Proposed format for recording the invariant characterisation of exact solutions", p. 301.

Miller, J.C.: 1983, "Non-axisymmetric instabilities in gravitational collapse", p. 443.

Williams, R.M., and Ellis, G.F.R.: 1983, "Observational results from Regge calculus", p. 451.

Other references

Brill, D.R.: 1959, "On the Positive Definite Mass of

the Bondi-Weber-Wheeler Time-Symmetric Gravitational Waves", *Ann.Phys.*, 7, 466.

Centrella, J., and Wilson, J.R.: 1983a, "Planar numerical cosmology. I. The differential equations", *Astrophys.J.*, 273, 428.

Centrella, J., and Wilson, J.R.: 1983b, "Planar Cosmology. II. The difference equations and numerical tests", University of Illinois, Astronomy preprint 83-11.

Collins, P.A., and Williams, R.M.: 1973, "Dynamics of the Friedmann Universe Using Regge Calculus", *Phys. Rev.*, D7, 965.

Eardley, D.M.: 1979, "Global problems in numerical relativity", in L.Smarr (ed.), *Sources of Gravitational Radiation*, Cambridge University Press, Cambridge, pp. 127-138.

Eardley, D.M., and Smarr, L.: 1979, "Time functions in numerical relativity: marginally bound dust collapse", *Phys. Rev.*, D19, 2239.

Eppley, K.: 1979, "Pure gravitational waves", in L. Smarr (ed.), *Sources of Gravitational Radiation*, Cambridge University Press, Cambridge, pp.275-292.

Estabrook, F., Wahlquist, H., Christensen, S., DeWitt, B., Smarr, L., and Tsiang, E.: 1973, "Maximally Slicing a Black Hole", *Phys. Rev.*, D7, 2814.

Evans, C.: 1984, "A method for numerical simulation of gravitational collapse and gravitational radiation generation", in J.Centrella, R.Bowers, J.LeBlanc, and M. LeBlanc (eds), *Numerical Astrophysics: A Festschrift in Honor of James R. Wilson.*

Evans, C., Smarr, L., and Wilson, J.R.: 1984, "Numerical relativistic collapse and collisions with spatial time slices", in K.-H.A. Winkler, and M.L.Norman (eds), *Radiation Hydrodynamics*, Reidel, Dordrecht.

Geroch, R.: 1971, "A Method of Generating Solutions of Einstein's Equations", *J.Math. Phys.*, 12, 918.

Haugan, M.P., Shapiro, S.L., and Wasserman, I.: 1982, "The soppression of gravitational radiation from finite-size stars falling into black holes", *Astrophys. J.*, 257, 283.

Hawley, J.F., Smarr, L., and Wilson, J.R.: 1983b, "A numerical study of nonspherical black hole accretion. I. Equations and test particles", University of Illinois Astronomy preprint 83-9, to appear in *Astrophys. J.*

Hawley, J.F., Smarr, L., and Wilson, J.R.: 1983c, "A numerical study of nonspherical black hole accretion. II. Finite differencing and code calibration", University of Illinois, Astronomy preprint, to appear in *Astrophys. J. Suppl.*

Isaacson, R.A., Wellings, J.S., and Winicour, J.: 1983, "Null cone computation of gravitational radiation", *J.Math. Phys.*, 24, 1824.

MacCallum, M.A.H.: 1983b, "Classifying metrics in theory and practice", in V.De Sabbata, and E.Schmutzer (eds), *Unified Field Theory of more than 4 Dimension Including Exact Solutions*, World Scientific Publishing, Singapore, p. 352.

May, M.M., and White, R.H.: 1967, "Stellar Dynamics and Gravitational Collapse", *Methods Comput. Phys.*, 7, 219.

Nakamura, T.: 1981, "General Relativistic Collapse of Axially Symmetric Stars Leading to the Formation of Rotating Black Holes", *Prog. Theor. Phys.*, 65, 1876.

Nakamura, T.: 1983, "General relativistic collapse of rotating stars", to appear in the proceedings of the 11th Texas Symposium on Relativistic Astrophysics.

Piran, T.: 1980, "Numerical codes for cylindrical general relativistic systems", *J.Comput. Phys.*, 35, 254.

Press, W.H.: 1971, "Long wave trains of gravitational waves from a vibrating black hole", *Astrophys.J. Lett.*, 170, L105.

Shapiro, S.L., and Teukolsky, S.A.: 1980, "Gravitational collapse to neutron stars and black holes: computers generation of spherical spacetimes",

Astrophys.J., 235, 199.

Smarr, L.: 1979, "Gauge condition, radiation formulae and the two black hole collision", in L.Smarr (ed.), *Sources of Gravitational Radiation*, Cambridge University Press, Cambridge, pp.245-274.

Smarr, L., and Hawley, J.F.: 1983, "General Relativistic Hydrodynamics and accretion physics: a numerical approach", to appear in the Proceedings of the Toulouse Workshop on Gravitational Collapse.

Stewart, J.M., and Friedrich, H.: 1982, "Numerical Relativity. I. The characteristic initial value problem", *Proc. Roy. Soc.*, A384, 427.

Wilson, J.R.: 1979, "A numerical method for relativistic hydrodynamics", in L.Smarr (ed.), *Sources of Gravitational Radiation*, Cambridge University Press, Cambridge, pp. 423-446.

Winkler, K.-H.A., and Norman, M.L.: 1984, "WH80's: Numerical Radiation Hydrodynamics", to appear in M.L.Norman, and K.-H.A.Winkler (eds),*Radiation Hydrodynamics*, Reidel, Dordrecht.

York, J.W., Jr.: 1979: "Kinematics and dynamics of general relativity", in L.Smarr (ed.), *Sources of Gravitational Radiation*, Cambridge University Press, Cambridge, pp. 83-126.

POSITIVE ENERGY AND KALUZA-KLEIN THEORY*

Edward Witten

Joseph Henry Laboratories
Princeton University
Princeton, New Jersey 08544, USA

In most field theories that enter physics, the energy is defined as the integral of a gauge invariant physically meaningful energy density $T_{00}(x)$:

$$E=\int d^3x\, T_{00}(x)\,.$$

Ordinarily (perhaps after addition of a suitable constant) $T_{00}(x)$ is a positive definite functional of the fields. For instance, in Maxwell's theory

$$T_{00}=\frac{1}{8\pi}(\vec{E}^2+\vec{B}^2)\,.$$

From the positivity of $T_{00}(x)$ it - generally - follows at once that the vacuum is stable, and that there are no excitations of negative energy.

The outstanding exception to this is general relativity. In general relativity, there is no good local measure of the energy content of the gravitational field. This essentially follows from general covariance. In theories like general relativity, with second order field equations, the energy density is usually a function of the fields and their first derivatives only. But in general relativity the values at any point of $g_{\mu\nu}$ and $\partial g_{\mu\nu}/\partial x^\alpha$ contain no information whatsoever. It is only the *total energy* of a gravitating system that can be defined; and this is possible only for an isolated system in an asymptotically flat world (or a world that is asymptotic to a

B. Bertotti et al. (eds.), General Relativity and Gravitation, 185–197.

suitable Kaluza-Klein vacuum state).

For such a system the energy can be defined as a surface integral. Remember, in electrodynamics there are two ways to define the total electric charge. It can be defined as a volume integral

$$Q=\int d^3x\,\rho(x)$$

where $\rho(x)$ is the charge density. Or, since the electric field is related to ρ by $\nabla\cdot E=4\pi\rho$, we can write the charge as a surface integral over a large sphere at infinity:

$$Q=\frac{1}{4\pi}\oint d\Sigma^i E_i \ .$$

In general relativity, there is a similar surface integral representation of the total energy, the ADM formula:

$$E=P_0=\frac{1}{16\pi G}\int d\Sigma^i(\partial_j g_{ij}-\partial_i g_{jj}) \ .$$

For instance, for a metric that is asymptotic to the Schwarzschild metric

$$g_{ij}\xrightarrow{|x|\to\infty}\delta_{ij}(1+2GM/r)$$

a simple calculation shows E=M, as expected. There is a similar surface integral expression for the total momentum $\vec{P}$ of an isolated system. The energy and momentum are *conserved*, essentially because they are surface integrals.

Of course, by using the divergence theorem we can convert the surface integral into a volume integral

$$E=\frac{1}{16\pi G}\int d^3x(\partial_i\partial_j g_{ij}-\nabla^2 g_{ii}) \ .$$

(Actually, one is here assuming that space is *topologically* Euclidean so that the Riemanniann metric can be represented by a collection of functions $g_{ij}(x)$ in some global coordinate system). This formula, however, is peculiarly unenlightnening. The "energy density" $\partial_i\partial_j g_{ij}-\nabla^2 g_{jj}$ is non-covariant and non-positive definite. It can be related by means of the Einstein equations to additional untransparent formulas.

Now the question arises: under what conditions can we assert that the energy of an isolated system is positive in general relativity?

In the case of pure gravity (no matter present), we can certainly assert that $E \geqq 0$ for very *weak* gravitational waves. The energy of linearized gravitational waves is positive, as the standard formulas show. But what happens for *strong* gravitational fields? Or in the presence of matter? These questions have been the subject of systematic study for about 25 years (Weber and Wheeler, 1957; Araki, 1959; Brill, 1959; Trautman, 1962).

A general statement that energy is positive is certainly possible only if the laws governing the matter fields are such that the energy content of matter is non-negative. Otherwise, one could consider a situation in which the gravitational contribution is negligible and the matter contribution is negative.

We assume, then that the properties of matter are such that $T_{00}(x) \geqq 0$ at each x and in each local Lorentz frame. With this restriction, we consider a general asymptotically Minkowskian world (non-singular, or with singularities hidden by horizons) with $G_{\mu\nu}=8\pi G T_{\mu\nu}$. Is the energy necessarily positive? For a flavor of the problem, consider the Schwarzschild solution:

$$g_{00}=-(1-2GM/r)$$

$$g_{ij}=\delta_{ij}\frac{1}{1-2GM/r}$$

$$g_{0i}=0.$$

There is a singularity at r=0. Two cases arise:

(i) If M>0, this singularity is harmless, being hidden by a horizon. What is more, as is well known, the singularity can be removed by matching the exterior Schwarzschild solution onto an allowed interior metric. For instance, one can take the Schwarzschild solution for $r \geqq r_0$ and match it to the metric of a spherically symmetric planet of radius r_0 and of allowed $T_{00}(x) \geqq 0$.

(ii) If M<0, the singularity at r=0 is a naked singularity. It corresponds roughly to a singular mass point with $T_{00}=-M\delta^3(\vec{x})$. We wish to show that - if $T_{00}(x) \geqq 0$ for all x - there is no way to eliminate this singularity while preserving the asymptotic behaviour of the negative mass Schwarzschild solution. Thus, the claim is that the negative M Schwarzschild metric can't be matched to an allowed interior metric of $T_{00} \geq 0$.

Now, since energy is conserved, it is really a property of the fields not on the whole space-time, but on a given space-like surface, which we may regard as the initial value surface. The problem is this. Given an asymptotically Euclidean space-like three-surface, embedded in space-time so that on this surface $G_{\mu\nu}=8\pi GT_{\mu\nu}$ (with $T_{00}(x) \geq 0$ everywhere and in each locally inertial frame) can we show

$$E=\frac{1}{16\pi G}\int d\Sigma^i(\partial_j g_{ij}-\partial_i g_{jj})>0?$$

This was first shown in a general way a few years ago by Schoen and Yau (1979a, b, c). I would like to sketch their method.

In discussions of the positive energy problem, there is a somewhat simpler special case - the case of "time symmetric" initial value data. This refers to a situation in which at t=0, $T_{0i}=0=\partial g_{0i}/\partial t$ so that the kinetic energy vanishes both for matter and for gravitational field. Plausibly, non-zero kinetic energy would only *increase* the energy, and Schoen and Yau proved this, following a suggestion by Jang (1978). I'll just sketch their treatment of the time symmetric case.

For time symmetric initial data the Einstein equations $G_{\mu\nu}=8\pi GT_{\mu\nu}$ split as follows on the t=0 surface. The equations $G_{0i}=8\pi GT_{0i}$ are obeyed trivially because of the $t\leftrightarrow -t$ symmetry. The equations $G_{ij}=8\pi GT_{ij}$ determine the future time development but do not restrict the initial data. Finally, there is one constraint on the initial data. It is the equation $G_{00}=8\pi GT_{00}$ with $G_{00}=R_{00}-\frac{1}{2}g_{00}R=\frac{1}{2}R^{(3)}$, where $R^{(3)}$ is the scalar curvature of the initial value three-surface. The constraint is $R^{(3)}=16\pi GT_{00}$ and (since $T_{00} \geq 0$) the restriction on the initial data is that $R^{(3)} \geq 0$.

What must be proved, then, is that given a three dimensional metric with $g_{ij}\to\delta_{ij}(1+2GM/r)$ and $R \geq 0$, we have necessarily $M \geq 0$.

Now, in three dimensions it is not obvious how to prove this. But in two dimensions the corresponding statement is much easier. The Schoen-Yau idea is based on an induction in the number of dimensions. They give a procedure which, given a counter example to the positive energy theorem in n dimensions, will produce

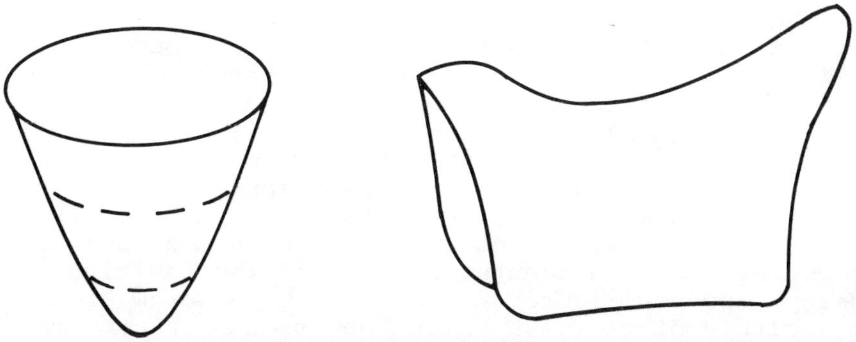

Fig. 1. Two dimensional spaces that are asymptotically flat need not be asymptotically Euclidean. They may be asymptotic to a cone or a saddle.

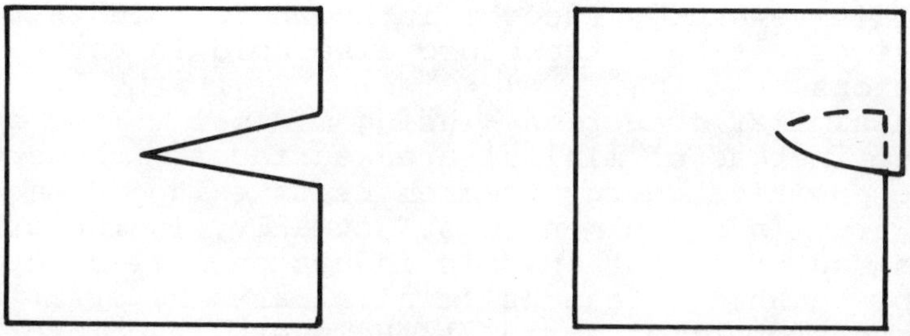

Fig. 2. Removing or adding a wedge from a plane to make a space of positive or negative deficit angle.

a counter example in n-1 dimensions.

In two dimensions a space that is asymptotically *flat* need not be asymptotically Euclidean. It may be asymptotic to a *cone* of positive or negative deficit angle (Figure 1). The cone consists of a plane that is cut open and has a wedge removed or added (Figure 2). The angle removed is called the deficit angle; if an angle is added, we call this a negative deficit angle. It turns out that the analogue of the ADM formula in two space dimensions is very simple. The mass is simply a constant times the deficit angle.

Now in two dimensions, $\sqrt{g}R$ is a total derivative whose integral can be expressed as a surface integral. And the Gauss-Bonnet theorem says that for an asymptotically Euclidean space

$$\text{Mass}=\text{deficit angle}=\text{constant}\cdot\int d^2x\sqrt{g}R.$$

So $M \geqq 0$ if $R=16\pi G T_{00} \geqq 0$.

How can this be generalized to three dimensions? The starting point is a theorem which is properly known as the solution of the Plateau problem but which we may refer to as the soap bubble theorem. The soap bubble theorem says that given any curve, there is a surface of least area spanning it. This surface can be realized physically (or at least approximated) by a soap bubble with the given curve as boundary.

Given a localized object in three dimensions that is claimed to be a counter example to the positive energy theorem (Figure 3), Schoen and Yau draw around it a circle of radius R which they span with a minimal surface. They show that if the object has negative mass this minimal surface converges as $R\to\infty$ to an asymptotically flat two manifold which violates the positive energy theorem in two dimensions if the original object violated it in three dimensions. Since the positive energy theorem is known to hold in two dimensions, it must therefore also hold in three dimensions.

A similar induction - using minimal volume hypersurfaces instead of minimal area surfaces - shows that if the positive energy theorem is true in d dimensions, it is true in d+1 dimensions. Actually, in the original Schoen-Yau presentation this induction failed beyond d=6 for technical reasons (minimal area hypersurfaces have bad properties for $d\geq 7$). However, I am informed (Yau, 1983) that this difficulty has been removed.

Several years ago, it was suggested by Deser and Teitelboim (1977) and by Grisaru (1978) that there might be another way to prove the positive energy

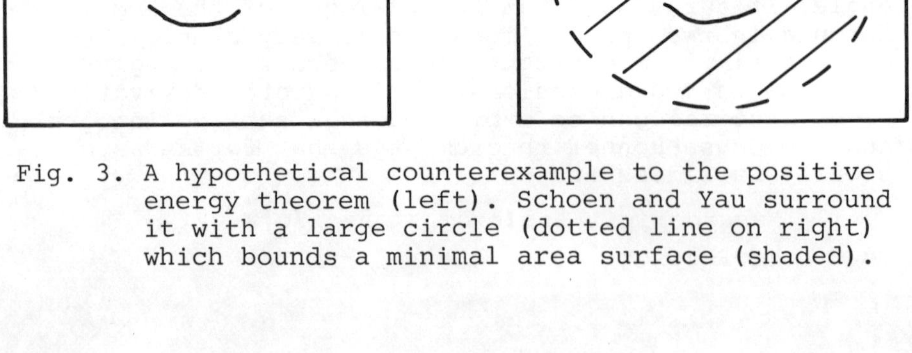

Fig. 3. A hypothetical counterexample to the positive energy theorem (left). Schoen and Yau surround it with a large circle (dotted line on right) which bounds a minimal area surface (shaded).

theorem, based on supergravity. Supergravity is a certain theory (actually a class of theories) of gravity coupled to spin 3/2 fermions. In this *quantum field theory* the Hamiltonian can formally be written as a sum of squares of Hermitian operators Q_α ("supercharges"):

$$H=\frac{1}{\hbar}\sum_{\alpha=1}^{4} Q_\alpha^2$$

so $H\geqq 0$ as an operator.

This one line proof of positivity of the energy is a formal argument in an unrenormalizable theory that may be nonsense. Perhaps the spin 3/2 fields do not exist in nature; supersymmetry may be wrong. We would like to understand the positivity of energy in standard, classical general relativity.

But the formal one line proof that $H\geqq 0$ in supergravity holds, formally, for any value of Planck's constant $\hbar$. So formally we have $H\geqq 0$ for any $\hbar$ and therefore also in the limit as $\hbar\to 0$. As $\hbar\to 0$, quantum field theory becomes classical field theory and fermions disappear. So the $\hbar\to 0$ limit of quantum supergravity is classical general relativity. If $H\geqq 0$ for any value of $\hbar$ in quantum supergravity, the positive energy theorem must hold in classical general relativity. And even if quantum supergravity is actually ill-defined because of the unrenormalizability, the $\hbar\to 0$ limit of the formal quantum argument may be a bona fide classical demonstration of the positive energy theorem. Let us see how this works.

To quantize supergravity about flat space, one expands the fields in a complete set of solutions of the linearized classical equations:

$$g_{\mu\nu}=\sum_i h_{\mu\nu}{}^{(i)}a^{(i)}+\text{h.c.}$$

$$\psi_{\mu\alpha}=\sum_j \chi_{\mu\alpha}{}^{(j)}b^{(j)}+\text{h.c.}$$

After quantization the $a^{(i)}, b^{(j)}$ and their hermitian conjugates become creation and annihilation operators for particles of spin 2 and spin 3/2. The supersymmetry operators Q^α turn bosons into fermions and fermions into bosons, so in the first approximation they are quadratic in creation and annihilation operators. Thus

$$Q_\alpha = \sum_i \lambda_i^\alpha (a^{(i)} b^{*(i)} + a^{*(i)} b^{(i)})$$

where the λ_i^α are constants depending on the energy and momentum of the ith mode. This equation is an expansion in normal modes of a formula that is roughly

$$Q = \int d^3x (\partial g/\partial x) \cdot \psi \ . \tag{1}$$

Here all indices have been suppressed.

Now, instead of quantizing about flat space, pick an arbitrary classical solution of the Einstein equations, which we'll call $g_{\mu\nu}^{cl}$, and quantize about it with the aim of showing that it describes a space-time of positive energy. So we write

$$g_{\mu\nu} = g_{\mu\nu}^{cl} + \sum_i (h_{\mu\nu}^{(i)} a^{(i)} + \text{h.c.})$$

$$\psi_{\alpha\mu} = \sum (\chi_{\mu\alpha}^{(j)} b^{(j)} + \text{h.c.}).$$

Now, it may appear that we are about to enter the thicket of quantum field theory, but this is not the case. The non-trivial background $g_{\mu\nu}^{cl}$ makes a big difference in equation (1).

With $g = g^{cl} + g^{qtm}$, where g^{qtm} refers to the quantum fluctuations, we have now

$$Q = \int d^3x (\partial g/\partial x) \cdot \psi = \int d^3x (\partial g^{cl}/\partial x) \cdot \psi + \int d^3x (\partial g^{qtm}/\partial x) \cdot \psi \tag{2}$$

Now, as $\hbar \to 0$, the quantum fluctuations are negligible compared to the classical background. So we drop the second term in equation (2), and keep only the first term. In the first term in equation (2), of the infinite number of normal modes of the ψ field, almost all are irrelevant since they are orthogonal to $\partial g^{cl}/\partial x$. For each Q, there is precisely one mode of the ψ field contributing to $\int d^3x (\partial g^{cl}/\partial x) \cdot \psi$. Since there are four Q_α's, $\alpha = 1 \ldots 4$, there are only four modes of the ψ field that are relevant. Thus, in the small $\hbar$ limit of the formal argument $H = (1/\hbar) \sum_\alpha Q_\alpha^2 \geq 0$, the positivity of energy should depend on the properties of just four modes of the Rarita-Schwinger field ψ.

In a convenient gauge, $\psi_0 = 0$, $\sum_{i=1}^{3} \gamma^i \psi_i = 0$, these modes may be expressed as $\psi_i = D_i \varepsilon$ where the spinor ε

must obey a hypersurface Dirac equation $\not{D}\varepsilon=0$ ($\not{D}=\sum_{i=1}^{3}\gamma^i D_i$). In addition ε must approach at infinity a constant ε_0. Essentially, we will be studying global supersymmetry transformations generated by the asymptotically constant spinor ε. (ε must be constant at infinity so that $\psi_i=D_i\varepsilon$ contributes to $\int d^3x(\partial g/\partial x)\cdot\psi$. Alternatively, if ε vanishes at infinity $\psi_i=D_i\varepsilon$ can be gauged away, while if ε diverges at infinity ψ_i is unnormalizable).

It is not difficult to prove that for given ε_0, there is a unique spinor field ε with $i\not{D}\varepsilon=0$ and $\varepsilon \xrightarrow{|x|\to\infty} \varepsilon_0$. Essentially, one writes (Witten, 1981) $\varepsilon=\varepsilon_0+\varepsilon'$, where ε' should vanish at infinity and obey $\not{D}\varepsilon'=-\not{D}\varepsilon_0$. Now, although ε_0 is a constant, $\not{D}\varepsilon_0$ is non-zero because of the Christoffel symbols in $\not{D}$. But in an asymptotically Euclidean initial value surface, the Christoffel symbols vanish fast enough so that $\not{D}\varepsilon_0$ is normalizable. We can solve for ε' as $\varepsilon'(x)= =-\int d^3y S(x,y)\not{D}\varepsilon_0$ where $S(x,y)$ is the Green's function of the hypersurface Dirac operator $\not{D}$. The hypersurface Dirac operator has no normalizable zero modes (Witten, 1981) and its Green's function has been rigorously shown to exist (Parker and Taubes, 1982).

By using ε, it is not difficult to prove the positivity of energy. For simplicity I will write the formulas only in the time symmetric case. We square the Dirac equation:

$$0=(i\not{D})^2\varepsilon=-\sum D_i D^i\varepsilon+\frac{1}{4}R\varepsilon$$

$$=-\sum D_i D^i\varepsilon+4\pi G T_{00}\varepsilon .$$

Then we multiply by ε^*, and integrate:

$$0=\int d^3x\sqrt{g}\,\varepsilon^*(-D_i D^i\varepsilon+4\pi G T_{00}\varepsilon) .$$

Next we integrate by parts:

$$\int d^3x\sqrt{g}(D_i\varepsilon^* D_i\varepsilon+4\pi G T_{00}\varepsilon^*\varepsilon)=\int d\Sigma^i\varepsilon^* D_i\varepsilon .$$

Since the left-hand side of this equation is clearly positive, we have proved the positivity of a certain surface integral, namely $\int d\Sigma^i\varepsilon^* D_i\varepsilon$. Only one surface integral is *expected* to be positive, namely the mass.

Indeed the two can be identified. With

$$i\not{D}\varepsilon=0,\quad \varepsilon\xrightarrow{|x|\to\infty}\varepsilon_0$$

$$g_{ij}\to\delta_{ij}(1+2GM/r)$$

one may readily show

$$\int d\Sigma^i\varepsilon^*D_i\varepsilon=4\pi G\varepsilon_0{}^*\varepsilon_0\cdot M$$

so we get

$$M=\frac{1}{4\pi G\varepsilon_0{}^*\varepsilon_0}\int d^3x(D_i\varepsilon^*D_i\varepsilon+4\pi GT_{00}\varepsilon^*\varepsilon)\geqq 0$$

as desired.

This reasoning was recently deduced from supergravity in a systematic way by several authors (Horowitz and Strominger, 1983; Deser, 1983; Hull, 1983).

I now want to discuss further the physical significance of the positivity of energy. Consider general relativity without matter. By dimensional analysis the energy E of any state is proportional to its size R:

$$E=\lambda R.$$

The constant λ depends on the structure of the object but not on its size R.

This type of scaling holds whether $\lambda>0$ or $\lambda<0$. The scaling shows that if there is a state of $\lambda<0$ the bottom falls out. It is energetically favored for this object to expand without limit and engulf everything. (Of course, since energy is conserved, while the negative energy object expands and gains additional negative energy, there would be a compensating production of positive energy material, perhaps gravitational radiation). The world as we know it would come to an end.

I now want to describe a theory in which this catastrophe occurs (Witten, 1982). The theory is the original Kaluza-Klein theory - five dimensional general relativity with the ground state assumed to be $M^4\times S^1$ - a product of four dimensional Minkowski space M^4 with a circle S^1. The metric of the vacuum is

$$ds^2=\eta_{\mu\nu}dx^\mu dx^\nu-R^2d\phi^2 \qquad 0\leqq\phi\leqq 2\pi.$$

Here ϕ is the angle parametrizing the Kaluza-Klein circle. As long as one considers initial value surface with the topology of the Kaluza-Klein vacuum - $R^3 \times S^1$ (R^3 is three dimensional Euclidean space) - the positive energy theorem holds in Kaluza-Klein theory. This may be shown by spinor methods.

Now, in standard four dimensional relativity, the positive energy theorem holds regardless of the topology of the initial value surface, even if "Wheeler wormholes" are present. It is believed, in fact, that classically the wormholes collapse into black holes. In Kaluza-Klein theory the situation is different (Witten, 1982).The positive energy theorem does not hold, in general, when the initial data set is topologically different from the vacuum. Here is an initial value set that is asymptotic to $R^3 \times S^1$ but is topologically different:

$$ds^2 = \frac{dr^2}{1-R^2/r^2} + r^2 d\Omega^2 + (1-R^2/r^2)R^2 d\phi^2 .$$

By studying the behavior near r=R one may see that this initial data set has a topological excitation, similar to a Euclidean black hole. It obeys the initial value constraint, but violates the positivity of energy; it has zero energy, and by methods of Brill and Deser (1968) one can see that there are nearby states of negative energy.

One can find explicitly the five-metric into which this initial value set develops:

$$ds^2 = -r^2 d\psi^2 + \frac{dr^2}{1-R^2/r^2} + r^2 \cosh^2\psi d\Omega^2 + (1-R^2/r^2)R^2 d\phi^2 .$$

It can be shown (Witten, 1982) that this solution describes an explosion - as one would expect for a case in which the positive energy theorem is wrong. The wormhole expands to infinity, asymptotically at the speed of light. Its radius at time t is roughly $R(t)=(R^2+t^2)^{\frac{1}{2}}$. All observers are accelerated to infinity by the expanding wormhole.

Of course, this is just a cautionary tale. There may be physical principles forbidding this clsss of topological excitations. In fact, (Witten, 1982) such a principle seems to emerge naturally in the theory just considered when spinor fields are included. I believe, though, that such catastrophes occur in many non-standard theories of gravity. For instance, I think they occur in the higher derivative ($\alpha R^2 + \beta R_{\mu\nu}^2$) theories if wormholes are considered and probably also otherwise.

For the future, I believe it is important to know for which Kaluza-Klein spaces $M^4 \times B$ (B being a compact manifold of dimension n) there is a positive energy theorem. I know only of certain partial results that can be obtained by spinor methods. For instance, it is enough for B to admit a covariantly constant spinor field - if wormholes of certain topological types are excluded.

To me, it is disappointing that the positive energy theorem is true in any number of dimensions. Had it been true in 3+1 dimensions but - say - wrong in 4+1 dimensions, this would have given a rationale for Kaluza-Klein theory, for the idea that the ground state of five dimensional general relativity is not five dimensional Minkowski space.

But the positive energy theorem is a remarkable feature of the Einstein $\sqrt{g}R$ theory, which has no obvious positivity property in Minkowski or Euclidean space.

NOTE

* Supported in part by the National Science Foundation under Grant N. PHY80-19754.

REFERENCES

Araki, H.: 1959, "On Weak Time-Symmetric Gravitational Waves", *Ann. Phys. (N.Y.)*, 7, 456.

Brill, D.: 1959, "On the Positive Definite Mass of the Bondi-Weber-Wheeler Time-Symmetric Gravitational Waves", *Ann. Phys.(N.Y.)*, 7, 466.

Brill, D., and Deser, S.: 1968, "Variational Methods and Positive Energy in General Relativity", *Ann. Phys. (N.Y.)*, 50, 548.

Deser, S.: 1983, "Energy in Gravitational Theories", Brandeis, Preprint.

Deser, S., and Teitelboim, C.: 1977, "Supergravity Has Positive Energy", *Phys. Rev. Lett.*, 39, 249.

Grisaru, M.: 1978, "Positivity of the Energy in Einstein

Theory", *Phys. Lett.*, 73B, 207.

Horowitz, G.T., and Strominger, A.: 1983, "On Witten's Expression for Gravitational Energy", IAS preprint.

Hull, C.M.: 1983, "The Positivity of Gravitational Energy and Global Supersymmetry", Cambridge University Preprint.

Jang, P.S.: 1978, "On the Positivity of Energy in General Relativity", *J.Math. Phys.*, 19, 1152.

Parker, T., and Taubes, C.: 1982, "On Witten's Proof of the Positive Energy Theorem", *Comm. Math. Phys.*, 84, 223.

Schoen, P., and Yau, S.T.: 1979a, "On the Proof of the Positive Mass Conjecture in General Relativity", *Comm. Math. Phys.*, 65, 45.

Schoen, P., and Yau, S.T.: 1979b, "Proof of Positive Action Conjecture in General Relativity", *Phys. Rev. Lett.*, 42, 547.

Schoen, P., and Yau, S.T.: 1979c, "Positivity of the Total Mass of a General Space-Time", *Phys. Rev. Lett.*, 43, 1457.

Trautman, A.: 1962, "Conservation Laws in General Relativity", in L.Witten (ed.), *Gravitation: An Introduction to Current Research*, Wiley, New York.

Weber, J., and Wheeler, J.A.: 1957, "Reality of the Cylindrical Gravitational waves of Einstein and Rosen", *Rev. Mod. Phys.*, 29, 509.

Witten, E.: 1981, "A New Proof of the Positive Energy Theorem", *Comm. Math. Phys.*, 80, 381.

Witten, E.: 1982, "Instability of the Kaluza-Klein Vacuum", *Nucl. Phys.*, B195, 481.

Yau, S.T.: 1983, Private Communication.

CLASSICAL AND QUANTUM ALTERNATIVES TO GRAVITATIONAL THEORIES
REPORT OF WORKSHOP A5

E.T. Newman

Department of Physics
University of Pittsburgh
Pittsburgh, PA 15260, USA

H. Goenner*

Institut für Theoretische Physik der Universität
Göttingen, FRG

Workshop A5 consisted of two parts, one on *classical* alternative approaches to theories of gravitation and the other on the fundamental use of *two-component spinors*. The 74 abstracts printed in part one of the GR 10 volume of contributed papers (Bertotti et al., 1983) obviously could not be presented in the four hours alotted to the workshop. The more so as the contents of the papers were scattered among different topics ranging, as far as the classical approaches are concerned, from gauge theories of gravitation within the Riemann-Cartan geometry to theories built on Finsler geometry; from exact solutions to the Hamiltonian formulation of alternative theories. Therefore it was thought best to review and/or summarize some interesting developments in the field discussed by a few selected speakers.

1. CLASSICAL ALTERNATIVE THEORIES OF GRAVITY

Hubert Goenner gave a brief review of classical field theories of gravitation on a space-time continuum. His interest in the subject stems from a systematic investigation into the foundations of general relativity. The main arguments for dealing with alter-

B. Bertotti et al. (eds.), General Relativity and Gravitation, 199–211.

native theories found in the literature are concerned with the "mending" of (real or apparent) weaknesses of general relativity (singularities, Mach's principle, local definition of energy and momentum, unified field theories).

On the real differential manifold underlying such theories three basic elements are given: i) an affine connection Γ (with torsion S) constituting affine geometry; ii) a non-singular bilinear form g (metric); iii) a 2-form f. The latter two are used mostly in the combination f+g describing a non-symmetric first fundamental form of space-time. The connection Γ, and f+g are the basic ingredients of the so-called "mixed geometry". The particular case f=0 is termed metric affine geometry. It can be reinterpreted as purely Riemannian geometry with two additional tensor fields (contorsion K and non-metricity Q). Within the metric--affine geometry all the geometries which have been the object of a lot of interest in the past can be found as particular subcases. If a complex structure is admitted, the case of a Hermitian manifold with the specific subcase of a Kähler manifold have obtained most attention in gravitational physics.

More recently, manifolds with the structure of a principal fibre bundle are studied in connection with gauge theories of gravitation and other interactions. In contrast to the "internal symmetry bundles" in which the connection on the bundle is irrelevant for space-time geometry, in the "external symmetry" - or "soldered" bundles occurring in gravitational physics the space-time connection is fixed by the Lie group acting on the fibres. For example, the bundle of linear frames gives a linear connection, the bundle of Lorentz-orthogonal frames a Lorentz connection compatible with a Lorentz metric.

Historically, three principles have helped the birth of general relativity, the covariance-, equivalence-, and Mach's principle. Today we know that all together are insufficient for deriving general relativity. Additional postulates are needed requiring, for example, the field equations to follow from a variational principle and to be partial differential equations of second order at most for the relevant dynamical variables (Cauchy problem, Newtonian limit). Also, the theory must by relativistic (light-cone structure).

The formulation of the three principles referred to above has been refined since they were put forward: there is a number of formulations of the equivalence principle of growing "strength" (weak, strong, superstrong). Concepts appearing in the wording as "inertial system" or "test body" are seen to be theory dependent. For example, in Riemannian geometry a particle with rest mass suffices as a test body while, in Riemann-Cartan geometry with torsion we need a particle with rest-mass and microscopic spin or massive spin fields to fully measure the connection. The situation is more complicated and even less clear in mixed geometry where additional internal parameters (such as electric charge) are required.

The covariance principle was interpreted by Goenner as saying that: i) all variables occurring in the theory must be *geometric objects* (i.e. transform in a prescribed way under the diffeomorphism group); ii) *absolute* variables must be excluded, i.e. those geometric objects of which the Lie-derivative for the generators of a *genuine* subgroup of the diffeomorphism group vanishes.

Moreover, there are twenty-odd formulations of Mach's principle which have been used either to show that general relativity is consistent with Mach's ideas or to construct alternative theories (Goenner, 1981).

Coming then to theories of gravitation built on the geometric structures described and obeying one or the other of the formulations of the three principles listed above, Goenner first dealt with unified field theories. He first pointed out that the ambiguity in relating the geometrical objects of such theories with physical variables had not been overcome since the times of Schrödinger and Einstein. In the framework of the affine geometry unified theories of gravitation and electromagnetism are mostly considered. Such a unification was also investigated within the mixed geometry by many workers (Einstein, Schrödinger, Einstein and Kaufman, Finzi, Bonnor, Tonnelat, Udeschini, Lichnerowicz, Hlavaty, Bertotti etc. and, more recently, by Klotz (1982) and Ponomariov and Obuchov (1982), for example). It seems, however, that, since 1946, real progress was made only in the field of exact solutions of the weak (real or complex) Einstein-Straus field equations, not in the interpretation of the theory. This applies also to the work of Ferraris and Kijowski (1982) claiming the equivalence of affine, metric-affine and pure metric (Riemannian) theories. This equivalence is shown only if: i) torsion is absent; ii) a certain prescription for the definition of the metric in the affine theory is given; iii) the light-cone structure of the metric is introduced ad hoc; iv) the matter Lagrangian does not depend on the connection.

Within the Hermitian geometry, pure gravitational

theories were suggested by Moffat and coworkers (Moffat and Boal, 1975; Moffat 1977, 1979; Kunstatter and Moffat, 1979) and by De Sabbata and Gasperini (1982). Treder (1957, 1980) suggested an interpretation including strong interactions besides the gravitational field. Indications along such lines also follow from an exact solution found by Antoci (1983a, 1983b). A similar interpretation, i.e. one taking into account some features of elementary particle physics (dilatation current) is attempted by theories constructed within metric-affine geometry (Hehl et al., 1977,1978; Hehl and Sijacki, 1980). As a source for the non-metricity tensor the so-called hypermomentum current is introduced. While Weyl geometry was revived as the framework for a pure gravitational theory by Dirac (1973) and Canuto, Adams, Hsieh and Tsiang (1977), also torsion is included in a theory of Papini and Gregorash (1981).

The center of current interest, however, is formed by alternative theories of gravitation built on Riemann-Cartan geometry. This is the natural geometry following from the application of the gauge principle to the homogeneous or inhomogeneous Lorentz group. The dynamical variables are a (Lorentz-orthonormal) tetrad field and a linear connection. While the interpretation of the connection as a gauge potential related to the homogeneous Lorentz group is undisputed, there are differing opinions concerning the role of the tetrad field (metric). While the tetrad components themselves are interpreted as gauge potentials by some scientists (Hehl, 1980; Cho, Hayashi and Shirafuji), others take the difference of two tetrad fields (one of which is fixed a priori) as gauge potentials (Kibble, Hayward). Moreover, Trautman (1980) calls the metric a *Higgs* field while Ivanenko and Sardanashvily (1983) prefer to interpret it as a *Goldstone* field. Since the corresponding gauge bosons (massive and massless) have not yet been found experimentally, this debate is surely academic. It all depends on the position given to the translations within the mathematical formalism. Nevertheless, one can safely say, that, in the gauge theories of gravitation, the Poincaré gauge invariance is always broken to Lorentz gauge invariance.

In contrast with the usual Yang-Mills' theories the gravitational free field Lagrangian is not uniquely determined. This is due to the soldering of the bundle of linear frames to the base manifold. The existence of a canonical one-form leads to torsion and to 9 different invariants quadratic in torsion or curvature plus one linear term (Ricci scalar). Thus, even in the form of a gauge theory, gravitation again plays an exceptional role among the fundamental interactions.

At the end, Goenner classified various generalizations of general relativity (changed dynamics, additional variables for the gravitational field, absolute elements etc.) in four-dimentional and higher-dimensional spaces (Kaluza-Klein, projective theory).

In the next talk, Edward Witten reviewed the revived interest in theories of the type of the Kaluza-Klein approach to a unification of gravitation and electromagnetism. From the present point of view the theory is interpreted as a theory with "spontaneously broken symmetry". This means that the diffeomorphism group in five dimensions is broken to the product of the 4-dimensional diffeomorphism group and a local U(1) gauge group.

On the road towards unification of other physically relevant gauge fields with the gravitational field, non-abelian generalizations of this local U(1) group come into play. One may start with the standard Einstein-Hilbert action in 4+n-dimensional space, possibly with additional matter fields. The "ground state" of this system is assumed to be a product space $M^4 \times B^n$ of space-time M^4 and a compact space of dimensions n accommodating a symmetry group, which is interpreted as the (non-abelian) gauge group in space--time. If one now considers small oscillations around the "ground state", the eigenvalues of the "internal" part of the operator appearing in the field equations for particles with spin (wave operator, Dirac-, Rarita-Schwinger-operator etc.) show up as a *mass term*, in the equations projected into space-time. As the internal operator acts on a compact space its spectrum is discrete with the eigenvalues being either zero or of the order of R^{-1} where R is the length dimension of the compact space. Since R^{-1} can be assumed to be of the order of magnitude of the Planck mass, the non-zero eigenvalues correspond to hitherto unobservable extremely massive particles. The observed particles should then relate with the zero-mass-modes.

For a unification of the electroweak and strong interactions with gravitation a gauge group $SU(3) \times SU(2) \times U(1)$ is needed.The minimal dimension of a compact manifold admitting this group as an isometry group turns out to be seven. Together with the four space-time dimensions the total dimensions of the manifold on which the unified field theory acts will be eleven. It so happens that supergravity theory does not exist in spaces with dimensions greater than eleven. (Otherwise, massless particles with spin greater than two would exist). There are serious problems to be solved, however, before a realistic model of such an eleven-dimensional supergravity unification can be worked out (Witten, 1981a).

The last talk in the first part of the workshop was given by Rafael Sorkin. His ideas ventured into space-time areas with an extension of the order of the Planck length where, conceivably, the continuum structure of the space-time manifold ceases to be meaningful. There, some more "finitistic" structure (not necessarily discrete) should appear. After discussing various hints from present day physics pointing in this direction Sorkin explained why the concept of simplicial complex used, for example, in the Regge calculus is not quite apt for replacing the continuum. The precise relation between the approximating simplicial complexes C_n and the limit space C_∞ as $n\to\infty$ is physically unacceptable. The "lattice" in the limit space corresponding to C_n may remain large, in some points, and not shrink to zero as it must.

This problem can be solved, however, if the simplicial complexes are replaced by partially ordered sets which are equivalent to topological spaces with only a finite number of points. Partially ordered sets are able to approximate arbitrary compact manifolds.

2. FUNDAMENTAL USE OF SPINORS IN GENERAL RELATIVITY

Among the many developments in general relativity over the past few years it appears to me that the most interesting and potentially most important one is the introduction of spinors into the classical theory of gravity at a fundamental level. Until recently spinors had been used solely as a powerful technical tool. All calculations done with spinors could have been translated and redone using standard tensorial methods. However over the past two years we have seen two very important cases (similar but apparently distinctly different) of the use of spinors in classical relativity which appear to be fundamental. The first case (Witten, 1981b; Horowitz and Tod, 1982)[(1)] was Witten's use of the Weyl neutrino equations "projected" onto an asymptotically flat space-like three-surface in order to prove the positivity of the ADM mass and the generalizations of this argument to the Bondi mass. The second case (Penrose, 1982; Tod, 1983) of the intrinsic use of spinors was in Penrose's quasi-local definition of mass-momentum and angular momentum. In this case, for an arbitrary two-surface in an arbitrary space-time, a pair of differential equations for a two-component spinor (defining 2-surface or "superficial" twistors) are given. The solutions are then used to weight 2-surface integrals of the Weyl tensor, which yield the mass-momentum and angular momentum enclosed by the surface.

In both cases, in spite of much effort, there has been no means of giving them a tensorial reformulation. There is thus the hint that perhaps a deeper structure underlies classical general relativity.

The fundamental use of two component spinors in general relativity was the basic theme of Session II of Workshop A5. There were three lectures; a review of twistor theory by R.Penrose, applications of Penrose's definition of quasi-local mass by P.Tod and a talk on an application of the Witten equation to the canonical formulation of general relativity by A.Ashtekar.

I will not say much about the Penrose talk as it was largely a review of material in the literature (Penrose and MacCallum, 1973). He did however briefly report on the basic ideas associated with the two-surface twistors as well as on new developments in twistor diagrams. A further very important idea was briefly alluded to; namely the possibility of producing "googly gravitons" and real vacuum solutions of the Einstein equations from twistor theory by means of the technical trick of "blowing up " $\mathscr{I}^+$.

As a personal comment, I would like to say that there always appears to be a difficulty with general talks on twistor theory. It seems as if most relativists (and certainly most physicists) have not shown much interest in twistor theory. Many find it too difficult, while others feel that its successes do not repay the effort to study it. (I do not share this view). This has the negative effect that every general talk must go over the basic ideas (at least 10-15 years old) for at least half the talk. It hence leaves essentially no time for presentation of new results. Unfortunately I do not know an efficient way of solving this problem.

Tod, after reviewing (Penrose, 1982; Tod, 1983) Penrose's quasi-local mass-momentum angular momentum complex, described a series of examples of the quasi--local mass construction for different two-surfaces in several different space-times. (Other than for two-surfaces on $\mathscr{I}$ there do not seem to be examples of the angular momentum worked out).

Some of the more important results reported by Tod are:

a) for Schwarzschild space-time, the Penrose mass, m_P, calculated for two-spheres in a t=const. surface is either (1) $m_p = m_s$, (m_s is the Schwarzschild mass) if the S^2 enclose the source, or (2) $m_p = 0$ if it does not enclose the source.

b) In Reissner-Nordström space-time for spherically symmetric spheres of radius r around the origin on a t=const. surface $m_p = m_s - \frac{1}{2}e^2/r$.

c) In Friedman-Robertson-Walker space-times for a spherically symmetric surface S, one has

$$m_p = \frac{4}{3}\pi \frac{r(R^3(t))}{f}\rho$$

where ρ is the matter density, R(t) is the scale factor and

$$f = 1 + \frac{1}{4}kr^2, \quad k = 0, \pm 1.$$

This is the product of the mass density with a volume enclosed by a surface in flat space with the same area as S.

d) For surfaces of time-symmetry in asymptotically flat vacuum space-times with worm holes one obtains the result that the total Penrose mass obtained by the two-surface enclosing all the worm holes is the ADM mass and the very beautiful result which relates the total Penrose mass to the Penrose mass associated with individual worm holes namely

$$M_T = \sum M_i - \sum_{i \neq j} \frac{GM_i M_j}{2d_{ij}}$$

where M_i are individual worm hole Penrose masses and d_{ij} are the separation between the worm holes.

e) Tod also discussed the Penrose mass for very small two-surfaces in general space-times. This calculation is however incomplete.

In the last lecture Ashtekar described a remarkable result in the canonical formalism obtained by adding Witten spinor ideas. As there is, of now, no written(2) account of this work I would like to devote slightly more time to it than to the other lectures.

He began with a review of the canonical formalism; consider a three manifold Σ, topologically R^3, with a positive definite metric g_{ab} which is asymptotically flat and a symmetric tensor density p^{ab}. $(q_{ab}, p^{ab}) \in \Gamma$ defines the phase space. The constraint surface is defined by $(q_{ab}, p^{ab}) \in \overline{\Gamma}$ satisfying

$$C^a \equiv -D_b p^{ab}=0,\quad C=q^{-\frac{1}{2}}(p^{ab}p_{ab}-\tfrac{1}{2}p^2)-q^{\frac{1}{2}}R=0.$$

The constraint functions

$$C_{N^a} \equiv \int_\Sigma N_a C^a \quad \text{and} \quad C_N \equiv \int_\Sigma NC,$$

with the lapse and shift function N_a and $N \to 0$ at infinity, are the generators of the diffeomorphisms of the space-time which is the identity transformation at infinity. From the Dirac point of view these are considered to be gauge transformations.

In order to obtain translations at infinity and generate the complete dynamics one needs a surface term added to the constraint functions. With the lapse and shift $(T, T^a) \to$ constant, at infinity, one defines

$$H_{T^a}=\frac{1}{8\pi}\left[\int_\Sigma T_a C^a+\oint_{r=\infty} T^a P_{ab} dS^b\right]$$

$$H_T=\frac{1}{16\pi}\left[\int_\Sigma TC + \oint_{r=0} T(\partial_a g_{bc}-\partial_b g_{ac})\,\varepsilon^{ac} dS^b\right].$$

The Hamiltonian $H=H_{T^a}+H_T$ generates the full dynamics on the constraint surface $\overline{\Gamma}$.

Ashtekar then pointed out some of the difficulties with canonical quantization; what polarization should be chosen, - are the quantum constraints compatible with the polarization, is there a problem of disentangling gauge transformations and dynamics, - how to face the problem of a very messy form of the Hamiltonian in terms of the q and p and finally is there a more natural choice of variables than q and p.

A partial answer to these questions comes from the use of spinors λ_A satisfying the Witten equation, $D_{AA'}\lambda^A=0$ on Σ, where $D_{AA'}=D_a=q_a{}^b\nabla_b$, $q_a{}^b=\delta_a^b-t_a t^b$ with t^a the unit normal to Σ and ∇_a the metric connection on the full four-space. One further requires that λ^A satisfy the boundary condition that $\lambda^A=\mathring{\lambda}^A$=const. at infinity.

The basic idea of Ashtekar was then to use a pair of these spinor fields to translate rigidly, throughout Σ, a constant vector at infinity. One can obtain in this manner four independent translations (relative to the choice of (q,p)). In particular one has a unique time translation, w^a, obtained from $\mathring{T}^a=0$, $\mathring{T}=1$ at infinity. If this is used in the Hamiltonian[(2)], one obtains H_w a generator of a "pure" time translation with no gauge freedom.

A final structure, namely a flat derivative operator on Σ (with torsion) is now defined. Let λ^A and μ^A be a Witten transported spinor dyad on Σ, then there exists a unique derivative operator D^0_a with the above properties such that $D^0_a\lambda^A=D^0_a\mu^A=0$. Ashtekar now obtains his *basic field* $\phi_{aA}{}^B$ by

$$(D_a-D^0_a)\alpha_A=\phi_{aA}{}^B\alpha_B \text{ for all } \alpha_A.$$

His main result is the remarkable simplification of the Hamiltonian H_w in terms of the new (non-local) variable $\phi_{aA}{}^B$, namely

$$H_w=\frac{1}{16\pi}\int_\Sigma \phi_{aBC}\bar{\phi}^a{}_{B'C'}W^{BB'}dS^{CC'}.$$

He ended the talk by pointing out that on the classical level, the new field ϕ_{aAB} could be thought of as a potential for the full Weyl tensor, i.e.

$$\Psi_{ABCD}=-\nabla_{A'(B}\phi^{A'}{}_{ACD)}+\phi_{a(AB}\phi^a{}_{CD)}$$

and hence it is intimately connected with H-space theory. On the quantum level it suggests that one should perhaps think of wave functionals of the ϕ_{aAB} rather than of the q_{ab} - they would then be something like helicity eigenstates.

In my own view and in the view of many in the audience this work was very impressive and injected new life into canonical quantization.

NOTES

* H.Goenner is author of 1. Classical Alternative Theories of Gravity, report of the first part of Workshop A5.

(1) See also the references cited in Horowitz and Tod, 1982.

(2) I have just been informed that there is an Ashte-

kar-Horowitz preprint that does describe some but not all of this work.

REFERENCES

Antoci, S.: 1983a, "The Meaning of the Hermitian Theory of Relativity Suggested by Exact Solutions", preprint, Dipartimento di Fisica A.Volta, Pavia.

Antoci, S.: 1983b, "Another Hermitian Generalization of the Weyl-Levi Civita Class of Vacuum Solutions and the Hypothesis of Quarks", preprint, Dipartimento di Fisica A.Volta, Pavia.

Bertotti,B., de Felice, F., and Pascolini,A. (eds): 1983, *10th International Conference on General Relativity and Gravitation. Contributed Papers*, Vols. 1-2, CNR, Roma.

Canuto, V., Adams, P.J., Hsieh, S. and Tsiang, E.: 1977, "Scale covariant theory of gravitation and astrophysical applications", *Phys. Rev.*, D16,1643.

De Sabbata, V. and Gasperini, M.: 1982, "The role of spin in a hermitian theory of gravity", *Lett. Nuovo Cimento*, 33, 177.

Dirac, P.A.M.: 1973, "Long range forces and broken symmetries", *Proc. R.Soc. London*, A333, 403.

Ferraris, M. and Kijowski, J.: 1982, "On the Equivalence of the Relativistic Theories of Gravitation", *Gen. Rel. Grav.*, 14, 37, 165.

Goenner, H.F.: 1981, "Mach'sches Prinzip und Theorie der Gravitation", in J.Nitsch, J.Pfarr, and E.-W. Stachow (eds), *Grundlagenprobleme der moderne Physik*, Bibliographisches Institut, Mannheim, p. 85.

Gregorash D., and Papini, G.: 1981, "Torsion in a multiply-connected Weyl-Dirac geometry", *Nuovo Cimento*, 64B, 55.

Hehl, F.W.: 1980, "Four lectures on Poincaré gauge field theory", in P.G.Bergmann and V. De Sabbata (eds), *Cosmology and Gravitation*, Plenum, New York, p. 5.

Hehl, F.W. and Sijacki, D.: 1980, "Towards a unified gauge theory of gravitational and strong interactions", *Gen. Rel. Grav.*, 12, 83.

Hehl, F.W., Lord, E.A., and Ne'eman, Y.: 1977, "Hadron dilation, shear and spin as components of the intrinsic hypermomentum current and metric-affine theory of gravitation", *Phys. Lett.*, 71B, 432.

Hehl, F.W., Lord, E.A., and Ne'eman, Y.: 1978, "Hypermomentum in hadron dynamics and in gravitation", *Phys. Rev.*, D17, 428.

Horowitz, G.T. and Tod, P.: 1982, "A Relation between Local and Total Energy in General Relativity", *Commun. Math. Phys.*, 85, 429.

Ivanenko, D. and Sardanashvily G.: 1983, "The gauge treatment of gravity", *Phys. Rep.*, 94, 1.

Klotz, A.H.: 1982, *Macrophysics and Geometry*, Cambridge University Press, Cambridge.

Kunstatter, G. and Moffat, J.W.: 1979, "Conservation laws in a generalized theory of gravitation", *Phys. Rev.*, D19, 1084.

Moffat, J.W.: 1977, "Space-time structure in a generalization of gravitation theory", *Phys. Rev.*, D15, 3520.

Moffat, J.W.: 1979, "New Theory of Gravitation", *Phys. Rev.*, D19, 355.

Moffat, J.W. and Boal, D.H.: 1975, "Solutions of the nonsymmetric unified field theory", *Phys. Rev.*, D11, 1375.

Penrose, R.: 1982, "Quasi-local mass and angular momentum in general relativity", *Proc. R.Soc. London*, A381, 53.

Penrose, R. and Mac Callum, M.: 1973, "Twistor theory: an approach to the quantization of fields and space-time", *Phys. Rep.*, 6C, 241.

Ponomariov, V.N. and Obuchov, Ju.: 1982, "The Generalized Einstein-Maxwell theory of gravitation", *Gen. Rel. Grav.*, 14, 309.

Tod, P.: 1983, "Some examples of Penrose's quasi-local mass construction", Preprint, Oxford University, Oxford.

Trautman, A.: 1980, Fibre bundles, gauge fields and gravitation, in A.Held (ed.), *General Relativity and Gravitation*, Volume 1, Plenum, New York,p.287.

Treder, H.J.: 1957, "Stromladungsdefinition und elektrische Kraft in der einheitlichen Feldtheorie", *Ann. Phys. Leipzig*, 19, 369.

Treder, H.J.: 1980, "Einstein's hermitesche Relativitätstheorie als Unifikation von Gravo- und Chronodynamik", *Ann. Phys. Leipzig*, 37, 250.

Witten, E.: 1981a,"Search for a realistic Kaluza-Klein theory", *Nucl. Phys.*, B186, 412.

Witten, E.: 1981b, "A New Proof of the Positive Energy Theorem", *Commun. Math. Phys.*, 80, 381.

Relativistic Astrophysics

RELATIVISTIC COSMOLOGY: ITS NATURE, AIMS AND PROBLEMS

G.F.R. Ellis

Max Planck Institüt für Astrophysik,
Garching bei München, FRG

and

Department of Applied Mathematics,
University of Cape Town+, South Africa

1. INTRODUCTION

1.1 Scope

This paper discusses *General Relativistic Aspects of Cosmology*. Thus it will make only passing reference to astrophysical aspects of cosmology, and such features as GUTS, quantum gravity, etc. (these are covered in depth in recent reviews, e.g. Balian et al., 1980; Gibbons, Hawking and Siklos, 1983; Royal Society 1983).

1.2 Aim

This paper intends to present the scope and aims of relativistic cosmology at present, and to show that there are still a series of significant and interesting problems that are presently unsolved, and in some cases have not even been seriously addressed(1).

1.3 Context

The standard Friedmann-Le Maitre-Robertson-Walker (FLRW) universe models have had considerable success in providing models of the geometry, dynamics and observational properties of the observed region of the

B. Bertotti et al. (eds.), General Relativity and Gravitation, 215–288.

universe (Ellis, 1971b; Weinberg, 1973; Gott et al., 1974; Gunn et al., 1978; and Part I of Balian et al., 1980). However,

(a) the relation between these models and real observations is complex, and is obscured in many analyses; in particular, these models are not uniquely indicated by observations. This issue of observational verification needs clarification;

(b) the real universe, unlike these models, is neither exactly spatially homogeneous nor isotropic. Present theory takes some cognisance of this, but does not fully come to grips with it; particularly the implications both for the field equations and for the geometrical models of cosmology need examination;

(c) by and large, present day theory is based on application of local physics to determining the behaviour, on a large scale, of geometry and the physical properties of matter; the converse relation - the effect of large-scale physics and geometry on local physical systems - is still relatively unexplored;

(d) underlying is the problem of the uniqueness of the universe, and what determines its particular form; this fundamental question (which is an aspect that cannot be fully answered without considering GUTS and quantum gravity) is still obscure;

(e) overall, we do not have a really satisfactory conceptual framework which provides an integrative view of cosmology, of the different philosophical approaches one may take to the subject, and of the problems in relating these various viewpoints to observational data.

This paper looks in turn at the nature of each of these issues (concentrating mainly on the first three).

2. THE DESCRIPTION ON A SMOOTHED-OUT SCALE OF THE LARGE-SCALE STRUCTURE OF THE UNIVERSE

2.1 The Standard Approach

The basis of the standard approach is to investigate the possibilities by,

(a) specifying a model of the smoothed-out geometry and matter content of the universe, determined up to some parameters (or possibly functions);

(b) determining the observational and dynamical properties of these models;

(c) comparing with astrophysical observations of idealised sources ('standard candles'), background ra-

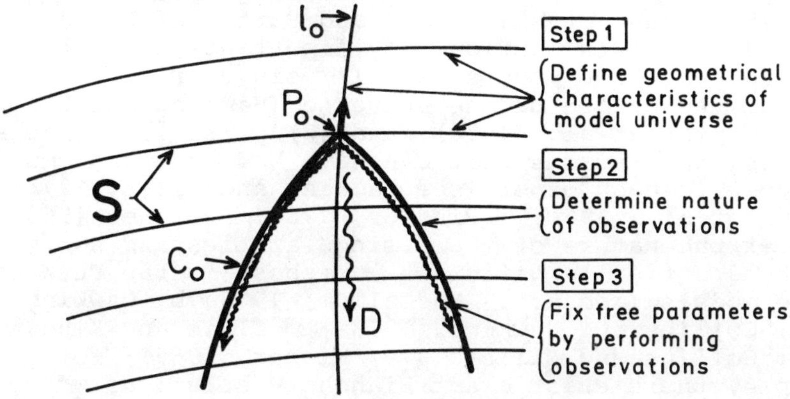

Fig. 1. Space-time diagram of the standard approach to Cosmology (time is indicated vertically), spatial distance horizontally). In this and the following space-time diagrams, l_0 is the world line of our galaxy; P_0 is the event "here and now" from which we make our observations; and C_0 is our past light cone. All astronomical observations are made on this light-cone. The surfaces S are surfaces of constant density.

diation, and element abundances, to fix the remaining parameters.

The aim is to obtain a "best-fit" model universe in this way (Figure 1). This has been done in great detail for the spatially homogeneous and isotropic FLRW universe models (see for example, Weinberg, 1973; Gott et al., 1974; Gunn et al., 1978; and Part I of Balian et al., 1980) and to a lesser extent for some more general universe models, particularly the spatially homogeneous but anisotropic Bianchi universe models (see MacCallum, 1979a,b; Hawking, 1969; Collins and Hawking, 1973a). Through these investigations we have obtained an understanding of the behaviour of broad classes of plausible universe models (Ellis, 1971b), and specifically of their thermal histories, their singularities, and their causal structure. These models expand from a hot initial phase; the observed (redshift, distance) relation is explained by this expansion, the light elements are synthesized in the 'hot big bang', and the observed microwave background radiation is understood to be relic radiation from this early era (Ellis, 1971b; Weinberg, 1973; Gott et al., 1974; Gunn et al., 1978; and Part I of Balian et al., 1980). It is particularly important that all "reasonable" universe models have (1) singularities in the past at the origin of the universe (Hawking and Ellis, 1973; Tipler et al., 1980), and (2) particle horizons limiting the regions that can have had causal communication with each other (see Hawking and Ellis, 1973; Tipler et al., 1980; Rindler, 1956; Penrose, 1963). However the nature of these singularities and horizons can be drastically different from those in the FLRW universe models (see e.g. MacCallum, 1979a,b; Hawking, 1969; Collins and Hawking, 1973a; Ellis and King, 1974; Collins and Ellis, 1979; Misner, 1969); for example, there exist exact Bianchi V solutions where both a non-scalar ('whimper') singularity and a conformal curvature singularity occur at the beginning of the universe, but the matter density is everywhere finite (Collins and Ellis, 1979).

Examination of, (a) the (redshift, distance) relation, where the distance is estimated from galaxy apparent magnitude, angular diameter, or other indicators (such as supernova apparent magnitudes); (b) galaxy number counts and projected correlation functions; (c) element abundances; (d) the ages of stars, star clusters, and elements; and (e) the cosmic background radiation over all wavelenghts, show reasonable agreement between the large-scale properties of the observed universe and the FLRW models, and give limits on the present density ρ_0 of matter, the Hubble constant H_0, and the deceleration parameter q_0 for the "best-fit" FLRW models (Ellis, 1971b; Weinberg, 1973; Gott et al., 1974; Gunn et al., 1978; and Part I of Balian et al., 1980). The present values of the shear σ_0 and rotation ω_0 (which distinguish the FLRW universe models, for which they vanish, from more complex models) are determined to have very low values because of the isotropy of the microwave background radiation (Hawking, 1969; Collins and Hawking, 1973a) and the observed helium and deuterium abundances (Barrow, 1976; Barrow and Matzner, 1977; Olson, 1978). However many problems arise in fitting the observations, and correspondingly the direct observational limits are relatively poor for ρ_0, H_0, and q_0. Reasonable estimates are 10^{-31}gm/cc$\leqq\rho_0\leqq10^{-29}$gm/cc, 25km/sec/Mpc$\leqq H_0\leqq$110km/sec/Mpc, $-1\leqq q_0\leqq 2$, $|\sigma_0|\leqq 10^{-3}H_0$, $|\omega_0|\leqq 10^{-3}H_0$; however these estimates could be substantially wrong, due to astrophysical uncertainties. In particular, we do not know if the density of matter and radiation in the universe is high enough to cause a recollapse in the future, or not.

The observational problems relate *firstly* to how well source properties are known. Problems arise here regarding (a) the source intrinsic magnitude and radius (there is a considerable statistical variation amongst individual sources in a class, and the average properties of a particular class of sources may change with time if there is source evolution (Tinsley, 1970, 1980; Lilly et al., 1983)); (b) the source spectra (these may change with source evolution, but in addition redshifting generally implies that we observe regions of a distant source's spectrum whose properties cannot easily be inferred from measurements of nearby sources, because of absorbtion); (c) the nature of the observed redshift (which may have an intrinsic component as well as local Doppler and overall cosmological components (Field et al., 1973; Roodt, 1981)). *Secondly*, problems arise from detection and measurement effects, which critically affect detection and selection probabilities (Segal, 1982; Soneira, 1979; Sandage et al., 1979). One must allow (d) for point spread effects in the observed image, (e) detection and recognition limits for images observed in a noisy background (Ellis, Fong and Phillips, 1977; Ellis, Perry and Sievers, 1983), and (f) aperture effects when measuring apparent magnitudes (Stock and Schücking, 1957; Ellis and Perry, 1979). *Thirdly*, problems arise from the location of sources relative to us; (g) there is a scarcity of suitable sources needed for adequate statistics nearby, and (h) the nature of possible intervening galactic and intergalactic matter is insufficiently well-determined to allow accurately for absorbtion and emission.

The outcome is that while we can fit various universe models to the observations, there is a considerable degree of uncertainty in this procedure. As well as the FLRW models, we can also fit various anisotropic Bianchi models (e.g. Bianchi I or V universes, which can approach FLRW model universes arbitrarily closely at late times but differ drastically from them at early times (MacCallum, 1979a,b; Hawking, 1969; Collins and Hawking, 1973a; Ellis and King, 1974; Collins and Ellis, 1979), or inhomogeneous models (even a static universe model, in which the redshifts are interpreted as cosmological gravitational shifts, can come close to explaining current observations (Ellis, Martens and Nel, 1978)).Thus the standard approach certainly succeeds in providing interesting, useful, and usable universe models; but leaves the question of *uniqueness* unresolved.

Part of the problem here is that the greatest effort has gone into investigating FLRW models only; thus leaving open the question of what alternative models could also explain the observational data. What in fact happens in the usual approach is that unverifiable *a priori* philosophical assumptions (in particular: a 'Cosmological Principle' (Ellis, 1975; Weinberg, 1973; Gunn et al., 1978; Bondi, 1960; McCrea,1953) are used as a basis for choosing particular universe models, rather than observational evidence (cf. Ellis, 1979). However a useful outcome of those investigations of alternatives to the FLRW models that have been carried out, is an indication as to the *degree of generality* of various of the models investigated (Collins and Hawking, 1973b; Siklos, 1981; Jantzen, 1979, 1983); this gives an idea of how probable they are if universe models are selected at random from the family of all possible universe models (Penrose, 1979, 1983). Clearly the FLRW models are extremely unlikely on this basis.

2.2. The Observational Cosmology Approach

In contrast to the usual approach outlined above, a direct observational approach was developed by Kristian and Sachs, 1966, following early work by McCrea, 1935, 1939.[(2)] This method has been further developed particularly by Dautcourt, 1977; Ellis, 1980a,b; Ellis, Martens and Nel, 1981; Martens, 1980; Martens et al., 1983; Nel, 1980; Nel et al., 1983.

The aim of this approach is to determine space--time geometry directly from observations. Thus (see Figure 2)),its basis is:

(a) determining the inner geometry of our past light cone and the matter distribution on it, direct from astronomical observations;

(b) using Einstein's field equations to determine the imbedding of the light cone;

(c) integrating Einstein's equations off the light cone to determine the space-time geometry and matter distribution elsewhere.

More precisely (Ellis, 1980a,b; Ellis, Martens and Nel, 1981; Martens, 1980; Nel, 1980; Nel et al., 1983), in following this method one needs to use *observational coordinates* $y^i=(w, y, \theta,\phi)$ centred on our world line (see Figure 3), where w is a null coordinate giving the time of observation; θ and ϕ are angles specifying the direction of observation; and y is a measure of distance down the null cone, which can conveniently be chosen as an affine parameter v down the past null geodesics. The matter 4-velocity

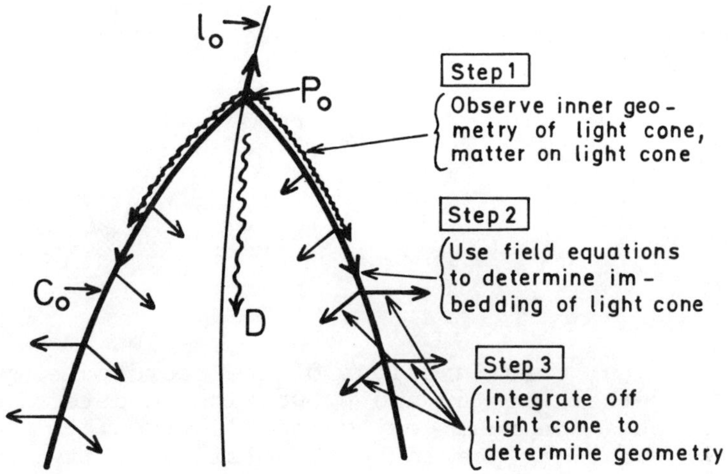

Fig. 2. Space-time diagram of the 'observational cosmology' approach to determining the structure of the universe. Both approaches utilise data D concerning conditions near our world-line at present and in the distant past; this is used to infer the nature of the distant sources observed.

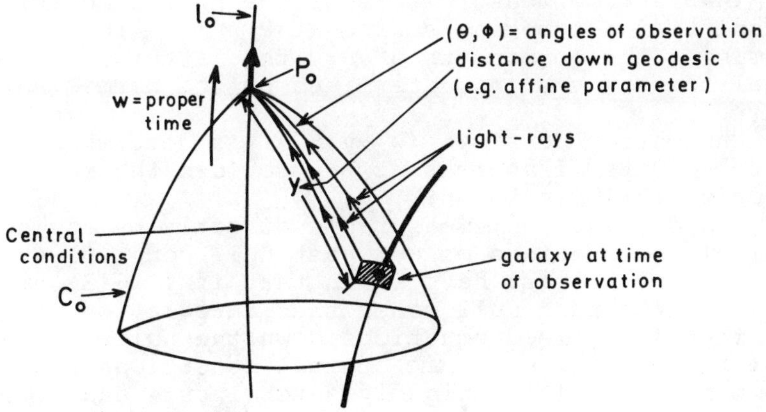

Fig. 3. The 'observational' approach utilizes observational coordinates (w, y, θ, ϕ) centred on our world line l_0.

and density will be $u^a(y^j)$, $\rho(y^j)$. Using these observational coordinates, the metric tensor components will take the form

$$g_{ab} = \begin{pmatrix} \alpha & 1 & v_1 & v_2 \\ 1 & 0 & 0 & 0 \\ v_1 & 0 & h_{11} & h_{12} \\ v_2 & 0 & h_{12} & h_{22} \end{pmatrix} \qquad (1)$$

where α, v_I, h_{IJ} are functions of the coordinates y^i which, together with u^a and ρ, we wish to determine by observations. One must very carefully stipulate the *central conditions* obeyed by the fluid velocity, null geodesic tangent vector, and metric components as $y \to 0$, expressing the regularity of space-time and the fluid flow at the origin and the relation between coordinate directions θ, ϕ at different times w (Ellis, Martens and Nel, 1981; Martens, 1980; Nel, 1980; Nel et al., 1983).

Now an analysis of ideal astronomical observations made at time w_0 shows that:
(1) the metric components $h_{IJ}(w_0, y, \theta, \phi)$ are directly measurable from image apparent angles, if the intrinsic size and shape of a source at the point P_1 with coordinates (w_0, y, θ, ϕ) are known;
(2) the matter 4-velocity $u^a(w_0, y, \theta, \phi)$ is directly determinable from measurements of the proper motion and redshift of a comoving source at P_1 (the former determining the components u^θ, u^ϕ, the latter u^w, the final component being determined by the normalisation $u^a u_a = -1$);
(3) the matter density $\rho(w_0, y, \theta, \phi)$ is determined by number counts of sources at P_1 provided the mass-to-luminosity ratio is known;
(4) α and v_1 are not measurable directly by any astronomical observations on our past null cone. However,
(5) Nel and Stoeger have shown that if (1)-(3) are known on the past null cone, then integration of four of Einstein's field equations down the null geodesics from the origin, using the central conditions and the known data (1)-(3), uniquely determines α and v_1 on the past light cone (one of the field equations determines the relation between the affine parameter v, which cannot be measured directly by observations, and redshift z, which is directly observable). Thus a remarkable feature of this approach is that *the data*

which is in principle observable by us on our past light cone is precisely sufficient to determine the space-time and its matter content on that light-cone when Einstein's field equations are used (Ellis, 1980a, b; Nel, 1980; Nel et al., 1983; *a priori*, the geometry could either have been over- or under-determined(3)).

Now standard results imply one can integrate the field equations to determine the geometry in the interior of our past light cone from this data. In principle we cannot predict to the future of our light cone at all. However in practice (as we have up to now successfully been able to predict the motions of the planets and their satellites, e.g. predicting the eclipses of the moon) we may assume that incoming radiation will not invalidate predictions made to the future from data on our past light cone, at least for short timescales. This 'non-interference' is basically due to the fact that for times a short way to the future of our past light cone, we know *nearly* all the data needed to make predictions; and the effects of uncertainties in the data are attenuated by the distance over which their influence must propagate, and by associated redshift effects (Figure 4).

Having integrated off the null cone, one should finally check the consistency of the results obtained with other data, e.g. that from element formation and the background radiation spectrum and isotropy. It is suggested that this last step (checking consistency with other data) should be postponed as long as possible, so that maximum information is obtained from the detailed null-cone observations (cf. Section 6).

As in the standard approach, a series of difficulties arise in carrying out this procedure. In fact all the problems (a)-(h) listed above for the standard approach will also occur here. In addition, this approach highlights three further problems (Ellis, 1980a, b), namely (i) the lack of standardised shapes for astronomical objects prevents one from accurately determining the distortion introduced into images by intervening gravitational lens effects; this prevents us from accurately determining the conformal geometry of our past light cone; (j) the distance scales involved are so large that we cannot obtain cosmologically useful estimates of galactic proper motions from observations; this prevents us from accurately determining the matter 4-velocity u^a, thereby limiting our ability to directly determine such features as the cosmological vorticity; and (k) the set of points on our past

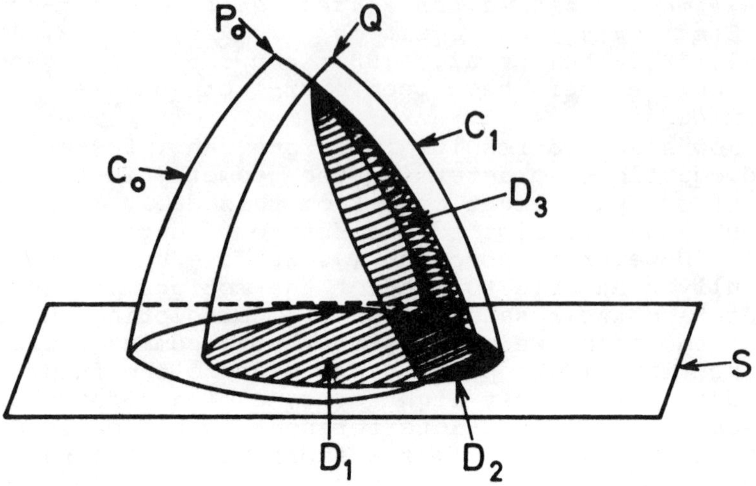

Fig. 4. Space-time diagram of prediction to the future of our past null cone C_0 from data on C_0. We wish to predict what will happen at the event Q. The causal past J_1 of Q intersects a surface S in $D=D_1 \cup D_2$; data on D is sufficient to determine what happens at Q. Here D_1 is the part of D lying in the causal past J_0 of P_0; therefore data on D_1 determines data on D_3, the segment of C_0 lying in J_1; in turn the data on D_1 is determined by data on C_0. Hence given data on C_0, we only lack data on D_2 before we can determine what happens at Q. However D_2 is 'much smaller' than D_1; in addition, D_2 is much further from Q than most of D_3, so the effect of data there is more attenuated by area distance and redshift effects. Hence if (1) Q is 'not too far to the future' of C_0; (2) S is sufficiently far to the past of P_0; and (3) data on D_2 is not too different from what we infer by simple extrapolation of conditions on D_1, we can predict conditions at Q to a good degree of accuracy.

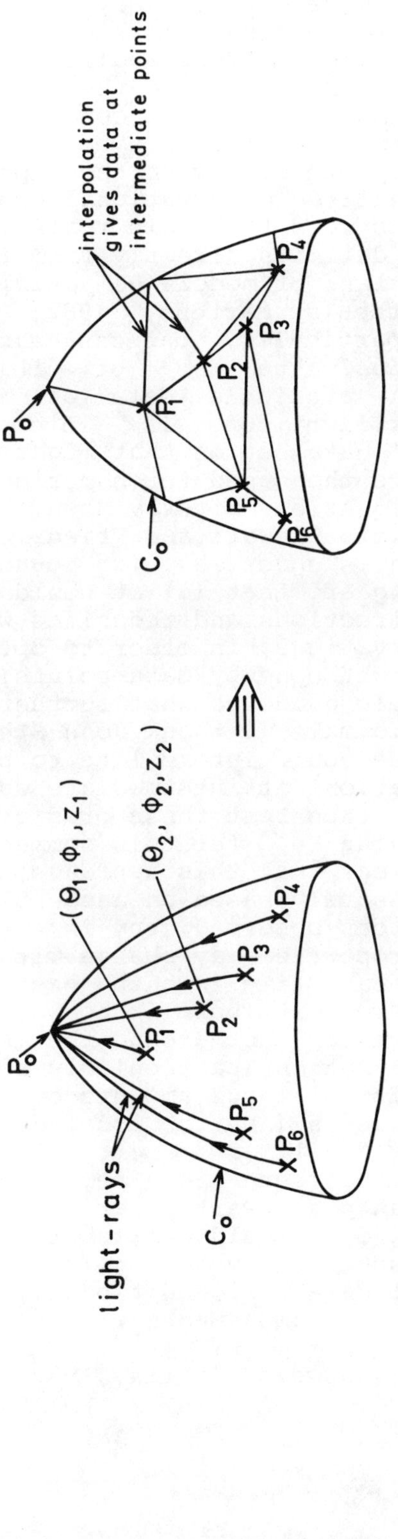

Fig. 5. We actually observe only at isolated points P_i on our past light cone C_0 (as each observation refers to a particular galaxy at distance z_i in direction θ_i, ϕ_i). In a practical observational cosmology approach, we interpolate data measured at the points P_i to the whole light-cone.

light cone for which we have data is very sparse (it can only be measured at isolated points where suitable galaxies occur, see Figure 5). Thus available or potential data does not nearly resemble the detailed information at every point of the light cone needed to carry out this programme in full.

In order to attempt to cope with these problems, one can modify the idealised observational cosmological approach in various ways to obtain realistic approaches in which one delimits directly from real observational data the class of models compatible with that data (see in particular Martens, 1980; Martens et al., 1983). In particular, one can aim to implement a *numerical version* of this approach, in which available data at isolated points on the past light cone is interpolated to obtain a 'best fit' data set on that light cone (Figure 5), and this is then used for numerical integration of the field equations firstly down the light cone and then off it (cf. Stewart and Friedrich, 1982). Study of this approach is under way. Two advantages this method would bring are that (i) it would allow one to specify what directions and redshifts would be best for further observations, in order to optimise the coverage of the light cone by data-points; and this would be a valuable guide to what further cosmological observations to make (I thank John Stewart for this comment); (ii) one could interpolate to predict the results of observations at intermediate data points on the light cone, and then test these predictions by observation (I thank Alan Held for this comment).

Overall, it is clear that this approach, like the standard one, is necessarily based on assumptions about the nature of the sources observed, and in particular how their intrinsic properties may change with redshift (i.e. with look-back time into the past and with radial distance away from us). There is no way round this fundamental limitation, nor around the issues of mass-to-luminosity estimation (particularly acute because of the possibility of large amounts of 'dark mass') or selection effects (and the model-dependence of associated 'corrections').

Given this approach, there are a series of significant questions one can pose. Particularly,
(1) what observational data would suffice to *prove* the observed universe is spherically symmetric about us?
(2) What observational data will additionally *prove* the universe is spatially homogeneous within our past light cone?
These questions can be answered (Ellis, 1980a): effec-

tively one must (1) find isotropy in all observable data, and (2) determine the number counts and the (redshift, distance) relation to have the FLRW form; together these suffice to directly prove spatial isotropy and homogeneity. However in practice the problems mentioned above (particularly, those posed by source evolution) make a real observational proof of homogeneity very difficult, if not impossible.

2.3. Observational Limitations

The fundamental question underlying the 'observational cosmology' approach outlined above is,

> QUESTION 1: What is the most that can be determined about the geometry of the universe directly from observations?

It may well be that ultimately use of this approach will confirm the validity of the FLRW description. If this is so, that would not mean the effort had been wasted: on the contrary, like Dr. Hellings, I must insist that science is tested by observations; and if cosmology really is an observational subject it is essential that this approach - which takes the direct observational method as far as it can go - be tried; a resulting confirmation of the FLRW models would be eminently worthwhile. It is always possible the answer would come out different.

In addition, the considerable benefit of this approach is that it greatly clarifies the nature of observational limitations in cosmology(4). Primarily, it clarifies the domains of predictability in cosmology, that is, the way conditions in different regions in space-time are related to possible observational verification (Ellis, 1975).

Three features are particularly emphasized by such an examination of observational verifiçation. These are:

(1) The loss of certainty with redshift z (cf. Metzner and Morrison, 1959; Whitrow and Yallup, 1964; Gudehus, 1972); as one attempts to determine conditions further and further down our past light cone, the quantity and quality of data rapidly declines due to (a) redshift effects, (b) increasing area distance(5), (c) absorbtion, (d) noise due to anisotropic lensing effects. Thus the precision with which we can determine conditions degrades greatly with distance.

(2) The interplay between uncertainty in physics and cosmology. We can use astrophysical data to determine conditions close to our world line in the very

distant past, and this is the basis of some of the best limits placed on the anisotropy of the universe at early times (e.g. limits on the distortion of the universe obtained by consideration of Helium and Deuterium abundances (Barrow, 1976; Barrow and Matzner, 1977; Olson, 1978)),so supporting the view that the universe is like a FLRW universe model then. However the physics involved is very uncertain, and an inverse approach uses the assumption that the universe is a FLRW universe at very early times to put constraints on high-energy physics, e.g. to limit the number of types and masses of possible neutrinos (see e.g. Olive et al., 1981; Turner and Schramm, 1979; Schramm, 1983). But one cannot have it both ways (Ellis, 1980a); one must decide whether to use the limited data available to constrain the cosmological model used, or the particle physics. To do both is illegitimate.

(3) The need to consider implications of our models outside our past light cone carefully. This analysis emphasizes that although we have some weak constraints on conditions outside our past light-cone, we have no direct observational data whatever for these conditions in distant regions (Ellis, 1980a,b); and so any statement we make about them are based on unverifiable philosophical assumptions (Ellis, 1979). In particular, if we use a strict 'Cosmological Principle' to determine space-time geometry, we inevitably predict that the universe is spatially homogeneous *everywhere*; in the case of a k=0 or k=-1 FLRW universe, this implies we are able to determine conditions 10^{10} Hubble radii away, from data within our past light cone whose future domain of predictability is strictly limited to lie within one Hubble radius; which is patently absurd. Even making the implication for say 5 Hubble radii seems scientifically unjustifiable; it is as if an ant in the middle of the Sahara desert were to declare that he had determined that the whole of the world was comprised of sand dunes (cf. Callan et al., 1965). If we consider a weakened version of the Cosmological Principle, e.g. some sort of simplicity principle where one takes implications of homogeneity seriously only inside our past light cone, the question then is: how far outside our past light-cone does this homogeneity extend? And if conditions are different outside our past light cone, why are we situated in a spatially homogeneous region?(6)

In Section 6 of this paper, we shall briefly consider some of the alternative philosophical approaches that might resolve this kind of problem.

3. IMPLICATIONS FOR THE GRAVITATIONAL FIELD EQUATIONS OF THE SMOOTHING SCALE USED

So far, we have looked at the description of the universe on a smoothed-out scale. Next we take a look at some of the implications of inhomogeneity.

3.1. Smoothing Procedures

Consider a given inhomogeneous universe, with the metric and matter properties represented at various scales of detail. Thus (Figure 6), let g_{1ab} and T_{1ab} represent the universe on a scale where detail down to the scale of stars is represented; call this *Scale 1*. This metric and stress tensor then represent both the large-scale and the small-scale structure of the universe. This is too detailed to handle easily (it will represent at least 10^{22} stars); so we let the metric g_{3ab} and matter tensor T_{3ab} represent the universe on a scale where the individual stars have been smoothed out, but all the galaxies are represented; call this *Scale 3*[(7)]. This is still too detailed for many purposes (it will represent at least 10^{11} galaxies), so we let g_{5ab} and T_{5ab} represent the universe on a scale where even the galaxies and clusters of galaxies have been smoothed out. This is the scale of description implied in the previous section, and envisaged in the standard use of the FLRW models. It represents only the very largest scale structure of the universe; we call this *Scale 5*[(7)]. One must emphasize here that these different matter tensors and metric tensors are intended to describe *the same* physical system and the same space-time, but at different scales of description.

The first major point is that the smoothing procedures mentioned in the above description are implicit in our models of cosmology, but while the matter smoothing procedures have been considered to some extent (e.g. in the transition from kinetic theory to a fluid description; Ehlers, 1971,1973; Israel and Stewart,1980), the metric smoothing has never been explicitly examined in detail. Thus there are implied smoothing operators S_{IJ}, S'_{IJ} such that

$$S_{31}(g_{1ab}) = g_{3ab}, \quad S'_{31}(T_{1ab}) = T_{3ab} \qquad (2a)$$

$$S_{53}(g_{3ab}) = g_{5ab}, \quad S'_{53}(T_{3ab}) = T_{5ab} \qquad (2b)$$

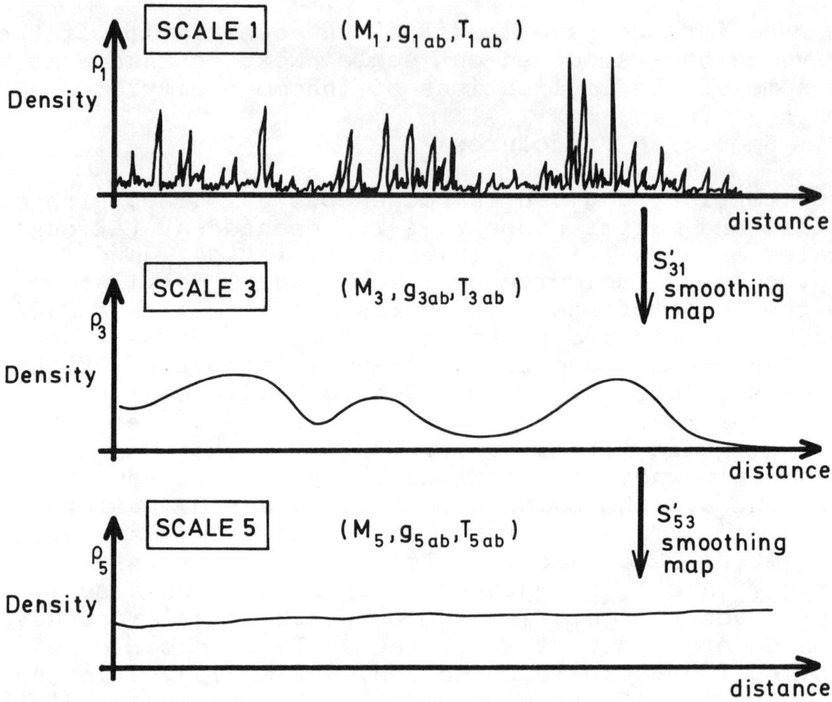

Fig. 6. A comparison of models of the same region of the universe on three scales showing different amounts of detail. Scale 1 represents all details down to stars. Scale 3 represents all details down to galaxies. Scale 5 represents only large scale features. The models at different scales are related by (1) identification of which points at the different scales correspond, through maps S*; (2) maps S determining the matter tensor and maps S' determining the metric tensor of the smoothed-out representation from the more detailed one. The usual models of cosmology are at Scale 5; but models for galaxy formation must be at Scale 3.

but we do not have a precise representation of *any* of these smoothing operations in the context of highly-curved space-times; much less a general prescription that would guarantee desirable properties, e.g. that

$$S_{55}=\mathrm{Id}, \qquad S'_{55}=\mathrm{Id}, \tag{2c}$$

$$S_{53}\circ S_{31}=S_{51}, \qquad S'_{53}\circ S'_{31}=S'_{51}, \tag{2d}$$

(where Id is the identity operator). These operators would have to involve some suitable kind of 4-dimensional volume integration to produce average metric and stress tensors on the chosen scale that made sense when considered physically. Their definition will be non-trivial particularly because description of the universe at the most detailed level (Scale 1) will involve high-curvature phenomena - there will be many cusps, caustics, and singularities (because we expect there to be many black holes in the universe at this scale). The volume integration will have to smooth these out in some suitable way. Problems may even arise from Scale 3 to Scale 5, because of the lumpiness of the distribution of galaxies (cf. the talk by Sato: it is possible that on a scale of 100 Mpc, most of space consists of large, empty voids between chains of galaxies). Before the tensor operations in eq. (2) can be properly defined one will have to define manifold projection operators S^*_{31}, S^*_{53}, S^*_{51} that make explicit which points in the different underlying manifolds are related to each other by the smoothing procedure (cf. Figure 6): that is,

$$S^*_{31}(p_1)=p_3, \quad S^*_{53}(p_3)=p_5, \quad S^*_{51}(p_1)=p_5 \tag{2e}$$

where p_1, p_3, p_5 are corresponding points at the different scales. Problems are likely to arise here too (cf. Section 5.3).

Thus, a significant problem at the foundation of cosmology is *to provide suitable definitions of averaged manifolds (defining projection operators as in (2e)), of metric and stress-tensor averaging and smoothing procedures (as in (2a,b)), and to show these have appropriate properties (as in (2c,d)).*

3.2. Implications for the Description of Matter

There are a series of questions related to the effect of the smoothing operations on the right-hand side of the field equations. Specifically, the usual procedure is to describe the matter content of the universe on a cosmological scale as a perfect fluid at almost all times. But the relation between this description and the kinetic theory description often used on smaller scales, leaves open questions about the validity of the perfect fluid description in situations where there is a large-magnitude shearing motion in the universe. Thus it is possible that a perfect-fluid matter tensor is only consistent with a kinetic-theory description if the shear σ of the fluid vanishes; and if so, then probably in almost all cases, the product $\omega\theta$ of the rotation ω and the expansion θ vanishes also (Treciokas and Ellis, 1971; Ellis, Matraves and Treciokas, 1983). Consistency of the small-scale and large-scale descriptions may then force one to use an imperfect-fluid description when significant shear is present; which could change dynamic properties of the solutions from those when the perfect fluid description is used. These consistency conditions need further investigation.

3.3. Implications for the Field Equations

Given suitable definitions of the smoothing operations, we will be in a position to tackle the next major problem arising, which is, What is the scale on which Einstein's field equations are to be taken to hold?

To make this issue clear, let us assume that Einstein's equations hold on Scale 1, the scale of stars (because the solar system tests described by Hellings, and the binary-pulsar data, refer to this scale). Then these equations will determine the metric g_{1ab}. In detail, this metric will determine a Ricci tensor R_{1ab} and associated Einstein tensor G_{1ab} in the usual way; then

$$G_{1ab} \equiv R_{1ab} - \tfrac{1}{2} R_1 g_{1ab} = T_{1ab}. \tag{3a}$$

Now g_{3ab} will similarly determine a Ricci tensor R_{3ab} and Einstein tensor G_{3ab}; but it is extremely unlikely that the field equations will hold on this scale, for in view of the non-linear nature of the field equations, in general the operations of smoothing will not commute with going to the field equations[8] (Figure 7; the Einstein tensor G_{3ab} determined by the smoothed out metric g_{3ab} will be related to G_{1ab} by a map S''_{31}, which will almost always not be the same as the smoothing operator S'_{31} acting on the matter tensor T_{1ab} to give T_{3ab}). Similar remarks apply in going from Scale 1 to the other scales (e.g. to Scale 5).

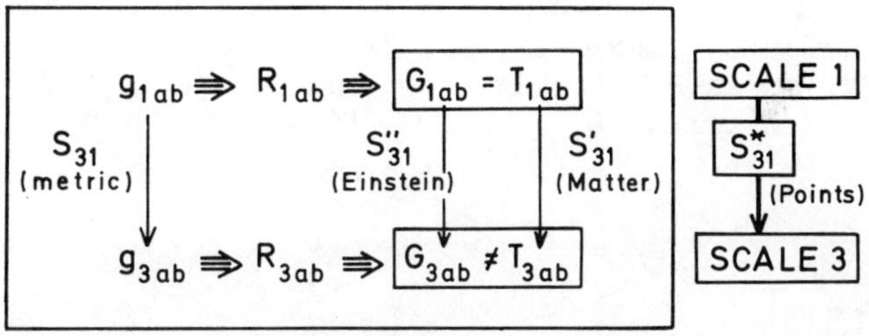

Fig. 7. The smoothing operations S_{31} on the metric tensor and S'_{31} on the matter tensor in general do not commute with taking the non-linear products of derivatives involved in the field equations; that is $S''_{31} \neq S'_{31}$. Therefore if the field equations hold at Scale 1, they will not hold at Scale 3.

What we can aim to do is to define a tensor P_{3ab} representing the difference between the Einstein tensor G_{3ab} defined from the smoothed-out metric g_{3ab}, and the smoothed-out matter tensor T_{3ab}. This is then the correction term to be added to the field equations at other scales than the scale at which they are verified (which we take to be Scale 1), and which is therefore presumably the scale on which these equations hold[9]. Thus at Scale 3 and 5 the field equations will become

$$G_{3ab} \equiv R_{3ab} - \tfrac{1}{2} R_3 g_{3ab} = T_{3ab} + P_{3ab}, \tag{3b}$$

$$G_{5ab} \equiv R_{5ab} - \tfrac{1}{2} R_5 g_{5ab} = T_{5ab} + P_{5ab} \tag{3c}$$

where the tensors P represent the effects of small-scale inhomogeneities in the universe on the dynamic behaviour at the smoothed-out scales. As usual these terms have been taken to the right hand side of the field equations, where they are effective source terms in the smoothed-out field equations.

The implicit assumption in the usual approach is

that at Scale 5, the term P_{5ab} vanishes. Clearly this may not be true; therefore the problem is to make good estimates of the terms P in eqs. (3), in the context of the scales associated with cosmological models.

Intriguing models have been given in the linear case (Szekeres, 1971; Bailey and Israel, 1975). For example, Szekeres (1971, eq. (4.4)) obtains the form

$$P_{ab}=Q'_{ab}{}^{cd}{}_{,cd} \tag{3d}$$

where P_{ab} is determined as the double-divergence of an effective quadrupole gravitational polarisation tensor Q'_{abcd} with suitable symmetries

$$Q'_{abcd}=Q'_{[ab]\,cd}=Q'_{ab\,[cd]}=Q'_{cdab} \tag{3e}$$

(Q'_{abcd} also incorporates any dipole polarisation effects that may occur). Such models might adequately cover the larger-scale transitions (e.g. from Scales 3 or 4 to Scale 5); but the difficulties of handling the full non-linear case have so far prevented any solution in the cases of real interest to us (from the smallest to the larger scales).

3.4. Further Implications

Let us now assume that the answer to the major question raised in this section, namely:

QUESTION 2: On what scale do Einstein's Field equations hold?

is "Scale 1". Then once the form of the tensors P in eqs. (3) is determined, the further issue is what effect they will have on the dynamics of the universe (Scale 5). Thus,

(1) What difference would inclusion of the term P_{5ab} in eq. (3c) make to the expansion of the universe? (It should lead to an extra term in the Friedmann equation, and so to different expansion rates). What difference would inclusion of the term P_{3ab} in eq. (3b) make to studies of galaxy formation?

(2) In particular, would the effective stress-energy tensor on the right hand sides of equations (3b,c), e.g. (3d,e), necessarily obey the usual 'energy conditions' (Hawking and Ellis, 1973; Tipler et al., 1980)? There seems to be no reason why they should do so(10).

(3) If effective energy violations are possible, could they be sufficient to allow avoidance of singularities in the smoothed out (Scale 5) model? Describing the universe at Scale 1, it would contain many singularities, because of the occurrence of small-scale black holes; but the smoothed-out Scale 5 structure would be locally singularity free. Its large-scale singularity structure might be like that of the FLRW or Bianchi universes, but also in principle singularities could be avoided at Scale 5, if sufficient violation of the energy-conditions could occur. This is perhaps unlikely, but it is certainly an intriguing possibility.

4. IMPLICATIONS FOR THE GEOMETRICAL MODELS OF COSMOLOGY OF THE DIFFERING SMOOTHING SCALES USED

There are a series of problems for the geometrical models of cosmology, which also have their roots in the fact that we do not yet have a good theory for the smoothing operations.

4.1. The Definition of Spatial Homogeneity and Isotropy

A first need is for a viable definition of spatial homogeneity, that is, one that could in principle be verified or disproved by detailed examination of the geometry and matter content of the universe. We do not currently have such a definition. In particular, we need definite criteria that should be satisfied in order that we can legitimately regard the universe as well-represented by a FLRW model. For example, (a) if we discover that the density of matter in the local supercluster of galaxies is four times that outside (Tamman et al., 1980; Yahil et al., 1980), is this sufficient to disprove FLRW or not? Or (b) if number counts of distant galaxies to a specified limiting magnitude differ by a factor 2 over an arc of 40° in the sky (Heidmann, 1980, pp.17-18), is that sufficient? Or (c) if number counts of radio sources were higher by 1% on a scale length of half a Hubble radius in one direction than in the opposite direction, in a way inconsistent with the micro-wave background radiation anisotropy (Ellis and Baldwin, 1984) would that be sufficient? We cannot tell, for we do not have usable criteria with which these statements can be compared (cf. Wesson, 1981; de Vaucouleurs, 1970).

Underlying this problem are two separate issues. The first is a statistical one, the question of what degree of divergence from the idealised model is acceptable; this calls for a decision about acceptable limits of tolerance, and is simply a matter of judgement; but one about which a decision is necessary before the concept of 'homogeneity' becomes testable. The second is the question how one can measure the degree of difference of one space-time (in this case, the real universe) from another (in this case, the idealised FLRW models).

This issue can be tackled in two ways. On the one hand, one can consider this problem in general for any pair of space-times, perhaps considering the use of suitable metrics on the 'space of space-times' to represent closeness of fit. On the other, with the particular application in mind, one can consider the problem of defining in a rigorous way 'almost isometries', so that we can tell when a given space-time nearly has the symmetries of some idealised comparison space-time. Either way, the essential question is,

> QUESTION 3: How do large-scale (Scale 5) symmetries manifest themselves in the geometry of the more detailed (Scale 3 or Scale 1) structure of the universe?

The theory of homogeneous turbulence of a fluid can give useful guide-lines as to what is meant by an almost-homogeneous matter distribution (or by 'physical homogeneity', suggesting rule-of-thumb definitions, cf. Batchelor, 1964; Wesson, 1981; de Vaucouleurs,1970) but does not tackle the basic problem of determining almost-symmetries of a curved space-time. Matzner (1968a,b) has proposed a way of finding *almost Killing-vectors* in a positive-definite space-time, by minimizing the overall degree to which Killing's equations are not solved for a vector field. Thus he looks for vector fields minimizing the quantity

$$\lambda(\zeta) = \left(\int \zeta^{(a;b)} \zeta_{(a;b)} dV\right) \left(\int \zeta^a \zeta_a dV\right)^{-1};$$

thereby he arrives at the eigenvalue problem

$$\zeta^{(a;b)}{}_{;b} + \lambda_{(n)} \zeta^a = 0. \tag{4a}$$

Taubes, (1978), looks for solutions of the "almost Killing-equation"

$$\zeta^{(a;b)}{}_{;b} = 0 \tag{4b}$$

(which is just (4a) with $\lambda_{(n)}=0$; Taubes calls eq. (4a) with $\lambda_{(n)}=\frac{1}{2}$, the "conformal almost-Killing equation").

These closely related approaches go a long way towards what we need and in particular permit a measure of deviation from homogeneity. However for our present purposes they omit a vital point: we need to determine if there is a *group* of almost-isometries transitive on spacelike surfaces in the space-time considered, and these procedures do not answer this question. Spero and Beierlein,(1977,1978) have proposed a method of finding if there is a simply-transitive group of almost-isometries on a 3-surface, by minimizing the volume-averaged deviations of the rotation coefficients of an orthonormal tetrad from those in a Bianchi universe; this permits investigation of spaces admitting simply-transitive groups of transformations that are *'almost-symmetry' groups*, and also allows a measure of deviation from homogeneity. On this basis, they are able to determine which Bianchi types are stable almost-symmetry groups under small but arbitrary metric perturbations. Further investigation is needed to see if this proposal is fully adequate for the purposes outlined here. In general, we will be able to describe a particular space-time in terms of several different almost-symmetry groups; but then presumably one will 'fit' the space-time better than the others.

In the approaches mentioned above, the essential feature is that we expect there to be exact continuous groups of symmetries, generated by Killing vector fields, in the smoothed out model (Scale 5), so that there are almost-Killing vector fields in the more detailed models (e.g. Scale 3). Remembering the smoothing operations (eqs. 2), one might hope that a suitable definition of almost-Killing fields would satisfy the requirement: when smoothing (say from Scale 3 to Scale 5) was performed, the almost-Killing field should be mapped into a Killing field; and a group of almost-isometries, into a group of isometries.

However the universe is certainly fairly inhomogeneous, on a detailed level, at Scale 1 (and this might also be true at Scale 3 or 4). The above approach might not be suitable then, in that it cannot describe universes that, while being approximately homogeneous in the large, are far from homogeneous in the small. Two further possibilities arise in this case. The first is to try to define a vector field which is not *almost* a Killing vector field, but rather is one *on average*; that is, one might define an average symmetry rather

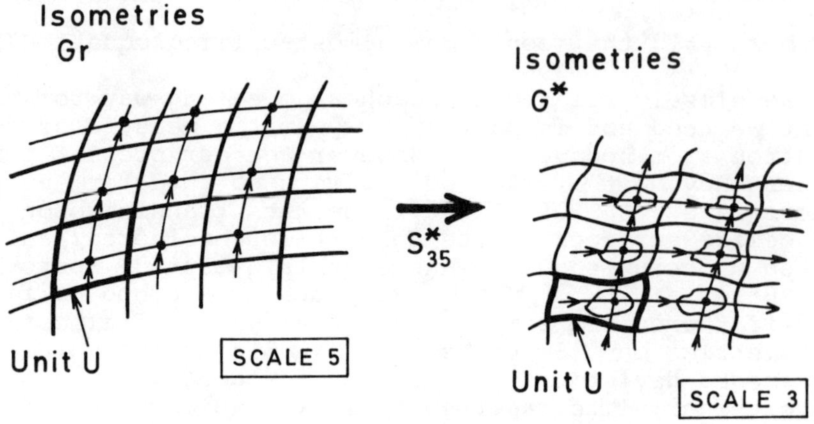

Fig. 8. A method of representing a universe which is inhomogeneous on a small scale but homogeneous on a large scale is via a discrete group of isometries G* which constructs a Scale-3 model by identical repetition of a small inhomogeneous unit U. The group G* should be a sub-group of the group of isometries G_r of the related Scale 5 description of the universe.

than an approximate symmetry. The same issues arise as in 'almost-symmetries'. One can test any proposed definition on an exact 'Swiss-Cheese' world model that is statistically homogeneous (cf. Dyer and Roeder, 1974, 1975; Kantowski, 1969; Bourassa and Kantowski, 1975), for this will be like the kind of universe we wish to describe.

A further possible approach is technically difficult but geometrically intriguing. This is that one could model the symmetry of the universe (at Scales 3 or 4) by the existence of a *discrete group of isometries* G*, which is a transitive subgroup of the continuous group of isometries G_r (acting at Scale 5). As there are no almost-Killing vector fields defined in this case, even determining what is meant by G* being 'transitive' becomes a problem. However the proposal has the merit of capturing the idea of a universe which on a large scale is uniform, but on a small-scale allows considerable deviations from homogeneity; it represents the universe as being built out of a repeating pattern of inhomogeneous units, thus generating the desired large-scale homogeneity (Figure 8). The two scales (say, 3 and 5) are then tied together by assuming that the group G* acts at both scales; at Scale 3, it is in general the largest group of isometries present, while at Scale 5, it is a sub-group of the full symmetry group. The group G* could be very complex (cf. Section 6.3 below).

Given this approach, we would still like a measure of the deviation from homogeneity within the fundamental 'cells'; this could presumably be done by a method based on searching for average-Killing vectors, and then measuring the deviation from satisfying Killing's equations; or by some method based on an examination of the variation of scalar invariants in these cells.

Such a model may sound a bit odd at first, with its precisely repeating patterns of structure. However reflection should show that it can give quite a good representation of the relation between smoothed-out FLRW universe models (at Scale 5) and more realistic models (at Scale 3 or 4); if the universe really is homogeneous in a reasonable sense, this is a good basic world-model on which perturbations may be imposed to represent details of the real distribution of matter. Indeed the size and nature of such perturbations could give an idea as to the degree of homogeneity of the universe. If such a model could not reasonably be fitted observationally, we would almost certainly have disproved the utility of the FLRW model at Scale 5.

An essentially similar set of questions arise in the context of *isotropies* (that is, symmetries generated by Killing vectors with fixed-points, which generate tangent-space isometries at each fixed-point and imply corresponding invariance of observational properties under subgroups of the rotation group (Ehlers, 1961; Ellis, 1967)).A usable definition is needed, i.e. one that can be contradicted by observation; similar geometrical questions to those discussed above, relating exact and approximate symmetries, underlie a realistic discussion of isotropy.

4.2. Geometrical and Dynamical Questions Arising.

Given suitable definitions of almost-homogeneity and almost-isotropy, a set of geometrical and dynamical questions arises.

The first is whether a generalization of the exact--symmetry results (Ehlers, 1961; Ellis, 1967) hold, namely, if almost-isotropy holds everywhere, does this imply almost-homogeneity? The answer certainly ought to be 'yes', but if the definitions are adequate this should be provable. This question underlies one of the major arguments used to give evidence for the homogeneity of the universe (see e.g. Hawking and Ellis, 1973, §10.1).

The second question is that of the stability of almost-homogeneity and almost-isotropy. If they hold at one time in an evolving universe, do they hold at other times? (In the case of exact symmetries, the result is true). Numerous studies of galaxy formation in a FLRW universe (Part V of Balian et al., 1980; Jones, 1976) suggest that almost-homogeneity is relatively stable at most stages of the evolution of the universe (and therefore it is difficult to form galaxies); such studies should be tied in to the precise definitions we have in mind here. Results by Collins and Hawking (1973b) suggest that almost-isotropy is unstable; it will not in general hold at all times. Again, this would need tying in to the precise kind of definition we have in mind here. If this is true, then observation of almost-isotropy suggests we live at a preferred epoch (unless some other principle intervenes to prevent anisotropies developing, cf. Section 6).

4.3. Observational Questions

Many authors (see e.g. Dyer and Roeder, 1974, 1975; Kantowski, 1969; Bourassa and Kantowski, 1975) have considered the effects on observations of being in a lumpy rather than a smoothed-out universe, that is, the effect of the transition from Scale 5 to Scale 3. Significant changes in observable relations are to be expected when this lumpiness is taken into account. Again, the theory suggests that gravitational lens effects will be important; and observations confirm they do indeed occur (Weedman et al., 1982; Canizares, 1981, 1982). Thus these observational implications of lumpiness have already been examined in detail. However various further issues arise from the discussion above.

First, just as in the case of the exact theory (cf. Section 2 above), one should determine from the geometrical definitions envisaged above, *real observational criteria* for almost-isotropy and almost-homogeneity. In the case of isotropy, the transition should be simple, for anisotropy of observations is directly observable. In the case of homogeneity, one may expect considerable problems in obtaining real observational criteria related to galaxy redshifts, number counts, magnitudes, etc. Nevertheless, it should be possible to state confidence limits that would be required for the desired fit of a FLRW universe model, if ideal observations were available. It is clear that such confidence limits must be stated before any test is done, and not after the fact! Unfortunately the observational difficulties discussed above (Section 2) that hinder verification of exact homogeneity, would make it very difficult to directly observationally verify approximate homogeneity.

In view of these difficulties, one should seriously consider the second issue, which underlies what we may really wish to do in practice (in a sense, this is the real question underlying the standard approach). Consider two space-times S_1 and S_2; suppose we have decided that we are going to model S_1 by S_2. What procedure should we use to obtain a *best-fit map* of space-time S_1 to space-time S_2? In particular, consider the universe described at Scale 1 or 3; what procedure will determine a best-fit of a FLRW model (at Scale 5)?

Note that this is essentially the *inverse* of the smoothing procedure discussed in the previous section: we wish to determine a point-map S^*_{35} and related maps of the matter tensor and metric tensor, from (smoothed-out) Scale 5 to (lumpy) Scale 3, in some optimal manner, for given space-times (g_{5ab}, T_{5ab}); (g_{3ab}, T_{3ab}) (Figure 9). Clearly one cannot simply try to minimize the difference between the components of corresponding tensors, as in general there will be no simple relation between the coordinate systems chosen in the two spaces.

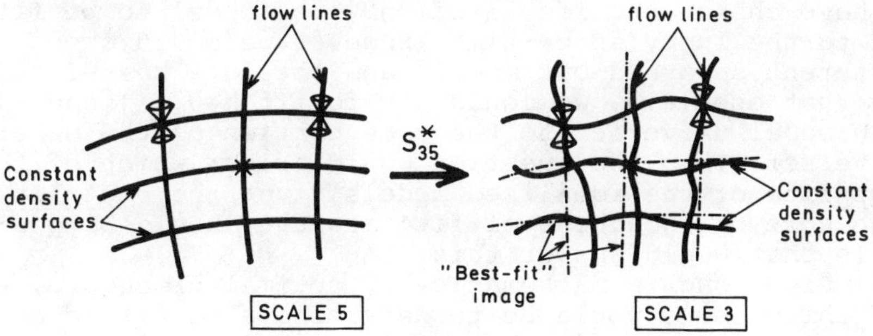

Fig. 9. We may wish to fit a specific smoothed-out (Scale 5) model to the real universe. The problem is that of determining a map from the Scale 5 model to a Scale 3 or Scale 1 model that gives a 'best-fit' of their metric properties and matter content. Thus one needs a measure of 'goodness-of-fit'.

Rather one would have to compare invariant quantities at corresponding points in the two spaces, for any particular correspondence proposed; and then choose the correspondence to minimize the difference, in some appropriate measure. The usual problems will arise from the fact that with an indefinite metric, invariants can vanish while the corresponding quantity does not. In fact the procedure will be intimately tied in with the equivalence problem and the theory of invariants (cf. Ehlers, 1981; Karlhede and MacCallum, 1982), for if the space-times are identical, then the procedure must prescribe that they be placed in precise alignment on each other. We also require that this process of obtaining a best-fit will in a suitable sense be the correct inverse of the smoothing-operation from Scale 3 to Scale 5; in particular, that if the smoothed model obtained from a more detailed description is indeed a FLRW universe, then the best-fit of the smoothed-out model will give the same correspondence from Scale 3 to 5 as the smoothing procedure. In a case where the smoothed-out universe exhibits symmetries, we may expect the procedure to throw light on the way the idealised symmetry is exhibited in the more detailed model, and hopefully on the degree to which the more detailed model approximates the idealised symmetries; and therefore should imply a criterion for when the attempt to fit the smoothed models is unreasonable.

This best-fitting procedure should be unique once we have chosen the idealised universe model to be fitted to the lumpy space-time. However we could choose a different smoothed-out model, and obtain a best-fit for that one (e.g. we could try to fit k=0, +1 and -1 FLRW model universes to the same portion of the observed universe). Then the question would arise; which of the best-fits of the idealised models,gives the best fit? Some sort of 'goodness-of-fit' measure should be possible that would answer this.

Given such a mathematically optimal procedure, the third issue would be to determine a realistic observational procedure that would *translate this optimal fitting process into practice*. In effect, observers are already engaged in this enterprise (e.g. by their efforts to determine the velocity of the local supercluster relative to a FLRW background, see e.g. Tammaunet et al., 1980; Yahil et al., 1980; here they are fitting observed velocity patterns to a hypothetical FLRW velocity field). They are doing so without the benefit of a sound theory as to how to obtain such a best-fit of idealised model universes to reality. We shall not be able to give this issue a serious answer until we have paid appropriate attention to the nature of the smoothing operations implied in our use of smoothed out universe models (cf. Section 3); and make a real attempt to translate our mathematical prescriptions into practicable observational procedures (cf. Section 2).

Clearly we will eventually wish to make such an attempt to fit smoothed-out universe models in the context of space-times obeying Einstein's field equations. Thus one needs a viable approach to the questions in the previous section, considering the form of the field equations in a lumpy universe, when making a comprehensive attempt to tackle the problems in this section, of observationally fitting smooth to lumpy universes. Particular problems may be expected to arise when the effective field equations (3) are substantially different at the two scales we are trying to relate to each other.

The above has referred to methods of implementing the 'standard approach' to cosmology. If one takes the 'observational cosmology' standpoint (Section 2), corresponding questions would arise: one would have to carefully choose on what scale one was trying to model the universe. Then actual observations would have to be 'smoothed out' in some way, to represent the metric and matter at the appropriate level of detail. Once a decision as to the representation scale had been taken, in principle this smoothing should be relatively easy; the essentials of the approach thereafter would be the same, whatever the scale of representation chosen (cf. Section 2). In practice, inevitably, problems would arise; they would particularly concern determining the average density and velocity of matter at a given distance down the past null cone. A clear approach to solving these problems will again depend on an explicit consideration of the different scales of representation involved in modelling the universe.

5. RELATION BETWEEN LOCAL PHYSICAL BEHAVIOUR AND BOUNDARY CONDITIONS IN THE UNIVERSE

Our understanding of the way local physical laws work is based on the concept of an "isolated system". However there are no exactly isolated systems in the real universe; and it seems likely that the effect of boundary conditions at the edge of space-time, and particularly at the beginning of the universe, may be crucial in determining such major features as the local direction of the arrow of time (cf. Penrose, 1979, 1983; Gold, 1967; Davies, 1978; this is not determined by fundamental local physical laws such as Maxwell's equations for electromagnetism and Einstein's gravitational field equations, for which both directions of time are completely equivalent). More fundamentally, in the context of the universe as a whole, the distinction between 'local physical laws' and 'boundary conditions' at the edge of space-time becomes questionable (cf. the closing remarks by Prof. Wheeler). Thus (see e.g. Ellis and Sciama, 1972), a significant question in cosmology is,

QUESTION 4: What is the nature of the interaction between local physical behaviour and boundary conditions in the expanding universe?

One wishes to understand what influence the universe has on an 'isolated system'; and what difference the use of realistic boundary conditions (i.e. the fact that the universe is not asymptotically flat) makes to those results that depend in an essential manner on asymptotic flatness (the 'peeling off' theorem, positive mass theorem, Black Hole uniqueness theorem, etc.).

5.1. Effective Infinity in the Expanding Universe

As discussed by Walker and Ashtekar, the analysis of isolated systems in general relativity is at present usually based on Penrose's concept of past and future null infinity ("Scri Plus" and "Scri Minus", each being a null surface), past and future timelike infinity (i^-, i^+, each being a point), and space-like infinity (i^0, a point). This extremely elegant formulation has led to many major results (see e.g. Hawking and Ellis, 1973; Tipler et al., 1980; Penrose, 1963; Ehlers, 1979), but is not really suitable for the analysis of isolated systems in the cosmological context.

What we want to describe is "an isolated system: a subsystem within the universe on which the effects of the rest of the universe are negligible" (Geroch, 1977). To put this in perspective, one should ask the question: *how far away is effective infinity for an isolated system in the expanding universe?*

To have a specific instance in mind, consider the solar system. A reasonable estimate for the distance one might regard as effectively being at infinity from the solar system, is one light-year out. This is far enough out, that distances within the solar system are relatively very small (the largest planetary orbit being approximately 0.0006 light-years); but it is not so far out that the gravitational effects of other systems have to be taken into account (if we chose a distance of more than two light-years, we should have to start worrying about the effects of neighbouring systems). Thus for purposes of analysing the solar system, we can regard a sphere at a distance of about one light-year as being effectively at infinity; and such a sphere 'isolates' this system, i.e. separates it from the rest of the universe.

In the space-time context, one can then visualise the world-line of the sun as being contained within a tube of radius one light year, representing the history of such an 'isolating' sphere. The interior of the tube ('Int') contains the solar system, which is the isolated system we wish to study; the region outside ('Ext') is the remainder of the universe, which - in this context - we do not wish to investigate (but we do wish to understand the influence of this outside region on the physics inside). We shall label the dividing surface (a timelike cylinder of radius one light year, see Figure 10) as 'Fi' (for *Finite Infinity*). The crucial feature to note is that the life-time of the solar-system as an isolated system may be about 10^9 years. Thus *Fi will be an extremely long, thin timelike tube*. Even if we restrict our attention to the period of reasonable quality astronomical observations (say, 10^2 years) the corresponding Fi is still a long, thin tube. Thus in the cosmological context, this is the correct surface on which to impose incoming and outgoing radiation conditions. More precisely, we need to consider the tangent bundle to Fi, and define from this the family of incoming null vectors 'In' (which pass from the exterior to the interior, in a future-directed timelike sense) and outgoing null vectors 'Out' (which pass from the interior to the exterior). Measurements on In and Out will specify the amounts of incoming and outgoing radiation; these correspond to Scri Plus and Scri Minus. A spacelike section of the tube will be a spherical surface 'Sp', which is effectively at infinity and enables one to measure the Coulomb field of the system (at any given time corresponding to the spatial section chosen); thus this family of surfaces is the analogue of i^0. Finally past and future timelike infinity i^-, i^+ correspond to spacelike surfaces 'Sur_-', 'Sur_+' with boundaries Sp_-, Sp_+, that bound Int to the past and future respectively and mark the beginning and end of the time-period of interest in the study; one can examine the 'initial' and 'final' states of the system on 'Sur_-' and 'Sur_+'. In principle one could consider the limit as these sections go to the past and future infinity of time respectively; but this would be against the spirit of the proposal made here, for we are

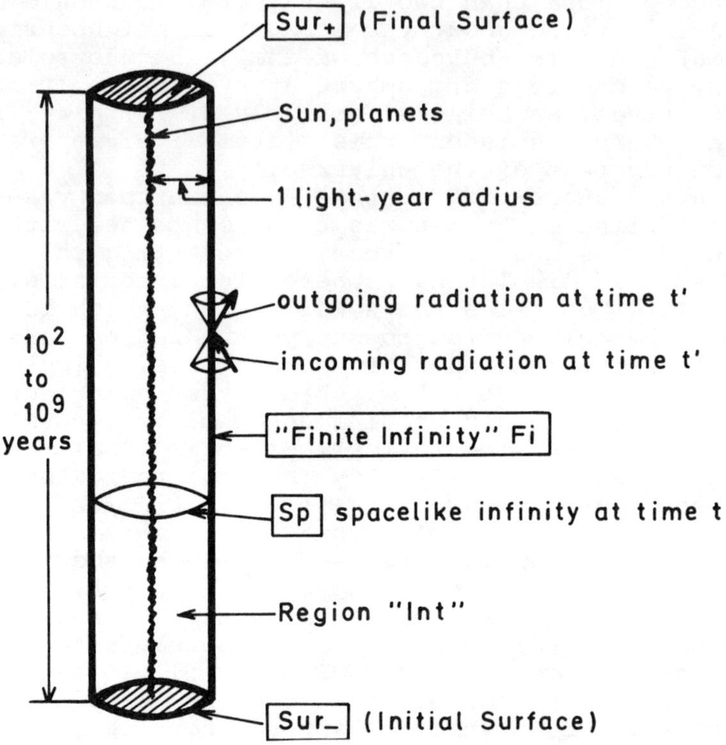

Fig. 10. Space-time diagram of 'Finite Infinity' Fi for the Solar System. This is the tube representing a sphere of radius about 1 light year, surrounding our world line l_0 from an initial surface Sur_- to a final surface Sur_+. The isolated system considered is the interior region Int bounded by Fi. Incoming matter and radiation conditions can be imposed on Fi, and outgoing matter and radiation measured there also.

actually usually interested in the behaviour of the system for large but finite times. Thus for the purposes of investigating the behaviour of astrophysical systems, finite time bounds will be appropriate. (This is emphasized by the fact that in a k=+1 FLRW universe, one could not extend Sp_-, Sp_+ to infinite times because the universe would come to an end first!). Finally, incoming and outgoing massive particles can easily be examined on appropriate sections of the tangent bundle

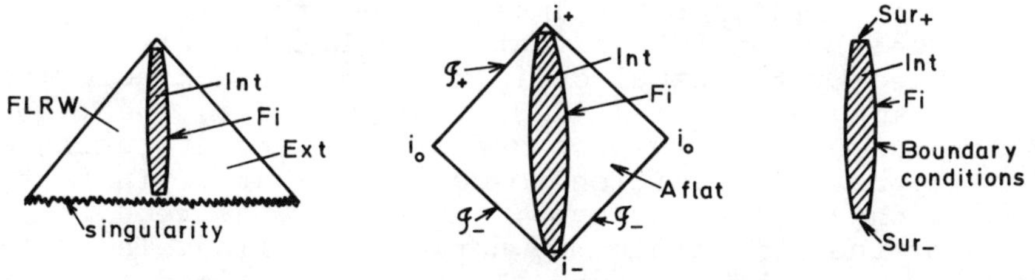

(a) (b) (c)

Fig. 11. (a) In a FLRW universe, Fi is surrounded by a region Ext which is matter-filled and is not asymptotically flat. Note that we cannot take the limit as $Sur_- \to -\infty$, because the universe started a finite time ago.
(b) In the asymptotically flat representation, we surround Fi by an exterior empty space-time region Aflat, with properties at infinity similar to those of the Schwarzschild solution.
(c) In the finite boundary representation, we simply consider the region Int bounded by Fi. Boundary conditions are stated as conditions on Sur_- and on Fi.

of Fi.

Comparing Fi and Scri, it becomes clear that Scri is too far out for many purposes. Consider the region of a realistic cosmological model isolated by Fi (Figure 11(a)). When we use Scri, we replace the region Out of the real universe outside Fi by an asymptotically flat space-time region 'Aflat' which goes to infinity (Figure 11(b)); and try to do so in a way compatible with Einstein's vacuum field equations. But the universe is almost certainly not asymptotically flat. Thus by doing so, we are including in our space--time model an inaccurate description of an infinite system we do not really wish to describe.

Various problems result. Firstly, our model now represents an exactly isolated system; thus it cannot be used to consider the kinds of questions we wish to consider here concerning the relation of the isolated system to the rest of the universe.

Secondly, because Scri is too far out, it may not really describe the physical situation we wish to investigate. For example, it is unrealistic when considering local gravitational collapse to worry about observers with infinite life-times because they (and their descendents) cannot live that long; but this is what is implied by the usual definition of a black hole (see e.g. Hawking and Ellis, 1973; Tipler et al., 1980; DeWitt and DeWitt, 1973) based on Scri. We are using a globally based definition to describe what is in fact a local physical phenomenon; but the global requirements of the definition may not be satisfied. The problem is highlighted by the sitation (mentioned above) of a k=+1 FLRW universe, where even the definition of a black hole is a considerable problem: there is no future infinity to use as a basis for the definition (as the universe recollapses in a finite time), so the usual definition is not applicable; special proposals have to be made to deal with this situation (Tipler, 1977; Krolak, 1983). Similar problems arise with other concepts determined by the use of Scri. For example, Hawking remarks in the context of a discussion of the BMS group, "if the asymptotic group has any physical significance... this significance presumably would attach to the asymptotic group corresponding to the real universe and not the one for asymptotically flat, empty space" (Hawking, 1968).

Thirdly, as emphasized by Ashtekar in his talk, problems arise with the existence of Scri; although major steps have been made in this regard recently, it is still not known if any asymptotically flat space--times exist which represent realistic isolated radiating systems and fulfil all the conditions usually demanded for infinity (cf. Ehlers, 1980). In a sense this is not surprising, for one is extending data on Fi for an infinite distance, using Einstein's vacuum field equations in the process, and it is not clear why a good asymptotically flat space-time should result from this process, particularly as this is required to happen both to the past and the future.

Finally, this infinite distance may even lead to problems in using Scri to fulfil its role of representing incoming and out-going null fields in a completely isolated system (cf. Geroch, 1978), for essentially the same reason; unlike the situation in the real world, these fields are depicted as propagating in from an infinite distance. This may lead to problems with divergences at Scri (Porrill and Stewart, 1981). Consequently, for example, a full interpretation of the Newmann-Penrose 'conserved quantities' may need consideration of spheres at a finite rather than an infinite radius (Press and Bardeen, 1971; Goldberg, 1972). Certainly Scri does not give a good description of incoming and outgoing massive fields (which is the reason for the search for projective and other space-time boundaries, cf. Schmidt, 1974).

We have considered these issues on the scale of stars (i.e. Scale 1); but similar considerations will hold at the scale of galaxies (Scale 3). The proposal, then, is that *in the cosmological context, incoming and outgoing matter and radiation affecting isolated systems should be analysed on an effective 'Finite Infinity' Fi, rather than Scri* (Figure 11(c)). Many of the standard results and methods could be taken over virtually unchanged; but the aim would be to obtain bounds on the remainder terms in the expansions (cf. the illuminating paper by Kerr and Goldberg, 1964 for the case of the electromagnetic field in flat space-time), so that they could be evaluated and bounded on Fi (at a finite distance from the source; cf. Figure 10).

5.2. Definition and Existence of Finite Infinity Fi

In pursuing this idea, the first issue one would have to consider is a usable *definition of Fi*. The requirements would be that it was a surface of topology $R\times S_2$ that

(Fi:a) surrounds the system of interest at a distance large compared with its spacelike length-scales,

which

(Fi:b) lies in an asymptotically flat region of space-time,

and on which

(Fi:c) incoming matter and radiation is appropriately bounded.

The first condition embodies the idea of looking at the system from the far zone, where outgoing radiation conditions may be expected to hold. The second contains the idea of 'gravitational isolation'. Determining if Fi lies in an asymptotically flat region again leads back either to the concept of obtaining a 'best-fit' of regions of two space-times, and a measure of lack of fit; or to the concept of almost-Killing vectors and approximate symmetry groups considered in the previous section. Either way , one will need to specify what is an acceptable deviation from flat space-time for which one will still accept that Fi lies in an asymptotically flat region (implying upper limits on the Weyl and Ricci tensors there). In practice one will be able to use asymptotically flat coordinates in these regions, which satisfy all the usual weak field conditions, and these coordinates (determined up to an asymptotic symmetry group) can usefully be used in analysis of the questions of interest. One might note here that at a distance of one ligth - year from the solar system, the expression (1-2m/r) takes a value of approximately $1-10^{-12}$, so the Schwarzschild solution there will indeed closely approximate flat space-time.

The third condition embodies the idea that one can (for physical purposes) regard the system as free from major interference from outside (Sachs, 1961; Penrose, 1965; Ehlers, 1973); such 'non-interference' conditions are needed in order to predict the future of a system in the real universe, as was emphasized in Section 2. However, it is unrealistic to set these limits to zero (for example, our observations of distant regions of the universe rely on this radiation being non-zero; thus we wish there to be a measurable interference term!) Note that it is proposed here to put limits on *incoming* fields, but not on those *outgoing*. Thus the system could conceivably satisfy these conditions and not be 'isolated', in that it could be emitting major quantities of matter or radiation. Given the isolation of the system from incoming disturbances, this would be a question to determine by investigating the system behaviour; and the proposal is that one should not prejudice the result of the investigation by also imposing outgoing radiation conditions (other than those implied by requiring asymptotic flatness).

Fi is not uniquely determined by these requirements: so there is considerable *arbitrariness* in its choice. Thus the position of Fi can be varied significantly without violating these conditions (e.g. its distance from the system considered might be varied by a factor of between 2 and 4). Assigning a characteristic 4-velocity to the 'walls' of Fi at each point, there will also be limitations on the change in this 4-velocity when making such variations; for if this change is too drastic, the incoming-radiation conditions will no longer be satisfied (cf. the next sub-section). One should note here that these restrictions on the choice of Fi are independent of the freedom of choice of appropriate asymptotically flat coordinates (which, if one was perverse, could be chosen in a way that did not suit the choice of Fi).

The question of *existence* theorems would come up somewhat differently in the context of Fi, than in the case of Scri. In effect the question from the cosmological viewpoint is not, why is space-time asymptotically flat at great distances away from the system considered, but rather, why is space-time asymptotically flat as one shrinks down to the line Fi in the universe? (cf. Figure 11(a)). There are basically two ways to approach this. One is to say that in the cosmological context, the time-scales we are considering are such that most systems we wish to analyse can be considered as occurring at a space-time point in the smoothed-out (Scale 5) cosmological model; then the basic result is effectively the standard theorem that in a small enough region, *any* space-time looks flat (and one can use normal coordinates to express this flatness). Fi can then be defined as a bounding surface in such a small asymptotically flat region in an expanding universe (there are of course still non-trivial problems regarding the junction of the 'interior' and 'exterior' solutions; but they do not involve problems due to infinities in the distances considered).

However when we wish to consider systems with life-times long compared with the life of the universe (e.g. the long-term history of the solar system), this approach is inadequate. Two possibilities arise. On the one hand, the system we consider may be a gravitationally bound system, effectively isolated for a long time period from the rest of the universe; in this case it should be surrounded by a suitable asymptotically flat region with almost-Killing vectors, where we can suitably choose Fi. On the other hand, this may not be true. In this case we have two choices. One is to abandon the attempt to discuss an isolated system. The other is to consider weakening the definition of Fi above. This might be usefully done in various ways. Particularly we might drop the asymptotically flat condition (Fi:b), for example instead choosing comoving (expanding) boundaries at a large distance from the system in a FLRW universe model. In this case, the spherical symmetry of the FLRW model would still guarantee 'gravitational isolation', and (Fi:c) that the system was otherwise isolated. This could be useful in examining an issue which has explicitly to be considered in the context of the expanding universe. Again, we could loosen (Fi:c), replacing the condition of isolation by a condition of equilibrium (or statistical balance); then instead of requiring that exterior regions do not interact with the system, we require that *on average* this interference is zero. While such extensions of the concept can well prove useful (cf. e.g. Harrison's discussion of thermodynamics in an expanding universe in Harrison, 1968, we will not pursue them further here.

5.3. Applications of Fi

There are two distinct but related uses of Fi. First, it could be used as the boundary for the study of isolated systems, e.g. when examining the radiation emitted by the system (cf. the lecture by Walker; it could be used in a variant of the 'initial value' formulation he mentions). In this case (see Figure 11(c)), one would set the incoming matter and radiation terms to zero, or (assuming they are of arbitrary form) put very small bounds on them. As mentioned above, the aim would be to bound appropriate quantities on Fi and take advantage of the approximate symmetry groups in the asymptotically flat region; one should then e.g. be able to prove that energy and momentum are conserved up to some specified limit of accuracy. The view taken would be that the essential physics of the situation is expressed in these finite, non-zero bounds. In this way one should be able to obtain suitable variants of the energy-momentum conservation laws (Horowitz and Schmidt, 1981; Penrose, 1982; Schattner and Streubel, 1980), positive mass theorem (Parker and Taubes, 1982; Gibbons et al., 1983), etc., in this context. Particularly, one would aim to give a definition of a black hole in this framework and then develop suitable versions of the black hole uniqueness theorems (DeWitt and DeWitt, 1973) in this context.

Secondly, Fi can be used in the study of the interaction of 'isolated systems' with the universe (cf. Figure 12(a)). Here one would specifically aim to examine the effect of non-zero interaction terms appropriate to the expanding universe context, where one would either know the form of these terms or try to determine this form as one of the main objects of the study. Thus one would relate initial conditions at the beginning of the universe with incoming radiation conditions on Fi, representing the immediate effect of distant regions in the universe on the isolated system; and outgoing radiation conditions on Fi to outgoing radiation conditions at the boundary of the universe. The outgoing radiation will have a different peeling-off behaviour in the asymptotically flat region than in the limit at infinity in the expanding universe (Hawking, 1968). One should be able to obtain total energy-momentum balances involving all the incoming and outgoing matter and radiation, and analogues of theorems such as the positive mass theorem (Parker and Taubes, 1982; Gibbons et al., 1983) in this context.

As an example, consider a change of velocity of an 'isolated system' to relativistic speeds (Figure

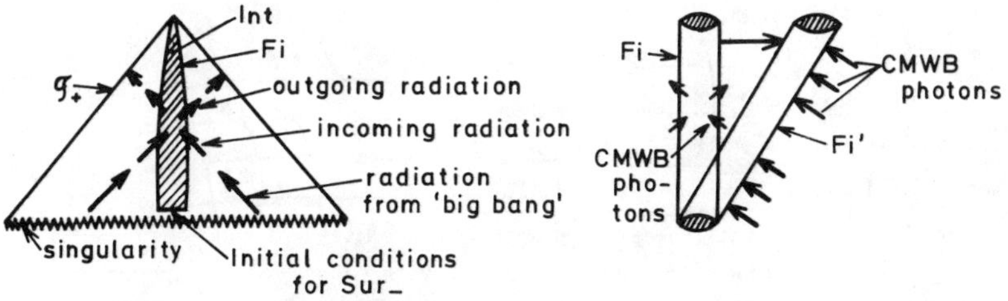

(a) (b)

Fig. 12. (a) The universe initially sets up conditions within Fi by determining conditions on Sur_- from those at the 'Big Bang'; thereafter, it interferes with conditions inside by sending matter and radiation in across Fi. The system responds by returning radiation and matter across Fi to the universe. Boundary conditions at the beginning and end of the universe are therefore related to the matter and radiation crossing Fi.
(b) The velocity of Fi cannot be changed too much, or the cosmic microwave background radiation will be experienced as high-energy radiation on Fi, exerting appreciable radiation drag; and the system will no longer be effectively isolated.

12(b)), resulting in a new boundary Fi'. In the new frame, the cosmic microwave background radiation will be experienced as high energy photons (X-rays and γ-rays) that will significantly interfere with the system, e.g. by exerting radiation pressure on it. Thus in this frame, the system will no longer be effectively isolated; this will be manifest in a significant amount of radiation crossing the surface Fi'. Hence the velocity of an 'isolated system' relative to the rest-frame defined by the cosmic background radiation, is limited to small values. One should note that this velocity limitation, needed to preserve the defining condition (Fi:c), is physically essential.

As a second example, consider the Solar System. A simple application to the Earth-Sun distance (500 light-seconds) of the Hubble law $v=H_0 r$ relating relative radial velocity v to relative distance r, with H_0=100 km/Sec/Mpc, leads to the conclusion that the orbit of the Earth expands by 1.5 km per century. Should

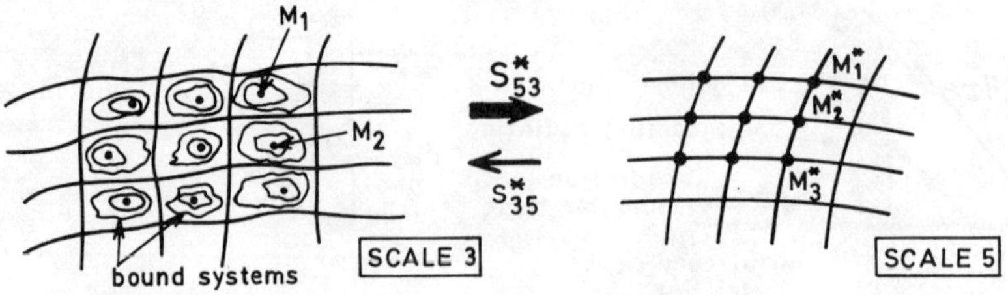

Fig. 13. When gravitationally bound systems exist at Scale 3, the smoothed-out model at Scale 5 refers to the motions only of the centres of gravity of these systems. Thus the representation from Scale 5 to Scale 3 maps isolated points M_i^* at Scale 5 to these centres of gravity M_i; they move with the Hubble flow. The motions of points between the M_i^* in the smoothed-out model do not correspond to the motion of particles in the corresponding positions in the real universe.

we attempt to measure such an effect, or not?

In fact, we are considering a gravitationally bound system; the initial kinetic energy of expansion has been countered by gravitational attraction, and the present expansion of the universe has no 'handle' with which it can now make this system expand (cf. Callan et al., 1965; Noerdlinger and Petrosian, 1971); so we do not expect this effect to occur. However the example raises serious issues. If the Hubble law does not apply at the scale of the solar system, at what scale, then, do we expect it to become relevant - Galaxies? Clusters of Galaxies? Superclusters? Below the scale chosen, the Hubble law will be inapplicable, because the matter is in bounds units not taking part in the 'Hubble Flow'. We need clearly stated criteria for gravitationally bound objects defining the 'units' comprising the expanding universe. This determination is essential, but it is non-trivial, as it is closely tied in to a solution to the 'missing mass' problem.

What is at issue is the nature of the map from Scale 5 to Scale 1 and Scale 3 in our models of the universe. The implication is that this is not really a continuous map at all; but rather that isolated points M_i in the smoothed-out universe model are mapped to whole regions in the more realistic models, these points representing the centres of gravity of gravitationally bound systems (Figure 13). The relative motion of these centres of gravity is determined by the Hubble law, and corresponds to the description at Scale 5. Other points of the smoothed-out model do not correspond to motion of matter in the real universe. Thus the Scale 5 FLRW models are tacked on to reality at the isolated points M_i (the bound systems and the detailed motions of their components are represented only in the more detailed descriptions). This has substantial implications for our attempts to determine if the FLRW model is a good fit or not, and for determining the best-fit procedure (cf. Section 4.3); it again points to the idea of using discrete symmetries to model lumpy universes (cf. Section 4.1).

Gravitational binding of these systems may be the process leading to the existence of approximate time-like Killing vectors, and so to the corresponding energy conservation laws (Horowitz and Schmidt, 1981; Penrose, 1982; Schattner and Streubel, 1980). Thus an essential feature of existence theorems for Fi will be a determination of conditions for formation of such bound systems exhibiting almost-symmetries. Closely related is the nature of the effective domain of dependence associated with such bound systems (see Figure 14). Provided the bound system remains isolated, i.e. a suitable 'no-interference' condition is satisfied, the effective data really determining conditions at a point Q at late times is bounded to within a relatively small region of the past world-line of the centre of mass M_i; it is much less than indicated by the past-light cone of Q. This is the reason we can predict conditions in the solar system to the future of our past null cone (as discussed in Section 2).

One can explore the relation between the bounded systems and the expanding universe by considering what happens as Fi is enlarged to encompass a number of systems (which together can be regarded as a single, more complex system; this is closely related to the transition from a Scale 1 to a Scale 3 description and from a Scale 3 to a Scale 5 description). If one considers moving a surface F surrounding an isolated system away from that system, the geometry of F will initially be highly curved in general (reflecting the near-field nature of the system); it will gradually get more and more like an asymptotically flat space-time, until one reaches an optimal 'best-correspondence' distance D_0; and will then become less like a flat space-time for larger distances as either other neighbouring systems, or the expanding universe, become predominant. This behaviour will be reflected in many pro-

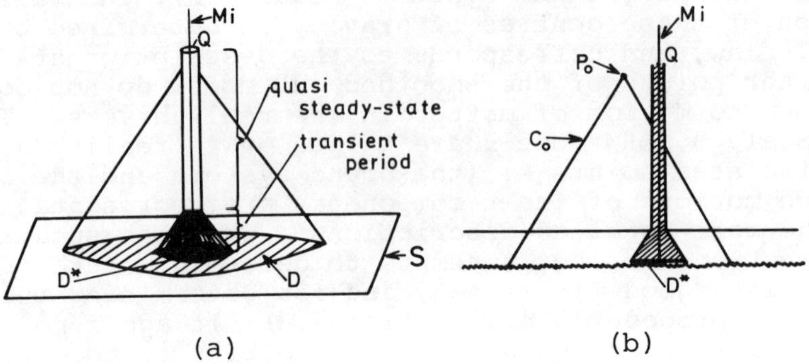

Fig. 14. (a) When quasi-stationary gravitationally--bound systems form in the universe, there is an initial transient period but thereafter the system is largely independent of conditions elsewhere. Thus providing the system remains isolated, the effective initial data D* is as shown; this is much less than the full null-cone data D relevant if unbound high-energy or zero-rest mass particles dominate.
(b) It is for this reason that we can predict the behaviour of bound systems at the point Q to the future of our past null cone C_0; for the data at D* is determined by data on C_0.

perties measured on F. For example, a solution of the massive or massless scalar wave equation will be different depending if F is large enough to encompass the expansion of a region of the universe; and this is related to the issue of particle creation in an expanding universe (Waylen, 1978; Gibbons and Hawking, 1977; Parker, 1982; Hawking and Page, 1983; Hawking and Moss, 1982; Mottola and Lapedes, 1983). This again points to the importance of determining where the boundary between 'bound' and 'expanding' systems will occur. At late times in the universe there will be *hierarchies* of bound systems (for they will occur at each of the Scales 1 to 4 considered in Section 3), and the description proposed here may be useful in considering their relation to each other, and hence the description of matter tensors at the various scales (I am indepted to David Matravers for discussions on this point).

Generally, in the context we have in mind one will aim to use bounding methods rather than infinite series, and to use inequalities rather than approximate equalities. However in some situations one may profitably use exact jump conditions instead, joining two regions together across a boundary where discontinuities occur; and perhaps use measures of the magnitude of the jump (e.g. that from the space considered to flat space-time, or to a FLRW universe) to estimate features of physical importance such as the degree of deviation from flatness.

Within the general framework outlined, with a 'local infinity' Fi, one will be able to return to consider the major issues underlying our concern for the relation between local physics and global physical behaviour; i.e. to consider the effect on local physics of the expanding universe (Ellis and Sciama, 1972). Thus the existence of event horizons in the universe may be important for the inflationary scenario (Hawking and Moss, 1982; Mottola and Lapedes, 1983), and there can be a minimum temperature for the radiation emitted by a Black Hole in some universe models (Hawking and Page, 1983), both reflecting the influence of the universe on local physics; one would wish to look at such effects in this framework. Particularly the problem of the arrow of time mentioned at the beginning of this section, and more generally of the consistency of local physical behaviour with global properties of the universe, are significant issues. A description of isolated systems such as suggested here may be helpful in pursuing these concepts.

6. SYNTHESIS: TOWARDS COSMOLOGICAL THEORY

The major problem in Cosmology is in justifying a claim of *Uniqueness* for a proposed universe model. One can use some combination of philosophical principles and observational data to obtain a unique model (it is necessary to introduce philosophical principles because a purely observational approach is in practice unable to determine a unique universe model, cf. Ellis, 1979, 1980a,b); but there are various competing such approaches. Because of the unique nature of the cosmological problem, these alternatives should be clearly distinguished and their relative merits considered with some care. Underlying is the question of why particular initial conditions have occurred in the universe, resulting in the actual situation we observe at the present.

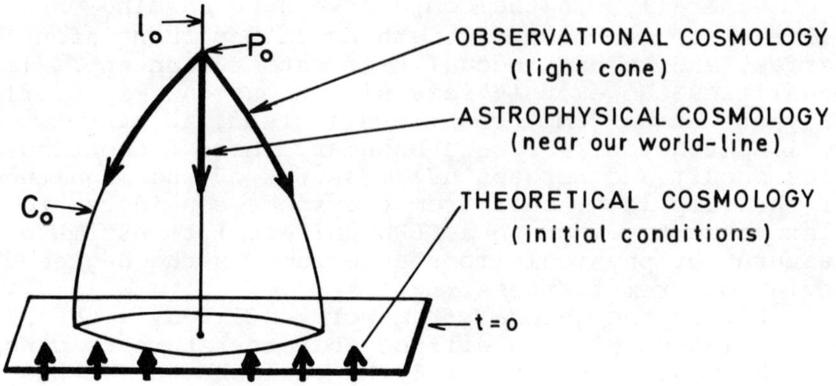

Fig. 15. Three conceptually different approaches to cosmology. Observational Cosmology attempts to determine conditions on our past light--cone, and to deduce conditions elsewhere from this data. Astrophysical Cosmology attempts to determine conditions near our world line (to very early times). Theoretical Cosmology attempts to prescribe initial conditions from which we can deduce later conditions.

6.1. Different Approaches

One possible approach is to just observe the geometry of the universe, and accept without further question whatever initial conditions are deduced. However this is ultimately unsatisfying (McCrea, 1970); rather than accepting this, the general tendency is to look for some underlying explanatory principle.

Different approaches to cosmology emphasize the various aspects of the problem. It is useful to distinguish these aspects by considering the way the terms 'Physical Cosmology', 'Observational Cosmology', and 'Theoretical Cosmology' are used. One may conceptualise this difference as follows (Figure 15): *Physical Cosmology* refers to investigation of processes such as star formation, galaxy formation, and nucleosynthesis near our world line in the past, up to very early times. *Observational Cosmology* refers to determination of conditions on our light cone (cf. Section 2) up to very large distances from us, and thereby (through use of the field equations) to estimation of conditions in the interior of our light cone. *Theoretical Cosmology* refers to proposing on the basis of theoretical considerations what initial conditions must have been, and working out the consequences of these conditions. These terms will be used in the sense made explicit here, in the rest of this article.

To satisfactorily reconcile theory and observation, we need to pursue all three of these approaches. It is suggested that in order to make maximum use of the limited data available to us, we should follow each approach as far as possible independently of the others, and only consider compatibility between them when each has been taken to its limits. Particularly useful are attempts at reconciling the different approaches far out from our world line. For example, determining element abundances as far out as possible along our light cone (by using high red-shift data to analyse element abundances), and then using the physical cosmology approach to determine conditions in the interior of the light cone from this information (Figure 16), probably gives the best method available for investigating very early conditions far out from our world-line. The results of this process can then be compared with implications of the observational and theoretical approaches for the same region (Figure 17), which in principle can give independent estimates of condition there. Needless to say there will be practical difficulties in doing this, but in principle this is the way to extract maximum information from the data. This kind of coherence check is obscured if one initially assumes a FLRW geometry, for then one simply looks for consistency with that geometry of the data analysed by each of the approaches, rather than pushing the data as far as it will go in each case before making comparisons with the other approaches.

6.2. Cosmological Principles

While the Theoretical Cosmology approach is an essential component of any satisfactory theory, the major problem arising is that various of the plausibility or legitimising principles we may use are in conflict with each other. This is obscured in the usual approach (see e.g. Hawking and Ellis, 1973; Weinberg, 1973; Gunn et al., 1978; Part I of Balian et al., 1980), where a particular one of various possible philosophical principles is picked out and emphasized as *the* 'Cosmological Principle', to the detriment of the others.

To clarify this situation, one needs a list of the various philosophical principles that can underly plausible cosmological models; and then to weigh up

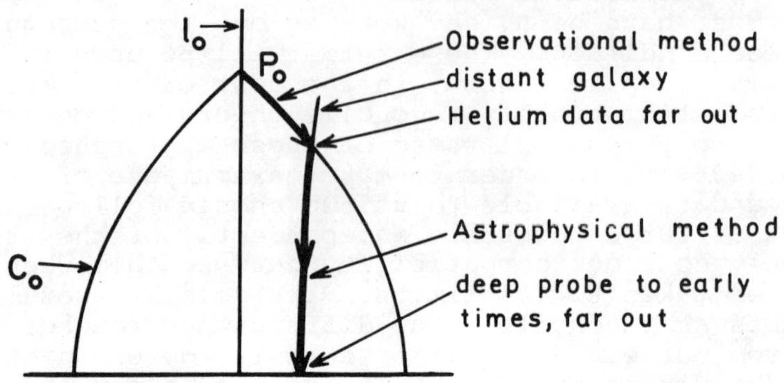

Fig. 16. If we can observationally determine astrophysically important data, e.g. element abundances, far out on our light cone, then we can use astrophysical methods to determine very early conditions, far out from our world line.

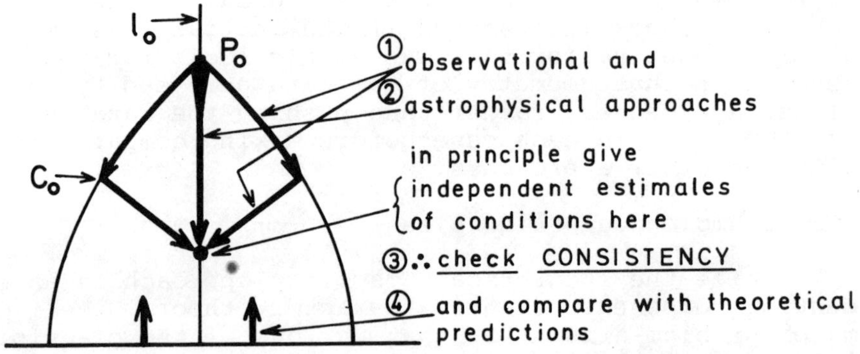

Fig. 17. In principle we can independently use each of these approaches to determine conditions at points in the interior of our past light-cone, and then check consistency of the predictions of the different approaches. This is probably the best way to extract maximum information from the available data.

their relative merits in the light of their consistencies and inconsistencies with each other. This will be briefly considered in the remainder of this section. One may hope that various of the inconsistencies can be removed by introduction of a unifying concept, or meta-principle, providing a unity to the various plausible principles. This will be considered in the following section.

All sensible approaches rely on the *Uniformity Assumption* U, that the laws of physics are the same everywhere in the universe, and accept the principle of *verification* V, that the ultimate test of proposed models lies in their compatibility with observations. A particular set of further cosmological principles, implicit or explicitly discussed in the literature (Ellis and Harrison, 1974; Harrison, 1974), are as follows[11]:

(a) The *Copernican* or *Spatial Homogeneity* Principle, SH. Combined with the assumption of *Spatial Isotropy*, I, this gives the usual *Cosmological Principle*, CP, underlying the standard FLRW universe models.

(b) A *Simplicity Assumption* S, stating that one should use the simplest possible universe model compatible with observations. This is the cosmological version of Occam's Razor.

(c) A *Probability Principle* P, stating that the universe model should be one that is a probable model within the set of all universe models (Penrose, 1979, 1983). This is closely related to a *Stability Assumption*, St, stating that the universe should be stable to perturbations.

(d) *Mach's Principle* M1, interpreted in the Einstein-Wheeler sense (Einstein, 1950; Wheeler, 1964, 1968; Isenberg and Wheeler, 1979) that viable universe models must have compact space-sections so that no boundary conditions have to be posed at spatial infinity.

(e) The *Anthropic Principle* A1, interpreted in the weak sense that the laws of physics and their boundary conditions in a cognisable universe must be compatible with the existence of intelligent life (Dicke, 1961; Carter, 1974, 1983; Carr and Rees, 1979); and so for example must allow heavy element formation (and probably star and planet formation).

Each of these has merit in its own right; not all are as well defined as the others, and some may be given various interpretations. Nevertheless it is clear they are in general not consistent with each other. Particularly, the Probability Principle P and Stability Assumption St are in conflict with the other principles. One way to resolve this is to reject these two principles, maintaining that in the cosmological context they are meaningless (because of the uniqueness of the Universe). However they are very powerful concepts that most physicists will be very unwilling to discard, even in the cosmological context (see e.g. Collins and Hawking, 1973; Ellis, 1979; Siklos, 1981; Penrose, 1979, 1983; Misner, 1968, 1969; Peebles, 1972; McCrea, 1970; Zeldovich, 1973a; Barrow, 1977; Eddington, 1930). Alternatives are to keep these two principles and abandon some of the others; or to seek a unifying concept that removes the apparent irreconciliability. Similar problems arise in the relation between various other of the cosmological principles. Priorities amongst them as desirable features in a cosmological theory, and ways of resolving their inconsistencies, need careful consideration.

Finally one may note that use of the 'cosmological principle' SH often almost casually introduces a spatial infinity, as in the cases of the k=0 and k=-1 FLRW universes. This has implications which, while not contradictions, are certainly awkward; namely (a) the existence of infinite numbers of 'identical twins' to each one of us (Ellis and Brundrit, 1979); and (b) the physical creation instantaneously of an *infinite* amount of homogeneous matter and radiation. One might perhaps wish the approach adopted to avoid such philosophically awkward situations, e.g. by adopting the Einstein-Wheeler suggestion of a spatially-compact universe (cf. the next section).

6.3. Unifying Principles

The basic issue we are concerned with can be summarised:

> QUESTION 5: How can we best synthesize the sometimes conflicting requirements of the different approaches to cosmology, observational verification, and the various philosphical principles of theoretical cosmology?

Various synthesizing approaches can be introduced for reconciling the different philosophical principles involved. We look briefly at some of these.

(1) A *Creation Theory* aims to describe the creation process, or at least the immediate results of that process, thereby necessarily implying the nature of the universe and relating previously unrelated features. Examples are the theories presented at this conference

by Hawking and Vilenkin. While this approach offers hope of a 'final solution' to the problems considered, conceptual difficulties arise with such approaches; particularly, it is unclear how one could hope to prove *uniqueness* for any posited theory (i.e. show that alternatives could not occur). It is particularly difficult obtaining a theory that explains observed features on *all* the scales considered here (Scales 1 to 5).

(2) Application of *Physical Principles* can imply relations between aspects that otherwise seem unrelated, so providing the desired unification; particularly Misner's *Chaos Theory* (Misner, 1968, 1969; Peebles, 1972; McCrea, 1970; Zeldovich, 1973a; Barrow, 1977) tries to show that generic initial conditions evolve towards particular (FLRW) conditions. Thus it aims to show consistency between the Probability and Uniformity principles. This is only partially successful, as it can explain reduction of inhomogeneity and anisotropy only for a restricted set of initial conditions (MacCallum, 1979a, 1979b; Hawking, 1969; Collins and Hawking, 1973a; Stewart, 1968).

The recent applications of the Grand Unified Theories ('GUTS', Balian et al., 1980, part IV; Rothman and Matzner, 1982; Fry and Turner, 1983) and analyses of the 'Inflationary Universe' (Gibbons, Hawking and Siklos, 1983; Hawking and Moss, 1982; Guth, 1981; Linde, 1982; Albrecht and Steinhard, 1982; Barrow, 1983) in essence provide a new version of this program. In particular, the GUTS are claimed to explain the entropy-to-baryon ratio because of baryon non-conservation, and the inflationary scenario the 'flatness problem' and 'horizon problem'. Whether they prove the necessity of local homogeneity and isotropy is the subject of debate(12). Certainly the physics underlying these models is still uncertain. In addition, it is not completely clear that conditions at the 'Bubbles' envisaged in the new Inflationary theory are fully consistent with the jump conditions of General Relativity; at late times these models may run into 'missing matter' problems; and they also have difficulty in linking together the properties of the different scales of structure observed. Thus there are various problems to be sorted out in this important and interesting approach.

(3) A *Consistency Principle* aims to relate previously unrelated aspects as a matter of necessity, e.g. *Mach's Principle* as investigated by Sciama and Raine (Sciama, 1969; Raine, 1975, 1981, 1982), the *Strong Anthropic Principle* of Wheeler (1977), Collins and Hawking (1973b), and others, or Harrison's *Boot-*

strap Principle (Ellis and Harrison, 1974; Harrison, 1974). Each of these is difficult to state precisely, let alone develop to a usable principle which can demonstrate the desired necessity; but all have interesting potential. The essential question is what can one deduce from the requirement that local physical behaviour be consistent with the global geometry and physics of the Universe (which itself has a structure partly determined by these local physical laws)? Underlying is the Machian issue of the extent to which one can separate local physical laws and distant boundary conditions in the universe; that is, does the geometry and physics of the universe as a whole affect the *nature* of the local physical laws themselves? If so, the requirement of consistency of local and global conditions might lead to very restricted solutions, forcing the theoretically possible nature of the universe into a very narrow mould. This line of argument is tantalising, but very difficult to follow through. It certainly deserves further thought.

The approaches above aim either to eliminate the element of random probability, or to show that arbitrary initial data will tend to the same final result. A quite different kind of resolution is possible:

(4) A *Geometrical Principle* may relate otherwise unrelated properties. An example is the Einstein-Wheeler observation that if the spatial geometry of the universe is closed, many 'Mach's Principle' ideas will necessarily be taken into account (Einstein, 1950; Wheeler, 1964, 1968; Isenberg and Wheeler, 1979). This idea can be taken further to what I believe is one of the philosphically most attractive proposals one can

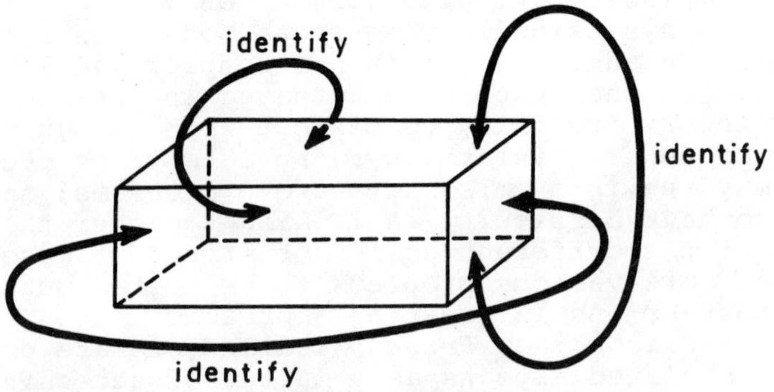

Fig. 18. The space-section of a small universe with a Torus topology.

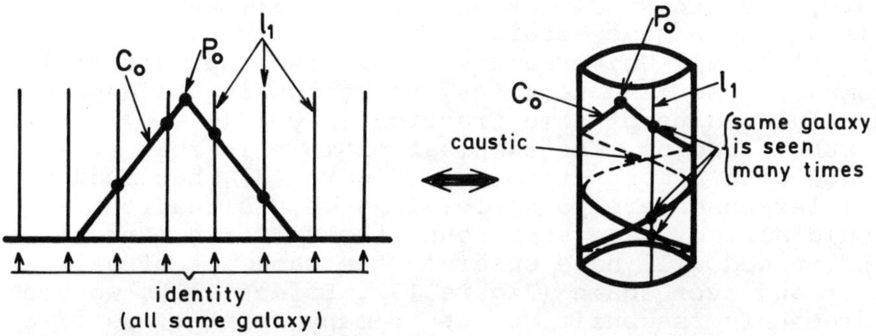

Fig. 19. Because of the small, compact space-sections, we can have seen the same galaxy many times as its world line will intersect our past light cone many times, as shown in these equivalent causal diagrams. Caustics will necessarily occur in our past light cone in this case.

Fig. 20. As we see the same spatial domains repeatedly, effectively we see a large universe built up out of repeating identical blocks. Thus if there are significant density inhomogeneities, we might detect resulting patterns of structures repeating with distance. However the light-travel time and apparent distance would be different for each image, making detection of this pattern difficult. A corresponding smoothed-out model should also permit the group G* of symmetries as a sub-group of its full group of symmetries G_r.

make, i.e. the *Small Universe* proposal, which will be discussed in some detail.

A 'Small Universe' is a universe that has small, compact spacelike sections, corresponding to a present spatial radius of some fraction (say 1/10 to 1/50) of a Hubble radius. The simplest example is that of a space with a Torus topology (Figure 18), but much more complex cases are possible. In such a situation, we would already have seen round the universe many times, and indeed will have observed the same set of galaxies over and over again (Figure 19). Effectively, we see a large universe built up out of many repetitions of a small building block (Figure 20)[(13)]. This provides a natural explanation of apparent homogeneity of the universe: namely it looks homogeneous on a large scale, even if the 'building block' is quite inhomogeneous, because we are seeing the same set of galaxies over and over again! Similarly, it can provide good reasons for the microwave background radiation isotropy (Gott, 1980).

If we live in such a universe it will in fact be difficult to verify that this is so (Ellis, 1971a), because we will observe each galaxy or cluster at different redshifts and observer area distances; effectively from different directions; and at different epochs in its history (when major evolution and proper motion can have taken place). Different distortion effects and absorbtion will affect the different images. Further the different kinds of identifications that can take place can be extremely complex (Gott, 1980; Ellis, 1971a; Milnor, 1979, 1982; Thurston, 1982; Heckmann and Schucking, 1962), so the geometry of the identifications could be quite non-obvious. These features imply it would in fact be quite non-trivial determining that one had indeed seen the same set of galaxies in different directions in the sky. One would have to focus on some identifiable non-uniformity in the distribution of galaxies, which in these universes would generate a repetitive pattern with distance; for example one might detect periodic features in redshift frequency distributions (Fang and Sato, 1983; note that the characteristic scale of repetition would vary with direction).Once one set of identifications had been proposed, it would then be relatively easy to check this by seeing if other inhomogeneities were multiply-imaged in a consistent way.

As the scale of identification is unknown, although observational data can limit the identification scale below, one can always model the observed universe by making this scale large enough that observational checks cannot disprove the hypothesis of a small universe; however one then loses the major advantage of this view, of explaining apparent homogeneity[14]. The hope would be to determine a relatively small scale of identification that agreed with observation. Attempts so far to determine if we live in a small universe by examining galaxy cluster catalogues (Gott, 1980; Sokolov and Shvartsmann, 1975) suggest lower bounds of the order of 200 to 400 Mpc for the scale of such identifications[15]. Fang and Sato (1983) have proposed that QSO redshift patterns confirm such a universe with an identification scale of about 600 Mpc; but Box and Roeder (1983) suggest these patterns are due to selection effects. General theorems on curvature (Frankel and Galloway, 1981; Galloway, 1982) imply that a large enough density of matter will force spatial closure (assuming the spatial sections are geodesically complete); however the density required to force a small universe in this way is much higher than we have yet detected[16]. Thus the question of whether we live in a small universe or not, is very open at present.

In terms of the scales of representation discussed earlier, we can have 'small universe' models at each scale, related by the smoothing operations discussed in Section 3. Some intriguing issues arise. There are a large family of Scale 5 (smoothed out) Small Universes that can be constructed from exact FLRW models (for any value of k (Heckmann and Schucking, 1962; Zeldovich, 1973b; Milnor, 1979)) by making identifications under an appropriate discrete subgroup G* of the full group of isometries G_r (Ellis, 1971; Milnor, 1982; Thurston, 1982; Gott, 1980). Although the global isotropy of the models would in general have been broken by the identifications made (Ellis, 1971; Milnor, 1982; Thurston, 1982), observationally such universes are indistinguishable from their exactly isotropic parent FLRW models (as there are no inhomogeneities present to be observed). These universes would not be simply connected; the possibilities in the cases k=+1 and k=0 are relatively simple, but in the case k=-1, the topology can be very complex (Ellis, 1971; Milnor, 1982; Thurston, 1982). The FLRW parent model would be recovered by going to the universal covering space of the Small Universe; clearly the group G* then acts as a discrete group of isometries.

Corresponding Scale 3 Small Universes could be constructed by appropriate perturbations of these compact, smooth Small Universe models; their universal covering spaces would correspond to FLRW universe models with perturbations imposed that are invariant under the discrete group of isometries G*. However one could also impose substantial irregularities (rather than just perturbations) to obtain very lumpy Scale 1 or Scale 3 Small-Universe models that also look like FLRW models at the large scale, being invariant under the group G* (which is a subgroup of the FLRW group G_r). Indeed these models would be observationally identical to the 'lumpy' approximately-homogeneous models proposed earlier in Section 4.1, invariant under discrete symmetry groups (one just compares those models with the universal covering spaces of the Small-Universe models discussed here; their observational properties are indistinguishable).

However we cannot necessarily get all the Small Universes in this way. In general, the only requirements for a *Small Universe model* are that it be a universe model which

(1) has compact spatial sections,

(2) is expanding, and

(3) is such that at late enough times,the world line of each galaxy intersects the past light cone of every other galaxy multiple times (i.e. the horizon size is much greater than the scale length of the compact space-sections)[17].

Thus the spatial topologies are restricted only by the requirement of compactness. There are a great many possibilities (Thurston, 1982), almost all being not simply connected, and some being exceedingly complex; in principle any of them could occur ("Why should nature not choose a non-trivial topology?", Zeldovich, 1973b). In general, the topology may be one that cannot be obtained by making identifications in a FLRW universe; correspondingly, if one goes to the universal covering space, in general the resulting discrete group G* of isometries will *not* be a subgroup of any of the FLRW symmetry groups G_r. In this case, it is difficult to see how the smoothing from Scale 1 or 3 to Scale 5 could produce a FLRW universe model. Thus one might be able to construct lumpy (Scale 1 or Scale 3) Small Universe models which *look* like perturbed FLRW universe models, but for which the smoothed-out (Scale 5) version is not a FLRW universe. They would *appear* approximately homogeneous, but their topology would be incompatible with the symmetry of exactly homogeneous universes.

To examine this, the first problem is to characterise which small universes correspond geometrically to 'identified' FLRW universe models, and which do not. This is a non-trivial question in geometry and group theory, for which some partial results are known

(Clarke, 1983)[18]. The second is to determine how anisotropic the models would appear. The third is to deduce a generalisation of the Friedmann equation governing the time-evolution of the volume of these space-sections, from Einstein's equations suitably averaged over the compact space-sections. This will determine when the average dynamical behaviour of these small universes is similar to that of the FLRW universes. The answers to these questions will give an idea of the similarity of general Small Universes to those obtained by identifications in FLRW models. In addition, we would want to examine the dynamics within the basic unit; particularly,
(i) examining the formation of bound systems, in these universes, and
(ii) seeing if the inhomogeneity would allow the initial singularities to be isolated singularities which could be avoided by much of the matter in the universe (cf. Hawking and Ellis, 1973).

If we were indeed living in a Small Universe, the implications would be considerable. One would have vastly simplified the concept of the universe, through explaining a great many images in terms of the properties of a much smaller set of objects. One could explain apparent homogeneity and isotropy of the universe in a simple way. The Einstein-Wheeler Machian conditions are satisfied, and interesting physical results might follow from the fact that all physical fields would at all times have to be periodic, with the maximum wavelength being the repetition scale of the universe model itself. And perhaps most important, the universe would in this case be quite exceptional in terms of its observational and causal properties; for in this case (and only in this case) we would already have had access to enough information to predict its future. There could only be a finite number of galaxies in the universe, and we would already have seen (or at least have had causal contact with) all the matter in it (Figure 19). In this situation the problem of how we can predict to the future of our past light cone (cf. Section 2) is fully resolved; and the observational limits restricting our knowledge of the universe are quite different from the usual case where we only have access to a tiny fraction of the matter in it (Ellis, 1975). Because of this different relation of the observer to the universe, there are no longer vast regions of space-time where conditions are completely beyond any relation to possible observations. The role of uncertainty in cosmology (Ellis, 1980a,b) is so diminished as to qualitatively change the nature of the subject. In these universes, and in them alone, there is a possibility of determining with reasonable confidence the structure of the universe in its entirety.

Sufficient will have been said to illustrate that this class of models, by virtue of their geometry, are able to integrate various features one might wish for in an intriguing way. It would be of major significance if we could show the Universe was like this. Thus it is well worth while on the one hand making considerable effort to put observational bounds on possible identification scales; and on the other, investigating further the relation of these models to the other unifying concepts considered here.

Overall, the essential point is that there are conflicts between various of the cosmological principles we might like to apply; the basic need is for a unifying principle that will relate some of them to others in a satisfactory way, and remove some of the apparent conflicts. There are various promising lines of investigation that can be explored, such as those discussed above. A mature cosmological theory will need to adopt some such coherent approach which also takes seriously the problem of observational verification in the face of abounding uncertainty. We do not yet have such a theory.

7. CONCLUSION

Relativistic Cosmology at present provides a fairly convincing picture of the structure of the observable universe, but does not answer many of the questions we would like to put; and various of its foundations need a surer footing than they have at present. This paper attempts to lay out clearly some of the relevant issues. Each of the main sections hinges on a central question which makes clear the concern of that section.

The principal observational suggestion is that we should seriously pursue the 'observational cosmology' approach initiated by Kristian and Sachs, developing this to the stage of a 'numerical observational cosmology' program whose results can be compared with those of the standard approach. This implies that we check more carefully than hitherto the limits of detection of galactic proper motions and image distortions, both of which are fundamental data needed in the observational approach. It opens the prospect of an algorithm to determine which new galactic observations would best increase our knowledge of the structure of the universe. This approach must be pursued in conjunction with a serious attempt to characterise detection and selection effects; these have not been discussed here (cf.Ellis, Fong and Phillips, 1977; Ellis, Perry and Sievers, 1983). Understanding of such effects is essential, for example, in searching to see if we live in a 'small universe'; it is suggested that this is an important issue worthy of considerable observational effort.

Little of what has been said here about theory is new. However I believe the attempt at synthesis is important, and particularly that one should make explicit the problems of reconciling the different scales of representation on the one hand, and contradictory cosmological principles on the other; these issues are not usually stated explicitly. Again the relation between local physics and the universe is not always regarded as a legitimate branch of cosmology; I believe it should be.

A theme running through the investigation of Cosmology is that of the relation between image and reality, between shadow and substance. Our problem is to sift the observable features and discern reality in the context of the uniqueness of the universe, where we cannot repeat experiments to see if the theory is correct (Bondi, 1960; McCrea, 1953), and uncertainty abounds (Ellis, 1980a,b). Our situation closely resembles that of the observers in Plato's cave (Plato), where only the blurred projection of reality onto a 2-dimensional surface is available for investigation. In Hubble's words,

> "From our home on the Earth, we look out into the distances and strive to imagine the sort of world into which we are born. Today we have reached far out into space. Our immediate neighbourhood we know rather intimately. But with increasing distance our knowledge fades, and fades rapidly, until at the last dim horizon we search amongst ghostly errors of observations for landmarks that are scarcely more substantial" (Hubble, 1958).

All these features compound the difficulties of achieving a satisfactory integration of cosmological theory and observations. This is why extra care and attention has to be given both to the nature of observational verification in cosmology on the one hand, and to the philosophical principles which inevitable play a large role in the cosmological models we use to understand the universe, on the other. Focussing on the five major questions discussed in this paper can significantly assist in this examination. We will ultimately have to take seriously each of these questions before we have a really satisfactory cosmological theory.

ACKNOWLEDGEMENTS. I should like to thank many colleagues, particularly Jurgen Ehlers and Bernd Schmidt, for enlightening comments. I thank the Max Planck Institute of Astrophysics, Munich, for hospitality, and the CSIR (RSA) and University of Cape Town for financial support of this research programme.

NOTES

+ Permanent address.

(1) Cf. Penrose's discussion (1980) of unsolved problems in General Relativity in the context of asymptotically flat space-times.

(2) Schucking was a proponent of this kind of approach at a Galileo Commemoration meeting in Padua in 1964.

(3) We remember here that in the cosmological context, we are effectively able to make observations down only *one* light cone: that of the event P_0, 'here and now' (Ellis, 1975, 1980a,b).

(4) These limits are obscured in the usual approach, where the symmetry assumptions made change the hyperbolic general form of Einstein's equations to that of a system of ordinary differential equations, with quite different predictive domains.

(5) But note that there will be a refocussing of null geodesics due to the Ricci tensor of matter and radiation present, and the area distance will *decrease* thereafter; this can happen at red-shifts as low as z=1.25 in a high density universe.

(6) It is intriguing that in the now fashionable 'new inflationary universe' model (Hawking and Moss, 1982; Linde, 1982; Albrecht and Steinhard, 1982; Barrow, 1983), the universe is viewed as composed of 'bubbles' with differing characteristics, and therefore is inhomogeneous on a very large scale. This is in agreement with one of the most plausi-

ble philosophical views concerning the homogeneity of the universe (Ellis, 1979), and in contradiction to some of the strongly stated versions of the 'Cosmological Principle'. We then *see* the universe as spatially homogeneous because there is a small probability of beiing situated so that our past light-cone intersects one of the bubble walls.

(7) One might wish to explicitly consider intermediate scales, e.g. Scale 2 (about 10^6 stars) and Scale 4 (clusters of galaxies); for simplicity we do not do so here, but have chosen the numbering to facilitate use of such further scales if desired.

(8) One conceivable exception would be if the model were *already* smooth at Scale 1. Then the operations would be the identity operations (if eq. (2c) holds). However this is clearly a degenerate case.

(9) One could also investigate the problems arising for Scale 1 if we assume the field equations hold on some smaller scale, say *Scale -2* (cf. the idea of 'space-time foam' advocated by Wheeler and Hawking).

(10) In a related study, Isaacson(1968a,b)proved appropriate positive energy properties; but the approximation was different, and does not apply here (he was describing the particle-approximation appropriate for gravitational radiation).

(11) Various others could be included; this discussion aims to indicate the possible lines of argument, rather than at completeness.

(12) See Rothman and Matzner, 1982; Ellis and Olive, 1983; Wald, 1983. The models of rotation and shear used so far in the discussions, e.g.(Ellis and Olive, 1983),are not generic. Wald's analysis (Wald, 1983) avoids this pitfall, but does not deal with rotation.

(13) The effect is that of looking into a box containing some millions of galaxies, with (expanding) walls made of mirrors. This would appear much like the universe we see.

(14) Early attempts by Refsdal and Petrosian and Sal-

peter to see if we lived in a universe we had already seen round, were restricted to the elliptic identification in Eddington-Le Maitre models; the scale of identification there is rather larger than we envisage here.

(15) The scales determined depend on various dynamical assumptions. A further crucial issue is the selection effects that determine when we regard a set of galaxy images as representing a 'cluster' of galaxies; these may need re-examination (cf. Segal, 1983).

(16) Correspondingly, we have not yet detected sufficient matter to imply the 'best-fit' FLRW universe must be a k=+1 universe.

(17) This implies the existence of cusps in the light cones as well as the caustics that will occur because of gravitational lens effects (cf. Rosquist, 1983). One of the observational issues is how one can detect these cusps and caustics.

(18) We do not even yet have a full characterisation of all possible Bianchi (spatially homogeneous) Small Universes.

REFERENCES

Albrecht, A., and Steinhard, P.: 1982, "Cosmology for grand unified theories with radiatively induced symmetry breaking", *Phys. Rev. Lett.*, 48, 1220.

Bailey, I., and Israel, W.: 1975, "Lagrangian dynamics of spinning particles and polarized media in general relativity", *Commun. Math. Phys.*, 42, 65.

Balian, R., Audage, J., and Schramm, D.N. (eds): 1980, *Physical Cosmology* (Les Houches, 1979), North Holland, Amsterdam.

Barrow, J.D.: 1976, "Light elements and the isotropy of the universe", *Mon. Not. R.Astron. Soc.*, 175, 359.

Barrow, J.D.: 1977, "A chaotic cosmology", *Nature*, 267, 117.

Barrow, J.D.: 1983, "Cosmology and elementary particle Physics", *Fundam. Cosmic. Phys.*, 8, 83.

Barrow, J.D., and Matzner, R.A.: 1977, "The homogeneity and isotropy of the universe", *Mon. Not. R.Astron. Soc.*, 181, 719.

Batchelor, G.K.: 1964, *Homogeneous Turbulence*, Cambridge University Press, Cambridge.

Bertotti, B., de Felice, F., and Pascolini, A. (eds): 1983, *10th International Conference on General Relativity and Gravitation. Contributed papers*, Vols. 1-2, CNR, Roma.

Bondi, H.: 1960, *Cosmology*, Cambridge University Press, Cambridge.

Bourassa, R.R., and Kantowski, R.: 1975, "The theory of transparent gravitational lenses", *Astrophys. J.*, 195, 13.

Box, T. and Roeder, R.C.: 1983, "An investigation of quasar redshifts and metric distance", in Bertotti et al., 1983, p. 795.

Callan C., Dicke, R.H., and Peebles, P.J.E.: 1965, "Cosmology and Newtonian mechanics", *Am.J.Phys.*, 33, 105.

Canizares, C.R.: 1981, "Gravitational focussing and the association of distant quasars with foreground galaxies", *Nature*, 291, 620.

Canizares, C.R.: 1982, "Manifestations of a cosmological density of compact objects in quasar light", *Astrophys. J.*, 263, 508.

Carr, B., and Rees, M.J.: 1979, "The anthropic principle and the structure of the physical world", *Nature*, 278, 605.

Carter, B.: 1974, "Large number coincidences and the anthropic principle in cosmology", in M.S.Longair (ed.), *Confrontation of Cosmological Theory with Observational Data*, D.Reidel, Dordrecht, p.291.

Carter, B.: 1983, "The anthropic principle and its implications for biological evolution", Talk given at Roy.Soc.Meeting (see Royal Society, 1983).

Clarke, C.J.S.: 1983, "Compact Cosmologies", Unpublished.

Collins, C.B., and Ellis, G.F.R.: 1979, "Singularities in Bianchi Cosmologies", *Phys. Rep.*, 56C, 65.

Collins, C.B., and Hawking, S.W.: 1973a, "The rotation and distortion of the universe", *Mon. Not. R. Astron. Soc.*, 162, 307.

Collins, C.B., and Hawking, S.W.: 1973b, "Why is the universe isotropic?", *Astrophys.J.*, 239, 317.

Dautcourt, G.: 1977, "The cosmological problem as an initial value problem on the observer's past light cone", in *GR8. Abstracts of the Contributed Papers*, University of Waterloo, Waterloo.

Davies, P.C.W.: 1978, *The Arrow of Time in Physics*, Cambridge University Press, Cambridge.

De Vaucouleurs, G.: 1970, "The case for a hierarchical cosmology", *Science*, 167, 1203.

DeWitt, B., and DeWitt, C.: 1973, *Black Holes*, Gordon and Breach, New York.

Dicke, R.H.: 1961, "Dirac's Principle and Cosmology", *Nature*, 192, 440.

Dyer, C.C., and Roeder, R.C.: 1974, "Observations in locally inhomogeneous cosmological models", *Astrophys. J.*, 189, 167.

Dyer, C.C., and Roeder, R.C.: 1975, "Apparent magnitudes, redshifts and inhomogeneities in the universe" *Astrophys. J.*, 196, 671.

Eddington, A.S.: 1930, "On the instability of Einstein's spherical world", *Mon. Not. R.Astron. Soc.*, 90, 668.

Ehlers, J.: 1961, "Beitrage zur relativistischen Mechanik Kontinuerlichen Medien", *Akad. Wiss. u Litt. Mainz Abh. Math-Naturwissens*, Kl, 11.

Ehlers, J.: 1971, "General Relativity and Kinetic Theory", in Sachs, 1971, p. 1.

Ehlers, J.: 1973, "Survey of General Relativity Theory",

in W.Israel (ed.), *Relativity, Astrophysics and Cosmology*, D.Reidel, Dordrecht.

Ehlers, J. (ed.): 1979, *Isolated Gravitating Systems in General Relativity*, Academic Press, New York.

Ehlers, J.: 1980, "Isolated systems in general relativity", *Ann. N.Y. Acad. Sci.*, 336, 279.

Ehlers, J.: 1981, "Christoffel's work on the equivalence problem and its importance for modern field theories of physics", in P.L.Butzer, and F.Feher (eds), *E.B.Christoffel*, Birkauser Basel, p. 526.

Einstein, A.: 1950, *The Meaning of Relativity*, Princeton University Press, 3rd Ed., Princeton, pp. 107-108.

Ellis, G.F.R.: 1967, "The dynamics of pressure free matter in relativity", *J. Math. Phys.*, 8, 1171.

Ellis, G.F.R.: 1971a, "Topology and Cosmology", *Gen. Rel. Grav.*, 2, 7.

Ellis, G.F.R.: 1971b, "Relativistic Cosmology", in Sachs, 1971.

Ellis, G.F.R.: 1975, "Cosmology and verifiability", *Q.J.R. Astron. Soc.*, 16, 245.

Ellis, G.F.R.: 1979, "The homogeneity of the universe", *Gen. Rel. Grav.*, 11, 281.

Ellis, G.F.R.: 1980a, "Limits to verification in cosmology", *Ann. N.Y. Acad. Sci.*, 336, 130.

Ellis, G.F.R.: 1980b, "Cosmology: Observational Verification, Certainty and Uncertainty", *South African J. of Sciences*, 76, 504.

Ellis, G.F.R., and Baldwin, J.E.: 1984, "On the expected anisotropy of radio source counts", *Mon. Not. R. Astron. Soc.*, 206, 377.

Ellis, G.F.R., and Brundrit, G.B.: 1979, "Life in the infinite universe", *Q.J.R. Astron. Soc.*, 20, 37.

Ellis, R.S., Fong, R., and Phillipps, S.: 1977, "Angular diameter counts of galaxies: a method for determining the selection criteria for deep

Schmidt plates", *Mon. Not. R.Astron. Soc.*, 181, 163.

Ellis, G.F.R., and Harrison, E.R.: 1974, "Cosmological Principles I. Symmetry Principles", *Comments Astrophys. & Space Phys.*, 6, 23.

Ellis, G.F.R., and King, A.R.: 1974, "Was the Big Bang a Whimper?", *Commun. Math. Phys.*, 38, 119.

Ellis, G.F.R., Maartens, R., and Nel, S.D.: 1978, "The expansion of the Universe", *Mon. Not. R.Astron. Soc.*, 154, 187.

Ellis, G.F.R., Maartens, R., and Nel, S.D.: 1981, "Observational Cosmology I: Ideal Cosmographic Observations" Preprint University of Cape Town.

Ellis, G.F.R., Matravers, D.R., and Treciokas, R.: 1983, "Anisotropic Solutions of the Einstein--Boltzmann Equations", *Annals of Physics*, 150, 487.

Ellis, G.F.R., and Perry, J.J.: 1979, "Towards a 'correctionless' observational cosmology", *Mon. Not. R.Astron. Soc.*, 187, 357.

Ellis, G.F.R., Perry, J.J., and Sievers, A.: 1983, "Cosmological observations of galaxies: the observational map", Preprint MPI Radioastronomie, Bonn.

Ellis, G.F.R., and Sciama, D.W.: 1972, "Global and non--global problems in cosmology", in L.O.'Raiffeartaigh (ed.), *General Relativity*, Oxford University Press, Oxford, p. 35.

Ellis, J., and Olive, K.A.: 1983, "Inflation can solve the rotation problem", *Nature*, 303, 679.

Fang, L.Z., and Sato, H.: 1983, "Is the periodicity in the distribution of quasar redshifts an evidence of multiply connected universe?", Preprint RFP-494, Kyoto, in Bertotti et al., 1983, p. 817.

Field, G.B., Arp, H., and Bahcall, J.N. (eds): 1973, *The Redshift Controversy*, Benjamin, New York.

Frankel, T., and Galloway, G.J.: 1981, "Energy density and spatial curvature in relativity", *J.Math.*

Phys., 22, 813.

Friedrich, H., and Stewart, J.M.: 1982, "Characteristic Initial Data and Wave Front Singularities in General Relativity", Preprint MPA 17, Munich.

Fry, J.N., and Turner, M.S.: 1983, "Baryogenesis and supersymmetric GUTS", *Phys. Lett.*, 125B, 379.

Galloway, G.J.: 1982, "Some global properties of closed spatially homogeneous spacetimes", *Gen. Rel. Grav.*, 14, 87.

Geroch, R.: 1977, "Asymptotic structure of space-time", in F.P.Esposito, and L.Witten (eds) *Asymptotic Structure of Space Time*, Plenum Press, New York, p. 1.

Geroch, R.: 1978, "Null infinity is not a good initial data surface", *J.Math. Phys.*, 19, 1300.

Gibbons, G.W., and Hawking, S.W.: 1977, "Cosmological event horizons, thermodynamics, and particle creation", *Phys. Rev.*, D15, 1738.

Gibbons, G.W., Hawking, S.W., and Siklos, S.T.C. (eds): 1983, *The Universe at Very Early Times*, Cambridge University Press, Cambridge.

Gibbons, G.W., Hawking, S.W., Horowitz, G. and Perry, M.: 1983, "Positive Mass Theorems for Black Holes", *Commun. Math. Phys.*, 88, 295.

Gold, T. (ed.): 1967, *The Nature of Time*, Cornell University Press, Ithaca.

Goldberg, J.: 1972, "Conservation of Newmann-Penrose conserved quantities", *Phys. Rev. Lett.*, 28,1400.

Gott, J.R.: 1980, "Chaotic cosmologies and the topology of the universe", *Mon. Not. R.Astron. Soc.*, 193, 153.

Gott, J.R., Gunn, J.E., Schramm, D.M., and Tinsley, B.M.: 1974, "An Unbound Universe?", *Astrophys. J.*, 194, 543.

Gudehus, D.: 1972, "Cosmology and Information", *Publ. Astron. Soc. Pac.*, 84, 818.

Gunn, J.E., Longair, M.S., and Rees, M.J.: 1978, *Observational Cosmology*, Swiss Society of Astronomy and Astrophysics, Geneva.

Guth, A.H.: 1981, "Inflationary universe: a possible solution to the horizon and flatness problems", *Phys. Rev.*, D23, 347.

Harrison, E.R.: 1968, "The Early Universe", *Phys. Today*, 21, 31.

Harrison, E.R.: 1974, "Cosmological Principles II. Physical Principles", *Comments Astrophys. & Space Phys.*, 6, 29.

Hawking, S.W.: 1968, "Gravitational radiation in an expanding universe", *J.Math. Phys.*, 9, 598.

Hawking, S.W.: 1969, "On the rotation of the universe", *Mon. Not. R.Astron. Soc.*, 142, 129.

Hawking, S.W., and Ellis, G.F.R.: 1973, *The Large Scale Structure of Space Time*, Cambridge University Press, Cambridge.

Hawking, S.W., and Israel, W. (eds): 1979, *General Relativity: An Einstein Centenary Survey*, Cambridge University Press, Cambridge.

Hawking, S.W., and Moss, I.G.: 1982, "Supercooled phase transitions in the very early universe", *Phys. Lett.*, 110B, 35.

Hawking, S.W., and Page, D.N.: 1983, "Thermodynamics of Black-Holes in Anti-de Sitter Space", *Commun. Math. Phys.*, 87, 577.

Heckmann, O., and Schucking, E.: 1962, "Relativistic Cosmology", in L.Witten (ed.), *Gravitation: An Introduction to Current Research*, Wiley, New York.

Heidmann, J.: 1980, *Relativistic Cosmology*, Springer, Berlin.

Held, A. (ed.): 1980, *General Relativity and Gravitation*, Vols. 1-2, Plenum Press, New York.

Horowitz, G.T., and Schmidt, B.G.: 1981, "Note on Gravitational Energy", Preprint, MPI-PAE/Astro

286, Munich.

Hubble, E.P.: 1958, "The Law of the Redshifts", George Darwin Lecture, *Mon. Not. R.Astron. Soc.*,113,666.

Isaacson, R.A.: 1968a, "Gravitational radiation in the limit of high frequency I", *Phys. Rev.*, 166,1263.

Isaacson, R.A.: 1968b, "Gravitational radiation in the limit of high frequency II", *Phys. Rev.*,166,1272.

Isenberg, J., and Wheeler, J.A.: 1979, "Inertia here is fixed by mass-energy there in every W model universe", in M.Pantaleo, and F. de Finis (eds), *Relativity, Quanta and Cosmology in the Development of the Scientific Thought of Albert Einstein*, Johnson Reprint Corp.,New York.

Israel, W., and Stewart, J.M.: 1980, "Progress in Relativistic Thermodynamics and Electrodynamics of Continuous Media", in Held, 1980, p. 491.

Jantzen, R.T.: 1979, "The dynamical degrees of freedom in spatially homogeneous cosmologies", *Commun. Math. Phys.*, 64, 211.

Jantzen, R.T.: 1983, "Perfect fluid sources for spatially homogeneous spacetimes", *Ann. Phys.*, 145, 378.

Jones, B.: 1976, "Galaxy formation", *Rev. Mod. Phys.*, 48,107.

Kantowski, R.: 1969, "Corrections in the luminosity--redshift relation of the homogeneous Friedmann models", *Astrophys. J.*, 155, 89.

Karlhede, A., and MacCallum, M.A.H.: 1982, "On determining the isometry group of a Riemannian space", *Gen. Rel. Grav.*, 14, 673.

Kerr, R., and Goldberg, J.: 1964, "Asymptotic Properties of the Electromagnetic Field", *J.Math.Phys.*, 5, 172.

Kristian, J., and Sachs, R.K.: 1966, "Observations in Cosmology", *Astrophys. J.*, 143, 379.

Krolak, K.: 1983, "Definitions of black holes without the use of the boundary at infinity", *Gen. Rel.*

Grav., 14, 793.

Lilly, S., Longair, M., and McLean, I.: 1983, "A radio galaxy at high redshift undergoing strong stellar evolution", *Nature*, 301, 488.

Linde, A.D.: 1982, "A new inflationary universe scenario", *Phys. Lett.*, 108B, 389.

Maartens, R.: 1980, "Idealized Observations in Relativistic Cosmology", Ph.D.Thesis, Univ. of Cape Town (unpublished).

Maartens, E., Ellis, G.F.R., Nel, S.D.: 1983, "Observational Cosmology III: Realistic Observations" in preparation.

MacCallum, M.A.H.: 1979a, "Anisotropic and inhomogeneous relativistic cosmologies", in Hawking and Israel, 1979.

MacCallum, M.A.H.: 1979b, "The Mathematics of anisotropic spatially homogeneous cosmologies", in M.Demianski (ed.), *Physics of the Expanding Universe*, Lectures Notes in Physics, Vol. 109, Springer, Berlin, p. 1.

Matzner, R.A.: 1968a, "3-Sphere 'backgrounds' for the space sections of the Taub cosmological solution", *J.Math. Phys.*, 9, 1063.

Matzner, R.A.: 1968b, "Almost symmetric spaces and gravitational radiation", *J.Math. Phys.*, 9, 1957.

McCrea, W.H.: 1935, "Observable relations in relativistic cosmology", *Z.Ap.*, 9, 290.

McCrea, W.H.: 1939, "Observable relations in relativistic cosmology", *Z.Ap.*, 18, 98.

McCrea, W.H.: 1953, "Cosmology", *Rep. Prog. Phys.*, 16, 321.

McCrea, W.H.: 1970, "A Philosophy for big-bang cosmology", *Nature*, 228, 21.

Metzner, A.W.K., and Morrison, P.: 1959, "The flow of information in cosmological models", *Mon. Not. R. Astron. Soc.*, 119, 657.

Milnor, J.: 1979, "Is space-time simply connected?", Preprint.

Milnor, J.: 1982, "Hyperbolic Geometry: The First 150 Years", *Bull. Am. Math. Soc.*, 6, 9.

Misner, C.W.: 1968, "The isotropy of the universe", *Astrophys. J.*, 151, 431.

Misner, C.W.: 1969, "Mixmaster universe", *Phys. Rev. Lett.*, 22, 1071.

Mottola, E., and Lapedes, A.: 1983, "Inflationary universe with gravity", *Phys. Rev.*, D27, 2285.

Nel, S.D.: 1980, "Observational Space-Times", Ph.D.Thesis, University of Cape Town (unpublished).

Nel, S.D., Stoeger, W.R., Ellis, G.F.R., Whitman, A.P., and Maartens, R.: 1983, "Observational cosmology II: ideal cosmological data", Preprint University of Cape Town (unpublished).

Noerdlinger, P.D., and Petrosian, V.: 1971, "The effect of cosmological expansion on self-gravitating ensembles of particles", *Astrophys. J.*, 168, 1.

Olive, K.A., Schramm, D.N., Steigman, G., Turner, M.S., and Yang, J.: 1981, "Big Bang nucleosynthesis as a probe of cosmology and particle physics", *Astrophys. J.*, 246, 557.

Olson, D.W.: 1978, "Helium production and limits on the anisotropy of the universe", *Astrophys. J.*,219,777.

Parker, L.: 1982, "Particle creation by the expanding universe", *Fundamentals of Cosmic Physics*, 7, 201.

Parker, T., and Taubes, C.H.: 1982, "On Witten's Proof of the Positive Energy Theorem", *Commun. Math. Phys.*, 84, 223.

Peebles, P.J.E.: 1972, "Light out of darkness vs order out of chaos", *Commun. Ast. Space Phys.*, 4, 53.

Penrose, R.: 1963, "Conformal treatment of infinity", in C. DeWitt, and B. DeWitt (eds), *Relativity, Groups and Topology*, Gordon and Breach, New York, p. 565.

Penrose, R.: 1965, "Zero rest-mass fields including gravitational radiation: asymptotic behavior", *Proc. R.Soc.*, A284, 159.

Penrose, R.: 1979, "Singularities and time asymmetry", in Hawking and Israel, 1979, p. 581.

Penrose, R.: 1980, "Some unsolved problems in classical relativity theory", Preprint, Mathematical Institute, Oxford.

Penrose, R.: 1982, "Quasi-local mass and angular momentum in general relativity", Preprint, Oxford University.

Penrose, R.: 1983, "Time asymmetry and quantum gravity", in *Quantum Gravity II, a Second Oxford Symposium*, Oxford University Press, Oxford.

Plato: *The Republic*, Book 7.

Porrill, J., and Stewart, J.: 1981, "Electromagnetic and Gravitational Fields, in Schwarzschild Space--Time", *Proc. R.Soc.*, A376, 451.

Press, W., and Bardeen, J.: 1971, "Non-conservation of Newmann-Penrose conserved quantities", *Phys. Rev. Lett.*, 27, 1303.

Raine, D.J.: 1975, "Mach's principle in general relativity", *Mon. Not. R. Astron. Soc.*, 171, 507.

Raine, D.J.: 1981, "Mach's principle and space-time structure", *Rep. Prog. Phys.*, 11, 1151.

Raine, D.J.: 1982, "Mach's principle and the microwave background", *Astrophys. Lett.*, 23, 37.

Rindler, W.: 1956, "Visual horizons in world models", *Mon. Not. R.Astron. Soc.*, 116, 662.

Roodt, H.J.: 1981, "Clusters of galaxies", *Rep. Prog. Phys.*, 44, 1077.

Rosquist, K.: 1983, "On the structure of space-time caustics", *Commun. Math. Phys.*, 88, 339.

Rothman, T., and Matzner, R.: 1982, "Grand unified reactions and dissipation in anisotropic cosmologies", *Astrophys. J.*, 263, 501.

Royal Society: 1983, "The Constants of Physics", Discussion Meeting, London, 25/26 May 1983, to be published in *Philos. Trans. R.Soc.*

Sachs, R.K.: 1961, "The outgoing radiation condition", *Proc. R.Soc.*, A264, 309.

Sachs, R.K. (ed.): 1971, *General Relativity and Cosmology*, Academic Press, New York.

Sandage, A., Tamman, G., and Yahil, A.: 1979, "The velocity field of nearby bright galaxies", *Astrophys. J.*, 232, 352.

Schattner, R., and Streubel, M.: 1980, "The connection between local and asymptotic structures for isolated gravitating systems with isometries", Preprint MPI-PAE/Astro 237, Munich.

Schmidt, B.G.: 1974, "A new definition of conformal and projective infinity", *Commun. Math. Phys.*, 36, 73.

Schramm, D.N.: 1983, "The early universe and elementary particle physics", *Physics Today*, 36, 27.

Sciama, D.W.: 1969, *Physical foundations of General Relativity*, Garden City, Doubleday.

Segal, I.E.: 1982, "A correction: the quadratic redshift distance law and the observational magnitude cut-off bias", *Astrophys. J.*, 252, 37.

Segal, I.E.: 1983, "Distance and model dependence of observational galaxy cluster concept", *Astron. Astrophys.*, 123, 151.

Siklos, S.T.C.: 1981, "Non-scalar singularities in spatially homogeneous cosmologies", *Gen. Rel. Grav.*, 13, 433.

Sokolov, D.D., and Shvartsmann, V.F.: 1975, "An estimate of the size of the universe from a topological point of view", *Sov. Phys.-JETP*, 39, 196.

Soneira, R.: 1979, "A cosmological redshift-distance square law?", *Astrophys. J.Lett.*, 230, L63.

Spero, A., and Baierlein, R.: 1977, "Approximate symmetry groups of inhomogeneous metrics", *J.Math.*

Phys., 18, 1330.

Spero, A., and Baierlein, R.: 1978, "Approximate symmetry groups of inhomogeneous metrics: examples", *J.Math. Phys.*, 19, 1324.

Stewart, J.M.: 1968, "Neutrino viscosity in cosmological models", *Astrophys. Lett.*, 2, 133.

Stewart, J.M., and Friedrich, H.: 1982, "Numerical relativity I. The characteristic initial value problem", *Proc. R.Soc.*, A384, 427.

Stock, J., and Schucking, E.: 1957, "Remarks on the magnitude of extragalactic nebulae", *Astrophys.J.*, 62, 98.

Szekeres, P.: 1971, "Linearized gravitational theory in macroscopic media", *Ann. Phys.*, 64, 599.

Tammann, G.A., Sandage, A., and Yahil, A.: 1980, "The determination of cosmological parameters", in Balian et al., 1980, p. 53.

Taubes, C.H.: 1978, "Solution of the almost Killing equation and conformal almost-Killing equation in the Kerr Space-Time", *J.Math. Phys.*, 19, 1515.

Thurston, W.P.: 1982, "Three Dimensional Manifolds, Kleinian Groups and Hyperbolic Geometry", *Bull. Am. Math. Soc.*, 6, 357.

Tinsley, B.M.: 1970, "On the interpretation of galaxy counts", *Astrophys. J.*, 241, 41.

Tinsley, B.M.: 1980, "Cosmology and Galactic Evolution", in Balian et al., 1980, p. 161.

Tipler, F.J.: 1977, "Black holes in closed universes", *Nature*, 270, 500.

Tipler, F.J., Clarke, C.J.S., and Ellis, G.F.R.: 1980, "Singularities and Horizons - a Review Article", in Held, 1980, vol. 2, p. 97.

Treciokas, R., and Ellis, G.F.R.: 1971, "Isotropic Solutions of the Einstein-Boltzmann Equation", *Commun. Math. Phys.*, 23, 1.

Turner, M.S., and Schramm, D.N.: 1979, "Cosmology and

elementary particle physics", *Physics Today*, September, p. 42.

Wald, R.: 1983, "Asymptotic behavior of homogeneous cosmological model in the presence of a positive cosmological constant", Preprint, MPI-PAE/PTh 38/83, Munich.

Waylen, P.C.: 1978, "Green's function in the early universe", *Proc. R.Soc.*, A362, 233.

Weedman, D., Weymann, R., Green, R., and Heckman, T.: 1982, "Discovery of a third gravitational lens", *Astrophys.J.*, 255, L5.

Weinberg, S.: 1973, *Gravitation and Cosmology*, Wiley, New York.

Wesson, P.: 1981, "How homogeneous is the universe?", *Astrophys. Lett.*, 21, 97.

Wheeler, J.A.: 1964, "Mach's principle as boundary conditions for Einstein's equations", in H.Y. Chiu, and B.Hoffmann (eds), *Gravitation and Relativity*, Benjamin, New York, p. 303.

Wheeler, J.A.: 1968, *Einsteins's Vision*, Springer Berlin.

Wheeler, J.A.: 1977, "Genesis and Observership", in R.E.Butts, and K.J.Hintikka (eds), *Foundational Problems in the Special Sciences*, D.Reidel, Dordrecht, p. 3.

Whitrow, G.J., and Yallup, B.D.: 1964, "Background radiation in isotropic world models. I.", *Mon. Not. R.Astron. Soc.*, 127, 301.

Yahil, A., Sandage, A., and Tammann, G.A.: 1980, "Deceleration of nearby galaxies", in Balian et al., 1980, p. 127.

Zeldovich, Ya.B.: 1973a, "A hypothesis unifying structure and entropy of the universe", *Mon. Not. R.Astron. Soc.*, 160, 1P.

Zeldovich, Ya.B.: 1973b, "Structure (topology) of the universe", *Comments on Astrophys. Sp. Sci.*, 5, 169.

NOTE ADDED IN PROOF

On the possible avoidance of singularities at Scale 5, if sufficient violation of the energy-condition could occur (cf. Section 3.4) see Marochnik, L.S., Pelikov,N.V., and Vereshkov,G.M.: 1975, "Turbulence in Cosmology", *Astrophys. Sp. Sci.*, 34, 249.

VOIDS IN THE EXPANDING UNIVERSE

Humitaka Sato

Research Institute for Fundamental Physics
Kyoto University, Kyoto 606, Japan

Enlargement of a spherical void in the expanding universe is discussed. In a nonlinear regime of density perturbation, the low-density region or a void behaves like an explosive source and the dense thin shell is formed by this explosion of the void. Israel's formalism for an infinitely thin shell is applied to derive a relativistic relation, which is analysed for various models of the universe. The motion of a void in the closed universe is also discussed.

Our result is also applicable for a blast wave propagation generated by an explosion in the early universe. These void-expansion phenomena may be relevant to a formation of the "large scale structure" in the galaxy distribution and the explosive galaxy formation theory.

1. INTRODUCTION

We investigate the expansion of a spherical shell surrounding a void in the universe. The shell propagates sweeping the ambient matter and its motion is decelerated as the shell mass increases. In the universe, such a shell may be generated by a rapid expansion of low density regions (voids) and by an explosion of pregalactic objects. The first process may relate to the formation of the recently observed "large scale structure" in the universe (Sato, 1982a, 1982b; Hoffman et al., 1983) and to the behavior of the large amplitude isothermal perturbation after the decoupling.

B. Bertotti et al. (eds.), General Relativity and Gravitation, 289–312.

The second process has been proposed in the explosive galaxy formation theory (Ikeuchi, 1981; Ostriker and Cowie, 1981).

The "large scale structure" is called either a "honeycomb" structure or "cellular" structure and it is a mosaic pattern which consists of large voids (holes) and superclusters of galaxies with a flattened or filamentary shape (Joeveer et al., 1978; Gregory and Thompson, 1978; Einasto et al., 1980; Krishner et al., 1981). Formation of this structure is related to the nonlinear behavior of density perturbation in a matter dominated universe (Zeldovich, 1970; Shandarin et al., 1983).

In a linear regime, a distribution of an over (positive) density perturbation and a low (negative) density perturbation has no meaning but, in a nonlinear regime, they behave in a quite different manner. Until recently, the "positive" case has been studied mainly (Tomita, 1969; White and Silk, 1979), but, in our series of papers (Sato, 1982a; Sato and Maeda, 1983; Maeda and Sato, 1983a, 1983b), a new feature of the "negative" case has been clarified; the formation of a dense thin shell surrounding the low density region. The thin-shell formation is merely an example of the "pancake" formation discussed by Zeldovich (1970) but our problem is concerned with the motion of "pancake".

The process of the cellular structure formation seen in the computer simulation (Doroshkevich et al., 1980; Melott, 1983; Centrella and Melott, 1983; Nakamura, 1983) may be interpreted not only in terms of the "positive" perturbation but also in terms of the "negative" perturbation. In the latter case, the dense shell formation is interpreted as the compression by a rapid expansion of the negatively perturbed region.

In this paper, we investigate a motion of a spherical shell in the universe using the method given by Israel (1966) for a singular layer. As opposed to the collapse case (White and Silk, 1979), the shape of the void tends to be spherically symmetric as the expansion proceeds (Fujimoto, 1983). By the Israel method, we can derive a relativistic relation, though the concerned processes can be treated in the Newtonian limit. We will also discuss a relativistic problem such as the behavior of the void in the closed universe, merely from theoretical curiosity.

In Section 2, we discuss the condition of shell crossing and the generation of dense thin shell by shock wave. In Section 3, Israel's formalism is generalized from a shell in the vacuum into a shell in the ambient matter. In Section 4, the formalism given in

Section 3 is applied to a spherical shell surrounding a void and the expansion of the shell is discussed in various models of the universe. In Section 5, the behavior of a void in the closed universe is discussed. In Section 6, our result in the Newtonian limit is compared with Sedov's relation for the expansion.

The main part of this paper is a summary of our previous papers (Sato, 1982a; Sato and Maeda, 1983; Maeda and Sato, 1983a, 1983b).

2. SHELL CROSSING AND FORMATION OF COMPRESSED DENSE SHELL

We denote the spherically symmetric spacetime metric as

$$ds^2=-e^{2\nu}dt^2+e^{2\lambda}d\chi^2+Y^2d\Omega^2. \qquad (1)$$

The dynamics is described by one of the Einstein equations (see for example Kramer et al., 1980):

$$D_t^2Y=-\frac{m+4\pi PY^3}{Y^2}-\{1-\frac{2m}{Y}+(D_tY)^2\}\frac{P'}{(\rho+P)Y'}, \qquad (2)$$

where $D_tY=e^{-\nu}\dot{Y}=e^{-\nu}\partial Y/\partial t$ and $Y'=\partial Y/\partial\chi$. In the dust model, we can put the pressure P=0 and eq. (1) is integrated to give

$$\dot{Y}^2-\frac{2m}{Y}=-\varepsilon f^2(\chi), \qquad (3)$$

where $m(\chi)$ and $\varepsilon f^2(\chi)$ are arbitrary functions of the comoving coordinates χ, $e^{2\lambda}=Y'^2/(1-\varepsilon f^2)$ and $\nu=0$. The density is computed from $m(\chi)$ and Y as $\rho=m'/4\pi Y^2Y'$ or

$$\rho=\rho_i\,\frac{Y_i^2Y_i'}{Y^2Y'}, \qquad (4)$$

where $Y_i=Y(\chi,t_i)$ and $\rho_i=\rho(\chi,t_i)$. Since $\rho_i>0$, the positivity of ρ requires the following condition (Sato and Maeda, 1983),

$$Y'/Y_i'>0. \qquad (5)$$

In the linear perturbation, condition (5) is satisfied but, in the nonlinear stage of the low density perturbation, this condition will be violated on the rim of the perturbed region for most cases. This phenomenon is known as "shell crossing". The "pancake formation" of Zeldovich (1970) is one example of this phenomenon. This was also discussed in the gravitational collapse which gives riso to a "naked" singularity (Müller zum Hagen et al., 1973). A general discussion about the development of singular density has been given for more complex topology (Arnold et al., 1982).

It is shown that shell crossing occurs if the density profile at the void's rim is steep enough (Sato, 1982a). For example, assume the profile is given in a form

$$\rho_i = \rho(t_i)\{1-\delta(\chi_v/\chi)^n\} \quad \text{for } \chi \geqq \chi_v \qquad (6)$$
$$= \rho(t_i)(1-\delta) \quad \text{for } \chi < \chi_v .$$

When the density contrast becomes nonlinear, the expansion in $\chi_v<\chi<\chi_c(t)$ is given for n<3 as

$$Y(\chi,t) \cong \frac{2}{3}\left(\frac{\delta\chi_v^n}{\chi^n}\right)^{\frac{1}{2}}\left(\frac{3}{3-n}\right)^{\frac{1}{2}}\left(\frac{t}{t_i}\right)Y(\chi,t_i) \qquad (7)$$

and

$$\chi_c(t) = \left(\frac{3}{3-n}\right)^{1/n}\delta^{1/n}\chi_v\left(\frac{t}{t_i}\right)^{2/3n} .$$

Then, condition (5) derives n<2, assuming that $Y(\chi,t_i)\propto\chi$. So, if the profile is steep enough as is n>2, shell crossing will occur. If the profile is dull, it does not occur (Occhionero et al., 1981).

In a real situation, shell crossing will be prevented by the effect of pressure. When two crossing shells collide, the pressure gradient force will increase just as the density and their relative velocity will be damped by shock dissipation. In this case, indeed, the "crossing" does not occur but a compressed thin shell is formed to move together. Thus, the shells swallowed in the compressed thin shell increase with time as shown in Figure 1 (Sato and Maeda, 1983).

The above process is described if we include the pressure gradient term in (2) in this compressed shell. By numerical computation (Suto et al., 1984), we can see how a thin shell is formed by shock compression as shown in Figure 2: as the density contrast becomes nonlinear, the low density region begins to enlarge and a thin shell is formed. The low density region

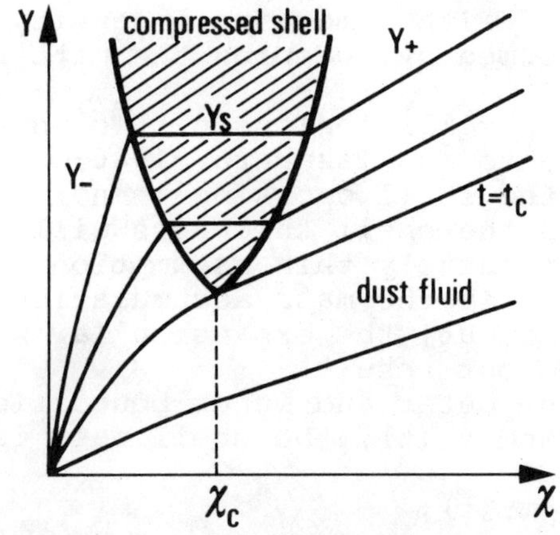

Fig. 1. A schematic diagram for the evolution of $Y(\chi,t)$. The shell-crossing starts to occur at t_c and χ_c where $Y'=0$. After that epoch, the mass in the hatched region has been compressed into a thin shell.

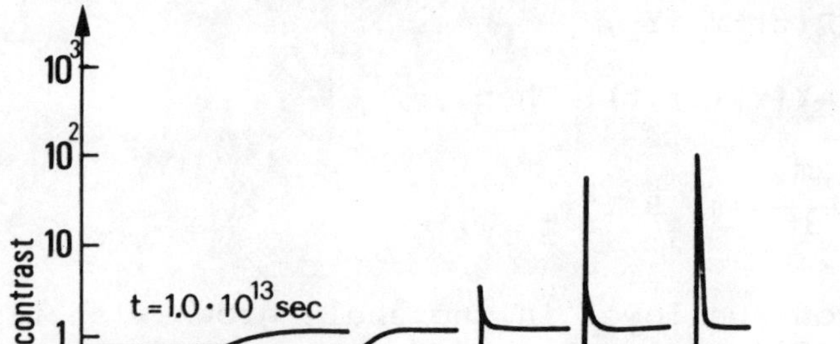

Fig. 2. As the negative density perturbation grows, the shock wave is generated surrounding this region and propagates outwards accumulating the expanding matter in the flat universe (Suto et al., 1984).

evolves into the void with almost zero density and into the massive shell formed by sweeping the matter inside the shell.

If the matter is collisionless particles such as massive neutrinos (Sato and Takahara, 1980; Sato, 1981), the dissipation at the shell crossing is not complete as assumed above and the shell thickness will be finite. Then, the infinitely thin assumption would not be verified. However, if the mass accumulation in the shell does actually occur, the expansion law would not be so different from our result.

If we denote the outer and inner boundaries of the shell by $\chi_+(t)$ and $\chi_-(t)$, the shell mass is

$$M_s(t)=m_+(t)-m_-(t), \tag{8}$$

where

$$m_\pm(t)=m(\chi_\pm(t)).$$

Approximating the shell thickness to be infinitely thin, the radius is given by

$$R(t)=Y_+=Y_-, \tag{9}$$

where $Y_\pm=Y(\chi_\pm(t),t)$. Then,

$$\frac{dm_\pm}{dt}=4\pi\rho_\pm R^2 Y'_\pm \dot{\chi}_\pm . \tag{10}$$

The region swallowed in the shell grows with time as shown in Figure 1.

If we neglect the internal energy in the shell, the energy conservation is derived from (3) as

$$\frac{d}{dt}\left\{M_s\left(\frac{dR}{dt}\right)^2-\frac{M_s^2}{R}\right\}=-\frac{dm_+}{dt}(\varepsilon f^2)_+ +\frac{dm_-}{dt}(\varepsilon f^2)_- . \tag{11}$$

Compared with the expansion equation used for the supernova explosion in the interstellar medium, equation (11) includes the self-gravity energy of the shell and the energy input from the surrounding "expanding" universe.

As an example, we consider the shell on the rim of the vacuum void in the flat universe. The energy conservation (11) is integrated under the initial condition of $(dR/dt)_i=\dot{Y}_i$ as

$$M_s\left(\frac{dR}{dt}\right)^2-\frac{M_s}{R}=\frac{M_{si}^2}{R_i}\ , \tag{12}$$

since $\dot{Y}_i^2-2M_{si}/Y_i=0$ and $R_i=Y_i$. As $M_s=4\pi\rho_+R^3/3$ and $\rho_+=1/6\pi t^2$, equation (11) becomes

$$\left(\frac{dy}{dx}\right)^2-\frac{2}{9}\,\frac{y^2}{x^2}-\frac{2}{9}\,\frac{x^2}{y^3}=0, \tag{13}$$

writing $x=t/t_i$ and $y=R/R_i$. We can integrate (13) as

$$x=\frac{2^{2/47}\left[(1+\sqrt{2})(\sqrt{1+u}-\sqrt{u})\right]^{15\sqrt{2}/47}}{\left[6\sqrt{2u}-5\sqrt{1+u}\,\right]^{36/47}} \tag{14}$$

with $u=y^5/x^4$, and

$$y\cong\left(\frac{25}{47}\right)^{1/5}x^{4/5}\cong 0.88x^{4/5}\quad \text{for } x\gg 1. \tag{15}$$

The relative velocity of the shell is estimated as

$$V\cong\frac{1}{5}\,V_H \qquad \text{for } x>>1, \tag{16}$$

where $V=dR/dt-V_H$ and $V_H=\dot{Y}_+$ (Sato and Maeda, 1983; Ikeuchi et al., 1983). The smallness of V/V_H will ensure the neglect of the internal energy in the shell, since it is of the order of ρ_+V^2.

3. MOTION OF A SINGULAR LAYER

The motion of the shell formed around the void can be treated by an analytic manner, if we assume the shell to be infinitely thin. Such a problem has been studied by several authors (Israel, 1966; Papapetrou and Hamoui, 1968; Raju, 1982; see also Misner et al., 1973). Israel (1966) formulated this problem considering the timelike hypersurface of discontinuity with two half-spaces. Papapetrou and Hamoui developed a four dimensional approach introducing δ-function and derived the surface stress-energy tensor introduced by Israel heuristically. These formalisms have been applied for the motion of a spherical shell in the vacuum and for a charged spherical shell (De La Cruz and Israel, 1967; Kuchar, 1968). Further details of the shell motion have been studied by Israel (1967),

Evans (1977), and Maeda (1980).

In this section, we apply Israel's formalism to the case where the singular layer moves in the matter instead of in the vacuum (Maeda and Sato, 1983). Our approximation to replace the blast wave by an infinitely thin layer will be verified for "isothermal shock" in the dust matter.

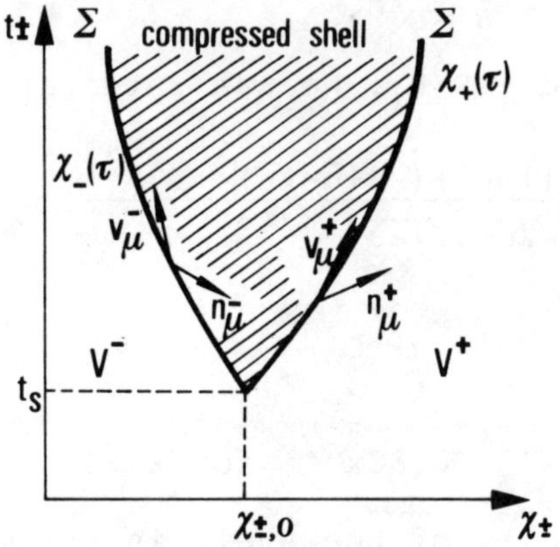

Fig. 3. Motion of the singular layer with radius R(τ). Spacetime V is divided into two regions V^- and V^+ by the compressed shell(Σ).

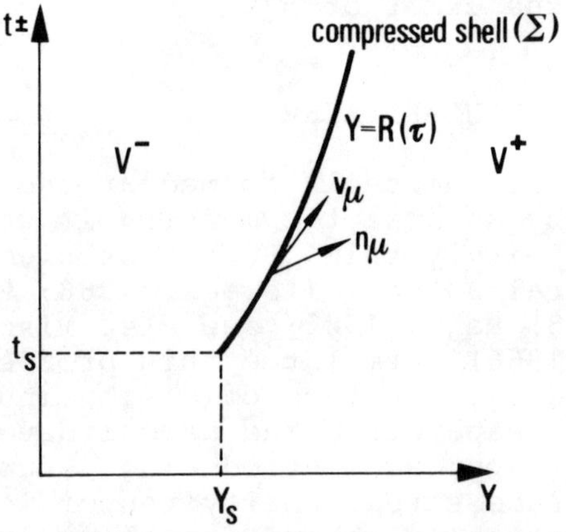

Fig. 4. The singular hypersurface in the t-χ diagram. The hatched region has been compressed into an infinitely thin shell.

Suppose the spacetime V is divided into two half regions V^+ and V^- by a timelike hypersurface Σ as shown in Figure 3 and Figure 4. Writing the unit spacelike normal to Σ by n_μ, the extrinsic curvature of Σ is defined as

$$K_{\mu\nu}=h^\rho_\mu h^\sigma_\nu n_{\rho;\sigma}, \tag{17}$$

where $h_{\mu\nu}=g_{\mu\nu}-n_\mu n_\nu$ (Misner et al., 1973). The Einstein tensor $G_{\mu\nu}$ is written in terms of $K_{\mu\nu}$ using the Gauss--Codazzi relation

$$2G_{\mu\nu}n^\mu n^\nu=K^2-K_{\mu\nu}K^{\mu\nu}-\overset{(3)}{R} \tag{18}$$

and

$$G_{\rho\sigma}h^\rho_\mu n^\sigma=K^\nu_{\mu//\nu}-K_{//\mu}\ , \tag{19}$$

where $K=K^\mu_\mu$, $\overset{(3)}{R}$ is the scalar curvature in Σ and // denotes the covariant derivative with respect to $h_{\mu\nu}$ in Σ.

Writing the extrinsic curvature of Σ associated with its imbedding in V^+ and V^- as $K^+_{\mu\nu}$ and $K^-_{\mu\nu}$ respectively, we define

$$\left[K_{\mu\nu}\right]_\pm=-8\pi(S_{\mu\nu}-\tfrac{1}{2}h_{\mu\nu}S) \tag{20}$$

and

$$\{K_{\mu\nu}\}_\pm=2\tilde{K}_{\mu\nu}, \tag{21}$$

where $[\psi]_\pm=\psi_+-\psi_-$ and $\{\psi\}_\pm=\psi_++\psi_-$. Here, $S_{\mu\nu}$ is interpreted as the stress-energy tensor of the shell such as

$$S_{\mu\nu}=\lim_{\varepsilon\to 0}\int_{-\varepsilon}^{\varepsilon}T_{\rho\sigma}h^\rho_\mu h^\sigma_\nu dx, \tag{22}$$

where x is the Gaussian coordinate in n_μ-direction.

The Einstein equation consists in two parts: the constraint equations in a "smoothed out" background spacetime $\tilde{V}$ such that Σ is regular and its extrinsic curvature is given by $\tilde{K}_{\mu\nu}$ and the evolution equations such that

$$\tilde{K}_{\mu\nu}S^{\mu\nu}=\left[T_{\mu\nu}n^{\mu}n^{\nu}\right]_{\pm} \tag{23}$$

and

$$S^{\nu}_{\mu//\nu}=-\left[T_{\rho\sigma}n^{\rho}h^{\sigma}_{\mu}\right]_{\pm}. \tag{24}$$

Now, we consider the dust shell as well as the dust background, i.e.

$$S_{\mu\nu}=\sigma v_{\mu}v_{\nu} \tag{25}$$

and

$$T^{\pm}_{\mu\nu}=\rho_{\pm}u^{\pm}_{\mu}u^{\pm}_{\nu}, \tag{26}$$

where σ is the surface density of the shell and $\rho_{\pm}$ and $u^{\pm}$ are density and four velocity in $V^{\pm}$ respectively.

Since v_{μ} is the unit timelike tangent vector to Σ, the equation (24) gives

$$S^{\nu}_{\mu//\nu}\,v^{\mu}=-\left[T_{\rho\sigma}n^{\rho}v^{\sigma}\right]_{\pm} \tag{27}$$

and

$$S^{\nu}_{\mu//\nu}\,n^{\mu}=0. \tag{28}$$

Substituting (25) and (26), the equations (27) and (28) derive a mass accumulation law

$$(\sigma v^{\mu})_{//\mu}=\left[\rho(u_{\sigma}n^{\sigma})^{2}\right]_{\pm} \tag{29}$$

and a geodesic equation in Σ

$$v^{\mu}_{//\nu}\,v^{\nu}=0, \tag{30}$$

respectively. Further, using (17) multiplied by $v^{\mu}v^{\nu}n^{\rho}$, $(v^{\rho}n_{\rho})_{;\sigma}=0$ and $\delta v^{\sigma}/\delta\tau=v^{\mu}v^{\sigma}_{;\mu}$, we can write the four acceleration as

$$\left.\frac{\delta v^{\alpha}}{\delta\tau}\right|^{\pm}=-v^{\mu}K_{\mu\nu}v^{\nu}n^{\alpha}\Big|^{\pm}, \tag{31}$$

where τ is a proper time on the shell. From (20), (21) and (31), we finally get

$$v_{\alpha}\left.\frac{\delta v^{\alpha}}{\delta\tau}\right|^{\pm}=0 \tag{32}$$

$$\left\{n_\alpha \frac{\delta v^\alpha}{\delta \tau}\right\}_\pm = -\frac{2}{\sigma}\left[\rho(u_\mu n^\mu)^2\right]_\pm \tag{33}$$

and

$$\left[n_\alpha \frac{\delta v^\alpha}{\delta \tau}\right]_\pm = 4\pi\sigma, \tag{34}$$

where (23) is used for $v^\mu v^\nu \tilde{K}_{\mu\nu} = S^{\mu\nu}\tilde{K}_{\mu\nu}/\sigma$.

4. SPHERICAL SHELL IN THE UNIVERSE

The result in Section 3 was applied to the case where Σ is a history of a 2-sphere and the background spacetimes $V^\pm$ are spherically symmetric (Maeda and Sato, 1983a, 1983b).

The metrics in $V_\pm$ are written from (1) as

$$ds_\pm^2 = -dt_\pm^2 + \frac{(Y_\pm')^2}{1-\varepsilon_\pm f^2(\chi)}\, d\chi_\pm^2 + Y_\pm^2 d\Omega^2. \tag{35}$$

Since the motion follows the θ=const. and ϕ=const. lines, we write the coordinates of the shell as $(t_\pm^s(\tau), \chi_\pm^s(\tau))$ in $V^\pm$ respectively. The (t,χ) components of $v_\pm^\alpha$, n_α and u^α are

$$v_\pm^\alpha = (N_\pm, \frac{d\chi_\pm^s}{dt}) \tag{36}$$

$$n_{\pm\alpha} = K_\pm(-\frac{d\chi_\pm^s}{dt}, N_\pm) \tag{37}$$

and

$$u^\alpha = (1,0), \tag{38}$$

where

$$N_\pm^2 = 1 + K_\pm^2\left(\frac{d\chi_\pm^s}{d\tau}\right)^2 \tag{39}$$

and

$$K_\pm^2 = \left.\frac{(Y')^2}{1-\varepsilon f^2}\right|_{\chi=\chi_\pm^s(t)}. \tag{40}$$

Then, the equations (32) and (33) reduce to

$$\left\{\frac{K}{N}\,\frac{\delta^2\chi}{\delta\tau^2}\right\}_{\pm}=-\frac{2}{\sigma}\left[\rho K^2\left(\frac{d\chi}{d\tau}\right)^2\right]_{\pm} \tag{41}$$

and

$$\frac{K}{N}\left[\frac{\delta^2\chi}{\delta\tau^2}\right]_{\pm}=4\pi\sigma. \tag{42}$$

Here, we introduce a radius of the shell $R(\tau)$ defined by

$$R(\tau)=Y_+(\chi_+(\tau),t_+(\tau))=Y_-(\chi_-(\tau),t_-(\tau)) \tag{43}$$

and, then,

$$\frac{K}{N}\,\frac{\delta^2\chi}{\delta\tau^2}\bigg|^{\pm}=\left\{\left[1+\left(\frac{dR}{d\tau}\right)^2-\frac{2m}{R}\right]^{-\frac{1}{2}}\left[\frac{d^2R}{d\tau^2}+\frac{m}{R^2}+4\pi\rho RK^2\left(\frac{d\chi}{d\tau}\right)^2\right]\right\}\bigg|^{\pm} \tag{44}$$

from (31).

If the background $V^{\pm}$ are uniform such as the universe model, the metric (35) is written as

$$ds_{\pm}^2=-dt_{\pm}^2+a_{\pm}^2(t_{\pm})(d\chi_{\pm}^2+f_{\pm}^2d\Omega^2), \tag{45}$$

where

$$f=\begin{cases}\chi & \varepsilon=0\\ \sin\chi & \varepsilon=1\\ \mathrm{sh}\chi & \varepsilon=-1\end{cases}\;. \tag{46}$$

Introducing the conformal time η such as $d\eta=dt/a(t)$, $a(t)$ is given as

$$\begin{aligned} a(\eta)&=\frac{a_0}{h'(\eta_0)}h'(\eta),\\ t(\eta)&=\frac{a_0}{h(\eta_0)}\,h(\eta) \end{aligned} \tag{47}$$

and

$$h(\eta)=\begin{cases}\frac{1}{6}\eta^3 & \varepsilon=0\\ \eta-\sin\eta & \varepsilon=1\\ \mathrm{sh}\eta-\eta & \varepsilon=-1.\end{cases}$$

Furthermore, in this case

$$K_\pm=a_\pm \text{ and } N_\pm^2=1+a_\pm^2\left(\frac{d\chi_\pm}{d\tau}\right)^2. \tag{48}$$

Then, substituting (44) into (41) and (42) and multiplying them, we can eliminate σ to obtain

$$\left[\left\{1+\left(\frac{dR}{d\tau}\right)^2-\frac{2m}{R}\right\}^{-1}\left\{\frac{d^2R}{d\tau^2}+\frac{m}{R^2}+4\pi\rho Ra^2\left(\frac{d\chi}{d\tau}\right)^2\right\}^2+8\pi\rho a^2\left(\frac{d\chi}{d\tau}\right)^2\right]_\pm=0. \tag{49}$$

For a vacuum void in the flat universe, i.e. $\varepsilon_-=-1, \varepsilon_+=0$, $m_-=0$ and $m_+=4\pi\rho_+R^3/3$, equation (49) can be written in terms of $R=a\chi$ as

$$\frac{d^2R}{d\tau^2}=\frac{1}{2R}\left[1+\left(\frac{dR}{d\tau}\right)^2\right]^{\frac{1}{2}}\left[1+\left(\frac{dR}{d\tau}\right)^2-\frac{2m}{R}\right]^{\frac{1}{2}}(1-3V_\tau^2) - \left[1+\left(\frac{dR}{d\tau}\right)^2\right](1+3V_\tau^2), \tag{50}$$

where $V_\tau=a\frac{d\chi}{d\tau}=\frac{dR}{d\tau}-\frac{da}{d\tau}\chi$. The mass of the shell is given by (42) as

$$M_s=4\pi\sigma R^2=\frac{2m(\chi)}{\left[1+\left(\frac{dR}{d\tau}\right)^2\right]^{\frac{1}{2}}+\left[1+\left(\frac{dR}{d\tau}\right)^2-\frac{2m}{R}\right]^{\frac{1}{2}}}. \tag{51}$$

If we use the time $t=t_+$ instead of τ, the equation of motion (50) is written as

$$\ddot{R}=\frac{1}{2R}\left[-(1+VV_H+2V^2+V_H^2)+(1-4V^2)(1+2VV_H+V_H^2)^{\frac{1}{2}}\right], \tag{52}$$

where $V_H=\dot{a}\chi=HR$, $H=\dot{a}/a$, $V=\dot{R}-V_H$ and we have used the relation $\ddot{R}=\dot{\tau}(d^2R/d\tau^2)+(\ddot{\tau}/\dot{\tau})R$. For a super-horizon case such as $R\gg H^{-1}$ or $V_H\gg 1$, the above equation reduces to

$$\ddot{R}=-\tfrac{1}{2}H^2R, \tag{53}$$

which coincides with the expansion law of the background space. In the Newtonian limit of $R\ll H^{-1}$ and $V\ll 1$, it reduces to

$$\ddot{R}=-\frac{1}{4}H^2R-\frac{3V^2}{R} . \tag{54}$$

In the Newtonian limit, the expansion law is rewritten from (49) and (51) as

$$\frac{d}{dt}(M_s\dot{R})+\frac{\bar{m}M_s}{R^2}=(\dot{m}V_H)_+-(\dot{m}V_H)_-, \tag{55}$$

where $\bar{m}=(m_++m_-)/2$. This relation may be interpreted as the momentum conservation law of the shell. Compared with the isothermal shell expansion law used for the supernova explosion, our expansion law (55) contains additionally a self-gravity term and a momentum supply term from the background medium, as in (11) for energy. For a vacuum void, i.e. $\varepsilon_-=-1$ and $m_-=0$, the equation (55) reduces to

$$\ddot{R}=-\frac{2\pi}{3}\rho R-\frac{3V^2}{R} , \tag{56}$$

which coincides with (54) for $\varepsilon_+=0$.

In case of $\varepsilon_+=0$, the expansion equation is integrated to give

$$R^4(t)=At^{\frac{15+\sqrt{17}}{6}}+Bt^{\frac{15-\sqrt{17}}{6}} \tag{57}$$

and the asymptotic form is

$$R(t)\propto t^{\frac{15+\sqrt{17}}{24}}\sim t^{0.797} \tag{58}$$

and $V/V_H\to(\sqrt{17}-1)/16$ (Maeda and Sato, 1983a; Ostriker and McKee, 1983).

For various universe models, the expansion of the shell around the vacuum void is integrated numerically and the results are shown in Figure 5. Since we are

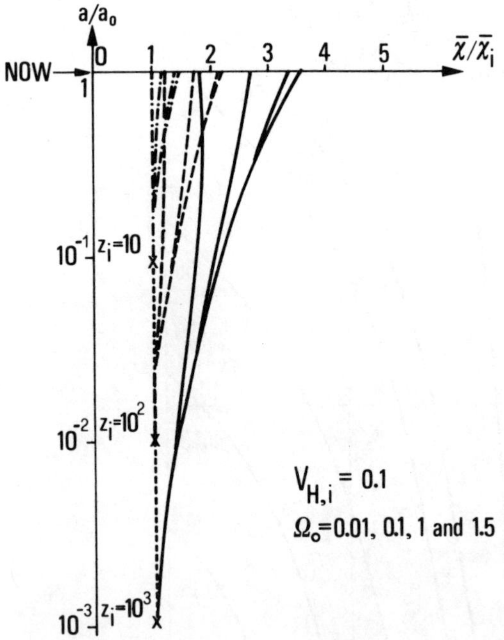

Fig. 5. The enlargement of voids is shown for various universe models Ω_0 and initial times t_i. The abscissa is a ratio of radius to the initial radius, $\bar{\chi}/\bar{\chi}_i$, where $\bar{\chi}$ is defined as $\bar{\chi}=R/a$. The initial state is taken to be the vacuum void. The ordinate represents the scale factor of the universe normalized by the present value a_0 (Maeda and Sato, 1983b).

interested in the nonlinear regime, we started the integration from the state such that $\rho_-=0$, $M_s=m_+(\chi_i)$ and $V=0$ at the initial epoch z_i. The fractional enlargement χ/χ_i is computed for four models. The enlargement is found to be more effective for the dense universe model with a large density parameter at the present Ω_0.

5. VOID IN THE CLOSED UNIVERSE

If $\Omega_0>1$, the space is closed and the background space collapses in the later phase. What happens for the expanding void in this collapsing closed space?

Using the conformal time η, the expansion equation (50) is written as

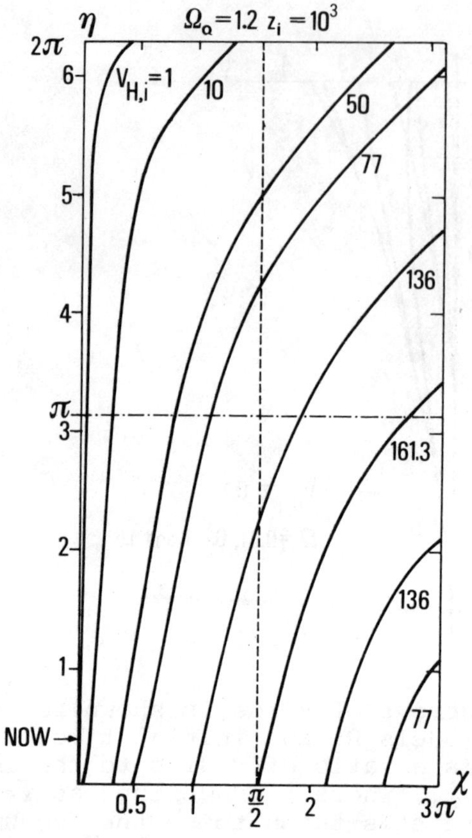

Fig. 6. The enlargement of the vacuum void in the closed universe with Ω_0=1.2. The voids begin to enlarge at $z_i=10^3$ with the V_i=0 and with various sizes, which are specified by $V_{H,i}$. The layer voids with $V_{H,i}>70$ occupy the whole space before the big crunch of the universe (Maeda and Sato, 1983b).

$$\frac{d^2\chi^s}{d\eta^2} = \frac{1}{2\sin\chi^s}\Big[(1+V_H^2-V^2\sin^2\chi^s+2\cos\chi^s V_H V)^{\frac{1}{2}}(1-4V^2)$$
$$-(1+2V^2)\cos\chi^s-3V_H V\Big], \qquad (59)$$

where $V=d\chi^s/d\eta$ and $V_H=(d(\ln a)/d\eta)\sin\chi^s$. In the case of Ω_0=1.2, the above equation is integrated numerically for voids with various sizes and V=0 at $z_i=10^3$, as shown in Figure 6. As seen from this figure, the fate of the void may be classified into two cases: (i) the enlarged void is yet smaller than the whole space at the big crunch when the background universe collapses to the future singularity; and (ii) the enlarged void has occupied the whole space before the big crunch. For the case of $\Omega_0=1.2$ and $z_i=10$ in Figure 6, the first case occurs for $V_{H,i}<70$ and the second case for $V_{H,i}>70$. The critical $V_{H,i}$ will depend on Ω_0 and z_i and $V_{H,i}$ will be slightly larger than 70 for $z_i>10^3$ and for $\Omega_0<1.2$.

Referring to the above result, we can guess the global spacetime structure as Figure 7. Our previous Penrose diagram given by Maeda et al. (1983) should be revised, because the positivity condition (5) is not satisfied in some regions. In Figure 7, such a negative density region has been removed. Further, the shell velocity V approaches light velocity at the final stage of $\eta=2\pi$ or $\chi=\pi$.

In the case of black hole formation in the closed universe (Qadiar and Wheeler, 1981), the spacetime can be sliced by a sequence of compact spacelike hypersurfaces over the whole space defined by the constant τ surface, where $\tau=(4/3)K^{\mu}_{\mu}$ is York time. Thus, the black hole singularity and the big crunch singularity are considered to be of the same nature. In the case of void, however, York time cannot foliate the whole space, since $\tau=-4\dot{a}/a$ is negative in the void but it is positive in the background space in the collapsing phase. As the void region crushes to singularity together with the big crunch, the constant t hypersurface can foliate the whole space. The inner boundary of the void $\chi_-(t_-)$ tends to zero as $\chi_- \propto (t_c-t_+)^{2/3}$ at the big crunch t_c, from (43) and $t_-=t_+$+const.

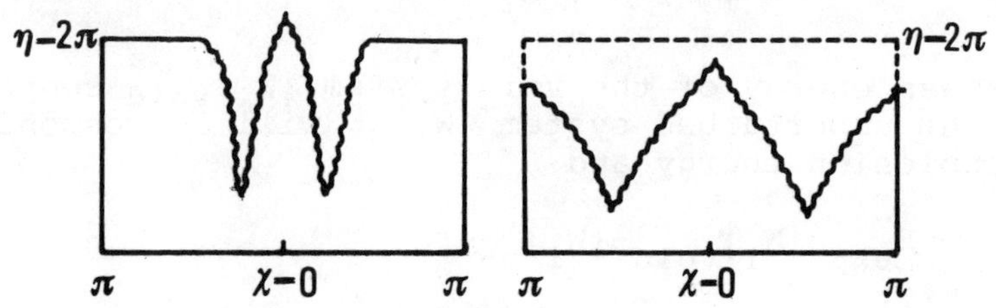

Fig. 7. A schematic Penrose diagram of the closed universe with a void in it. If the void is large enough, the void occupies the whole space before the big crunch.

6. VOIDS AND EXPLOSIONS

In this section, we restrict our discussion to the Newtonian limit.

The adiabatic expansion of the blast wave generated by an explosive energy E_{exp} is given by Sedov's relation

$$R(t)=1.2\left(\frac{E_{exp}}{\rho}T^2\right)^{1/5} \tag{60}$$

where R(t) is a radius of the expanding shell and T is a time after the explosion. In an expanding medium such as $\rho \propto t^{-2}$, the above relation gives $R \propto t^{4/5}$ for $T \sim t$, which coincides with our result (15). Compared with (15), and (60), an effective explosion energy for void is obtained as

$$E^v_{exp} \sim (\rho_i R_i^2 t_i^{-2}), \tag{61}$$

which is rewritten also in terms of the mass of the shell $M_{si} \sim \rho_i R_i^3$ as

$$E^v_{exp} \sim M_{si}\left(\frac{R_i}{H_i^{-1}}\right)^2 \sim \frac{M_{si}^2}{R_i} . \tag{62}$$

Thus, we can regard the void expansion as an explosion phenomenon (Sato and Maeda, 1983).

The total energy of a spherical region with mass m_i is given

$$W_i(<m)=\tfrac{1}{2}\int_0^m \dot{R}_i^2 dm-\int_0^m \frac{m}{R_i}\, dm. \tag{63}$$

An excess energy of the void system $\{W_i\}_{void}$ compared with the unperturbed system $\{W_i\}_{un}$ will correspond to the explosion energy and

$$E^v_{exp}=\{W_i\}_{void}-\{W_i\}_{un}, \tag{64}$$

where

$$\{W_i\}_{un}=-\varepsilon\int_0^m f^2(\chi)\,dm/2 \tag{65}$$

and

$$\{W_i\}_{void}=\frac{M_{si}^2}{2R_i} - \frac{M_{si}}{2}\varepsilon f^2(\chi_i^s). \tag{66}$$

Then, the explosive energy is

$$E_{exp}^v = \frac{M_{si}^2}{2R_i} -\Delta E, \tag{67}$$

where

$$\Delta E=\frac{\varepsilon}{2}(M_{si}f(\chi_i^s)^2-\int_0^{m_i} f^2dm)$$

$$\cong\frac{1}{5}\varepsilon M_{si}(\chi_i^s)^2$$

$$\cong\frac{1}{5}\varepsilon M_{si}\left(\frac{R_i}{H_i^{-1}}\right)^2\left(\frac{\Omega_i}{\Omega_0}\right)(\Omega_0-1)^2\frac{1}{(z_i+1)^2}, \tag{68}$$

assuming $\chi_i^s<<1$. Then, ΔE is always smaller than the first term in the right hand side of (67) for $z_i>>1$, and it is found to coincide with (62) in order of magnitude.

If the perturbation grows from a linear stage, the negatively perturbed region has an excess energy

$$E_{exp}^v \sim m_i\left[\left\{\tfrac{1}{2}\dot{Y}_i^2-\frac{m_i-\Delta m}{Y_i}\right\}_{void}-\left\{\tfrac{1}{2}\dot{Y}_i^2-\frac{m_i}{Y_i}\right\}_{un}\right]$$

$$\sim\frac{m_i\Delta m}{Y_i}, \tag{69}$$

where Δm is a mass deficiency in the perturbed region and the equal time hypersurface has been chosen so that $(\dot{Y})_{void}=(\dot{Y})_{un}$. In the linear stage, the perturbation grows as $\delta\rho/\rho\propto a$ and reaches the nonlinear stage at $a_{n\ell}\sim a_i(\rho/\delta\rho)_i=a_i(m/\Delta m)_i$. Then, E_{exp}^v of (69) is rewritten as

$$E_{exp}^v\sim M_{si}^2/Y_{n\ell} \tag{70}$$

where $Y_{n\ell} \sim Y_i(a_{n\ell}/a_i)$ is the radius at $a \sim a_{n\ell}$.

We can regard the explosion energy of the void as the built-in energy in our universe due to the inhomogeneous big bang energy or the inhomogeneous curvature term. The explosive process at their nonlinear stage is a dissipative process of these built-in inhomogeneities. In contrast with the explosion of quasars or population III stars in the early stage, the void explosion would be called "clean bomb" (Sato, 1983), since there is no pollution by heavy elements or by radiation.

If the blast wave is generated by some explosion, E_{exp} is an additional parameter to M_{si} different from the void case. Suppose that the mass of M_E is ejected with total kinetic energy E_E. Since the mass accumulation is negligible up to t_i defined by $M_E \sim 4\pi\rho(t_i)R(t_i)^3/3$, the expansion proceeds with a constant velocity $V_E=\sqrt{2E_E/M_E}$, where $\rho(t)$ is the density of the background medium. After t_i, the expansion obeys the relations discussed in this paper.

As an example, let us consider the isothermal shock in the flat universe. Then, the expansion obeys relation (57). The constants A and B are determined by E_E and $R(t_i)$. Instead of E_E, we use the relative velocity V of the shell at t_i: $V_i=\dot{R}_i-H_iR_i$. Then, using $y=R/R_i$ and $x=t/t_i$, it becomes

$$y^4=\frac{12}{\sqrt{17}}\left[\beta_i\left(1-x^{-\frac{\sqrt{17}}{3}}\right)+\frac{\sqrt{17}+1}{24}\left(1-\frac{(\sqrt{17}-1)^2}{18}x^{-\frac{\sqrt{17}}{3}}\right)x^{\frac{15+\sqrt{17}}{6}}\right], \quad (71)$$

where $\beta_i=V_it_i/R_i$.

In case of void, we have taken $V_i=0$, but, in the explosion case, V_i would be very large, i.e. $V_i \cong V_E >> V_{H,i}$. Then, the energy parameter β_i is written as

$$\beta_i=\frac{V_Et_i\rho_i^{1/3}}{M_E^{1/3}}\left(\frac{4\pi}{3}\right)^{1/3}$$

$$\cong 10^{4.3}\left(\frac{V_E}{c}\right)\left(\frac{10^9M_\odot}{M_E}\right)^{1/3}\frac{h^{-1/3}}{(z_i+1)^{\frac{1}{2}}}, \quad (72)$$

where $h=H_0/100\text{km}\cdot\text{s}^{-1}\cdot\text{Mpc}^{-1}$. The present radius is estimated as

$$R(t_0) \sim 10^{-1} h^{-2/3} \left(\frac{10^9 M_\odot}{M_E} \right)^{1/3} \left[0.21 + \beta_i \right]^{1/4} (z_i + 1)^{0.195} \text{Mpc}. \tag{73}$$

The explosion time t_E will be nearly equal to t_i if $\beta_i >> 1$ because $(t_i - t_E)/t_i = (1 - R_E/R_i)/\beta_i$, R_E being the radius at t_E. The relation (73) shows that $R(t_0)$ can be larger for stronger explosion or $\beta_i >> 1$ but it depends on β_i very weakly as $\beta_i^{1/4}$. Then, the enlargement factor of the shell would not be so different for the explosion and the void.

ACKNOWLEDGEMENTS. The author thanks K.Maeda for collaboration. This work is supported by the grants from the Ministry of Education, Science and Culture of Japan (58540212).

REFERENCES

Arnold, V.I., Shandarin, and Zeldovich, Ya.B.: 1982, "The large scale structure of the universe I. General properties, one-and two-dimensional models", *Geophys. Astrophys. Fluid Dyn.*, 20, 111.

Centrella, J., and Melott, A.: 1983, "Three dimensional simulation of large scale structures in the universe", Preprint of Illinois Astronomy, 83-6.

De La Cruz, V., and Israel, W.: 1967, "Gravitational bounce", *Nuovo Cimento*, 51A, 744.

Doroshkevich, A.G., Kotok, E.V., Novikov, I.D., Polyudov, A.N., Shandarin, S.E., and Sigov, Yu.S.: 1980, "Two-dimensional simulation of the gravitational system dynamics and formation of the large-scale structure of the universe", *Mon. Not. R.Astron. Soc.*, 192, 321.

Einasto, J., Joeveer, M., and Sarr, E.: 1980, "Structure of superclusters and supercluster-formation", *Mon. Not. R.Astron. Soc.*, 193, 353.

Evans, A.B.: 1977, "Relativistic dynamics of spherical counter-rotating dual bodies", *Gen. Rel. Grav.*, 8, 155.

Fujimoto, M.: 1983, "Dynamics of ellipsoidal void of matter in the expanding universe", *Publ. Astron. Soc. Jpn.*, 35, 159.

Gregory, S.A., and Thompson, L.A.: 1978, "The Coma/A 1367 supercluster and its environs", *Astrophys.J.*, 222, 784.

Hoffman, G.L., Salpeter, E.E., and Wasserman, I.: 1983, "Spherical simulations of holes and honey combs in the Friedmann universe", *Astrophys. J.*, 268, 527.

Ikeuchi, S.: 1981, "Theory of galaxy formation triggered by quasar explosions", *Publ. Astron. Soc. Jpn.*, 33, 221.

Ikeuchi, S., Tomisaka, K., and Ostriker, J.P.: 1983, "The structure and expansion law of a shock wave in the expanding universe", *Astrophys. J.*, 265, 583.

Israel, W.: 1966, "Singular hypersurfaces and thin shells in general relativity", *Nuovo Cimento*, 44B, 1; correction, ibid., 18B, 463 (1968).

Israel, W.: 1967, "Gravitational collapse and causality", *Phys. Rev.*, 153, 1388.

Joeveer, M., Einasto, J., and Tago, E.: 1978, "Spatial distribution of galaxies and of clusters of galaxies in the southern galactic hemisphere", *Mon. Not. R.Astron. Soc.*, 185, 357.

Kirshner, R., Oemler, A., Schechter, P., and Schechtman, S.: 1981, "A million cubic megaparsec void in Boötes?", *Astrophys. J.*, 248, 257.

Kramer, D., Stephani, H., Herlt, E., and MacCallum, M.: 1980, *Exact Solution of Einstein's Field Equations*, Cambridge U.P., Cambridge.

Kuchar, K.: 1968, "Charged shells in general relativity and their gravitational collapse", *Czech. J.Phys.*, B18, 435.

Maeda, K.: 1980, "Gravitational collapse of charged dust shell and maximal slicing condition", *Prog. Theor. Phys.*, 68, 425.

Maeda, K., and Sato, H.: 1983a, "Expansion of a thin

shell around a void in expanding universe", *Prog. Theor. Phys.*, 70, 772.

Maeda, K., and Sato, H.: 1983b, "Expansion of a thin shell around a void in expanding universe II", *Prog. Theor. Phys.*, 70, 1276.

Maeda, K., Sasaki, M., and Sato, H.: 1983, "Voids in the closed universe", *Prog. Theor. Phys.*, 69, 89.

Melott, A.: 1983, "Two-dimensional simulation of the gravitational superclustering of collisionless particles", *Mon. Not. R. Astron.Soc.*, 202, 595.

Misner, C.W., Thorne, K.S.,and Wheeler, J.A.: 1973, *Gravitation*, Freeman, San Francisco.

Müller zum Hagen, H., Seifert, H.J., and Yodzis, P.: 1973, "On the occurrence of naked singularities in general relativity", *Commun. Math. Phys.*, 34,135.

Nakamura, T.: 1983, unpublished.

Occhionero, F., Veccia-Scavali, L., and Vittorio, N.: 1981, "Cavities in high density universe", *Astron. Astrophys.*, 99, L12.

Ostriker, J.P., and Cowie, L.L.: 1981, "Galaxy formation in a intergalactic medium dominated by explosion", *Astrophys. J.*, 243, L127.

Ostriker, J.P., and McKee, C.F.: 1983, "Analytic models for blast wave in homogeneous or inhomogeneous and stationary or expanding media", *Rev. Mod. Phys.*, to be published.

Papapetrou, A., and Hamoui, A.: 1968, "Spherically symmetric surface layers in general relativity", *Ann. Inst. Henri Poincaré*, 9, 179; correction, *Gen. Rel. Grav.*, 10, 253, 1979.

Qadiar, A., and Wheeler, J.A.: 1981, "Black hole singularity as part of big crunch singularity", in R. Deveber, and M.Demeur (eds), *Memorial volume dedicated to J.Geheniau*, U.Brussel Press, Brussel.

Raju, C.K.: 1982, "Junction conditions in general relativity", *J.Phys.*, 15A, 1785.

Sato, H.: 1981, "The early universe and clustering of

the relict neutrinos", in R.Ramaty, and C.Jones (eds), *Proceeding 10th Texas Symposium*, Academy Press, New York, p. 43.

Sato, H.: 1982a, "Enlargement of a void in the expanding universe", *Prog. Theor. Phys.*, 68, 236.

Sato, H.: 1982b, "Cluster and void in the expanding universe", in Hu Ning (ed.), *Proceeding 3rd Marcel Grossmann Meeting*, Beijin.

Sato, H.: 1983, "Non linear density fluctuations in expanding universe", in *Proceeding Workshop "GUT and Early Universe"*, KEK 83-13, Tsukuba, p. 97.

Sato, H., and Maeda, K.: 1983, "The expansion law of the void in the expanding universe", *Prog. Theor. Phys.*, 70, 119.

Sato, H., and Takahara, F.: 1980, "Clustering of the relic neutrinos in the expanding universe",*Prog. Theor. Phys.*, 64, 2029.

Shandarin, S.F., Doroshkevich, A.G., and Zeldovich, Ya.B.: 1983, "The large-scale structure of the universe", *Usp. Fiz. Nauk*, 139, 83.

Suto, Y., Sato, K., and Sato, H.: 1984, "Expansion of voids in a matter-dominated universe", *Prog. Theor. Phys.*, 71, n. 5.

Tomita, K.: 1969, "Formation of gravitationally bound primordial gas clouds", *Prog. Theor. Phys.*, 42, 9.

White, S.D.M., and Silk, J.: 1979, "The growth of aspherical structures in the universe: is the local supercluster an unusual system?", *Astrophys. J.*, 231, 1.

Zeldovich, Ya.B.: 1970, "Gravitational instability: an approximate theory for large density perturbation", *Astron. Astrophys.*, 5, 84.

NONLINEAR WAVE PROPAGATION IN RELATIVISTIC HYDRODYNAMICS AND COSMOLOGY

A.M. Anile

Seminario Matematico, Città Universitaria
Viale A. Doria, 6 - 95125 - Catania, Italy

1. INTRODUCTION

Relativistic nonlinear wave propagation plays an important role in various areas of relativistic astrophysics and cosmology.

Usually nonlinear waves are encountered in situations where there is: i) a large release of energy within a small region, giving rise to blast waves, or ii) an already formed smooth wave of finite amplitude (as opposed to infinitesimal) which distorts owing to non-linearity.

Blast waves become relativistic when the amount of energy E released in a volume V of a gas of rest--mass density ρ is such that $E \gg \rho V c^2$. Relativistic blast waves have been invoked in connection with supernovae explosions and gravitational collpase (Chevalier, 1981), models for superluminal variations of radio-sources (Blandford and McKee, 1976) and the formation of jets (Shapiro, 1979, 1980), explosions of pregalactic objects and the formation of galaxies in the post-decoupling era (Ikeuchi et al., 1983).

The effects of general relativity can be important for the structure and the propagation of the blast wave. In the spherically symmetric case exact self-similar solutions of general relativistic hydrodynamics were found in the pioneering work of Cahill and Taub (1971). Successively Bogoyavlenskii (1977) made a deep study of general relativistic self-similar solutions containing a spherical shock wave. Numerical codes have also been developed in the spherically

B. Bertotti et al. (eds.), General Relativity and Gravitation, 313–335.

symmetric case, mainly in connection with gravitational collapse (May and White, 1967; Van Riper, 1979). Very little is known in the non-spherically symmetric case, apart from some rather crude approximate methods by Shapiro (1979) in the case of special relativity.

The case of the nonlinear distortion of a smooth wave of finite amplitude, its breaking and the formations of a shock wave within the profile, is thought to be important in connection with adiabatic perturbations in the cosmological radiation era (Peebles, 1980; Zeldovich and Novikov, 1982). Shock formation acts as a dissipative mechanism which adds to the usual linear dissipation caused by radiative viscosity and heat conduction. This nonlinear effect could be important in determining the spectrum of adiabatic perturbations at the time of decoupling (Barrow, 1980; Jones, 1976; Liang, 1977b). Theoretical analysis of this problem initiated with Peebles (1970) and Jones (1971). Successively it was reconsidered with more detail by Liang (1977a). All these authors treated the problem in special relativity for onedimensional hydrodynamics and special initial conditions. Geometrical effects, general relativity, general wave profiles as well as interaction among waves have not been considered in an appropriate way. The extreme case where wave-wave interaction is dominant with respect to the nonlinear distortion of the wave has been treated recently by Vishniac (1982).

Special relativity is important when the ambient medium is relativistic, as in the cosmological radiative era, or when the wave amplitude is relativistic. In this latter case a new interesting effect was discovered by Liang and Baker (1977) by using very crude approximate methods, an effect which was successively proved rigorously by Anile, Miller and Motta (1980). An assessment of this effect under a wide variety of conditions is warranted because nonlinear dissipation could affect the spectrum of the microwave background in a potentially observable way (Zeldovich and Novikov, 1982).

In this talk I shall report on some recent work on nonlinear wave propagation in relativistic fluid dynamics. I shall concentrate on the analytical methods and the qualitative results. In Sec. 2 I shall discuss the nonlinear steepening of a finite amplitude wave and introduce some of the basic mathemathical concepts. In Sec. 3 I shall treat shock waves with the method of singular hyperfurfaces. In Sec. 4 I shall discuss briefly the structure of shock waves and suggest that the present formulation of dissipative processes in relativistic fluid dynamics is not adequate to cope with these phenomena. Throughout this work units are such that c=1.

2. NONLINEAR STEEPENING OF A FINITE AMPLITUDE WAVE IN RELATIVISTIC HYDRODYNAMICS

The phenomenon of the nonlinear steepening of a finite amplitude wave is best understood by considering a family of exact solutions of the special relativistic hydrodynamic equations, called *simple waves*. These solutions were first obtained by Taub (1948) and are characterised by the constancy in space and time of two of the three Riemann invariants for one dimensional special relativistic fluid flow (Jeffrey, 1976). This is equivalent to saying that the specific entropy is constant and all the other fluid variables (density, velocity, temperature, etc.) are expressed as a function of one of them (e.g. the velocity). These solutions are essentially the nonlinear analogue of the plane waves of linearised theory.

In the case of a barotropic fluid with the state equation

$$p=(\gamma-1)\rho$$

where ρ is the total proper energy density, a right propagating simple wave has the following form, in an inertial frame,

$$\rho=\rho_0\left(\frac{1+v}{1-v}\right)^{\gamma/2(\gamma-1)^{\frac{1}{2}}} \qquad (1)$$

$$v=f\ x-\left(\frac{v+c_s}{1+vc_s}t\right) \qquad (2)$$

with v the x-component of the 3-velocity, f an arbitrary C^1 function, $c_s=(\gamma-1)^{\frac{1}{2}}$ the local sound speed and $\rho_0=\rho(v=0)$.

For a polytropic fluid explicit expressions have been found by Liang (1977a) and for a relativistic gas, described by the Synge equation of state, by Lanza, Miller and Motta (1983a,b). Recently simple waves have been found and studied in the more complicated case of relativistic magnetohydrodynamics (Anile and Muscato, 1983).

If the initial profile has a compressive part,

the regular solution (1), (2) will break down at some critical time t_c corresponding to the earliest time at which the slope of the waveform becomes infinite. Usually this is interpreted as the time of shock formation (Whitham, 1974), i.e. a discontinuity is inserted at some point in the waveform. The precise way by which the discontinuity,once originated, spreads over the waveform is rather complicated and, in general, can only be studied numerically (Whitham, 1974). An interesting phenomenon occurs for strongly nonlinear simple waves. In this case the density ratio in the simple wave must be smeared out considerably before the discontinuity can extend right across the profile (Anile et al., 1983). Simple waves, however, are rather special cases.

For applications to realistic problems in astrophysics and cosmology it is necessary to consider more general waveforms, allowing for non planar wavefront geometry, an arbitrary background space-time, the self-gravity of the wave, the interaction among wave modes. In one particular case it is possible to carry over the concept of simple waves from special relativistic to general relativistic test-hydrodynamics. This is a consequence of the following result which is very easy to prove (Anile and Greco, 1978).

Let $T^{\mu\nu}$ be a traceless energy-momentum tensor satisfying

$$\nabla_\alpha T^{\alpha\beta}=0$$

with respect to the conformally flat metric

$$g_{\alpha\beta}=\exp(\psi)\,\eta_{\alpha\beta}.$$

If one defines

$$\tilde{T}^{\alpha\beta}=\exp(3\psi)\,T^{\alpha\beta}$$

then $\tilde{T}^{\alpha\beta}$ satisfies

$$\partial_\alpha \tilde{T}^{\alpha\beta}=0.$$

In the cosmologically important example of a radiation fluid ($\gamma=4/3$) in the background spacetime of the spatially flat Robertson-Walker model with the metric

$$ds^2=R^2(-d\tau^2+\delta_{ij}dx^i dx^j) \tag{3}$$

this amounts to performing the transformation

$$\tilde{\rho}=R^4\rho; \qquad \tilde{u}^\alpha=Ru^\alpha$$

where u^α is the 4-velocity of the fluid flow and τ is the conformal time. If the fluid motion occurs in the x-direction only, and v denotes the peculiar velocity relative to the background comoving frame, then simple waves are characterized by

$$\rho=\rho_0\left(\frac{1+v}{1-v}\right)^{2\sqrt{3}/3} \tag{4}$$

$$v=f\left(x-\frac{v+1/\sqrt{3}}{1+v/\sqrt{3}}\tau\right), \tag{5}$$

where $\rho_0=\tilde{\rho}_0/R^4$ is the unperturbed background energy--density and f is an arbitrary C^1 function. These solutions had already been found by Liang (1977a) by an *ad hoc* procedure, not as a special case of the more general result sketched above.

Proceeding along these lines it should be possible to extend the concept of simple waves to the Robertson-Walker models with positive or negative spatial curvature. A more promising way of dealing with an arbitrary background space-time and at the same time allowing for arbitrary wavefront geometry,is to resort to *asymptotic waves*. These were first introduced by Choquet-Bruhat (1969) for quasi-linear hyperbolic systems. The idea of the method is the following. Any of the field variables, i.e. pressure p, 4-velocity u^α, metric $g_{\mu\nu}$, etc. is expanded in a formal series of the kind

$$p=\sum_{n=0}^{\infty}\omega^{-n}\underset{n}{p}(x^\mu,\xi), \quad \xi=\omega\boldsymbol{\varphi}, \tag{6}$$

etc. where $\underset{0}{p}$, $\underset{0}{u}^\mu$, $\underset{0}{g}_{\mu\nu}$, etc. represent the unperturbed solution of the field equations (which, according to the cases can be taken as those of test hydrodynamics or the full Einstein equations), ω is a real parameter (interpreted as frequency), $\boldsymbol{\varphi}(x)$ is a C^1 function (interpreted as phase) to be determined. Essentially the ansatz (6) resembles the standard high-frequency expansion (Kline and Kay, 1965) suitably extended to deal with nonlinear equations.

By substituting expansions such as (6) into the field equations, treating x^μ and ξ as independent variables, and equating to zero the coefficients of the resulting powers of ω, after various manipulations one obtains the following results:

i) $\varphi(x^\mu)$ satisfies the characteristic equation of the given system of field equations,

ii) if, for the sake of simplicity, $\varphi(x^\mu)$ is taken to be a simple root of the characteristic equation, then all the first-order perturbations $\underset{1}{p}$, $\underset{1}{u}^\mu$, etc. can be expressed as funtions of one of them, say $\underset{1}{p}$ (this property makes asymptotic waves, in some sense, a generalization of simple waves),

iii) by introducing the *rays* as the bicharacteristics of the characteristic equation (which form a congruence of time-like or null curves), $\underset{1}{p}$ satisfies a transport equation along them,

$$\partial \underset{1}{p}/\partial\sigma + N\underset{1}{p}\ \partial\underset{1}{p}/\partial\xi + M\underset{1}{p} = 0 \tag{7}$$

where σ is a parameter along the rays, N and M appropriate coefficients depending only on the unperturbed solution.

By this method all the non-linearity of the field equations is reduced to that of eq. (4) which is a simple, well known equation of the Hopf type and has been extensively studied (Whitham, 1974).

An interesting example is provided by a barotropic fluid in a background space-time given by the spatially flat Robertson-Walker metric (3) (Anile, 1978). In this case the characteristic equation for the acoustic mode reads

$$(\varphi_0)^2 - (\gamma-1)\delta^{ij}\varphi_i\varphi_j = 0 \tag{8}$$

If, for the sake of simplicity, one restricts oneself to right-progressive plane waves solutions of (8), one obtains

$$\varphi = -(\gamma-1)^{\frac{1}{2}}\ell\tau + \ell x^1,\quad \ell=\text{constant}.$$

The rays are then the straight lines (in the τ, x^i coordinates)

$$\frac{dx^1}{d\tau} = (\gamma-1)^{\frac{1}{2}},\quad \frac{dx^2}{d\tau} = \frac{dx^3}{d\tau} = 0,\quad \frac{d\tau}{d\sigma} = 1. \tag{9}$$

It is convenient to express all the perturbations as function of v, the peculiar velocity of the fluid along the x^1-axis relative to the background comoving frame. Then

$$\rho_1=\gamma(\gamma-1)^{-\frac{1}{2}}\rho_0 v \qquad (10)$$

which is identical to the relationship (1) for simple waves in the limit $v<<1$. The transport equation for $v(\sigma,\xi)$ reads

$$\frac{\partial v}{\partial \sigma}-\ell(\gamma-2)v\frac{\partial v}{\partial \xi} - \frac{3\gamma-4}{2}\frac{\dot{R}}{R}v=0. \qquad (11)$$

Notice that for $\gamma=2$ (the Zeldovich stiff fluid) the nonlinear term vanishes and this corresponds to the exceptional case (Boillat, 1975).

The solution of eq. (11) is easily obtained by the method of bicharacteristics. For $\gamma=4/3$ one recovers the exact solution for simple waves (5) in the limit $v<<1$. Application of this formalism to relativistic magneto-hydrodynamics (Anile and Greco, 1978) and to radiative gasdynamics (Anile and Greco, 1979a,b) have been made.

There are some difficulties with this method. The first is that, in its present formulation, it breaks down at caustics (Choquet-Bruhat, 1969). However this difficulty could perhaps be overcome by using a technique similar to that of Ludwig (1966) for the linear case (see also Choquet-Bruhat, 1982). Another difficulty is that the method does not seem capable of dealing with the interaction among waves and with the self-gravity of the wave. A possible way out could consist in adopting a two-timing technique (Whitham, 1974; MacCallum and Taub, 1973). Finally, at present there is no general proof of the asymptotic convergence of the formal expansions in a suitable topology. However one could perhaps try to apply some methods recently used by Eckhoff (1981) for the linear case, which yield asymptotic convergence in the Sobolev space H^s.

3. EVOLUTION OF SHOCK WAVES

At the shock front, treated as a discontinuity in the profile, the usual differential equations for the flow are no longer valid and one must resort to an integral formulation, treating the shocks as weak solutions (Whitham, 1974; Jeffrey, 1976). This implies that at the shock front the jump relations hold, which express the local conservation of mass, momentum and energy. Let $j^{\mu}=nu^{\mu}$ be the mass-flux vector, $T^{\mu\nu}$ the energy-momentum tensor, Σ the oriented hypersurface (time-like or null) representing the history in the space-time M of the shock front and denote the jump of a quantity f across Σ by $[f]$ (a precise definition will be given immediately below). Then the jump relations (or generalized Rankine-Hugoniot conditions) (Lichnerowicz, 1971) read

$$n_{\mu}[j^{\mu}]=0 \tag{12}$$

$$n_{\mu}[T^{\mu\nu}]=0 \tag{13}$$

where n_{μ} is the orthogonal form of Σ (the unit normal when Σ is time-like). To these one must add the Weyl condition expressing the entropy increase across a shock, (s being the specific entropy)

$$[s]>0$$

which selects the class of physically admissible shock waves (Lichnerowicz, 1970; Ruggeri and Strumia, 1981).

In relativistic fluid dynamics, under mild thermodynamical restrictions, it can be proved that, except in the extreme case of a stiff barotropic fluid ($\gamma=2$), the shock solutions are represented by time-like hypersurfaces (Lichnerowicz, 1970; Ruggeri and Strumia, 1981). Therefore, in the following, Σ will be assumed to be a time-like hypersurface of discontinuity. Also, only acoustic shocks will be considered here, i.e. those for which there is a net mass flow across Σ, $n_{\mu}[j^{\mu}]\neq 0$, because they are the only ones which can be produced by the nonlinear steepening of an acoustic wave of finite amplitude. Slip discontinuities, i.e. shocks for which $n_{\mu}[j^{\mu}]=0$, correspond to the so-called characteristic shocks and have been studied extensively by Boillat (1982).

In order to study the evolution of relativistic shock waves extensive use is made of the covariant compatibility relations (Maugin, 1976). These can be obtained by a suitable use of distribution theory (Lichnerowicz, 1970; Friedländer, 1975) or more simply by introducing the concept of regularly discontinuous tensor fields. Let Σ be orientable. Then it is easy to see (Modugno and Stefani, 1979) that there exist three differentiable submanifolds of space-time M, Ω, Ω_+, Ω_-, (Ω_+, Ω_- having boundary) such that

i) $\bar{\Omega} = \Omega_+ \cup \Omega_-$

ii) Ω is an open neighbourhood of Σ

iii) $\Omega_\pm \cap \Sigma = \partial\Omega_\pm$.

A tensor field T of type (r,s), smooth on $\Omega-\Sigma$, is said to be regularly discontinuous across Σ if there exist two tensor fields $\tilde{T}_+$, $\tilde{T}_-$, of type (r,s), smooth on Ω_+ and Ω_- respectively such that

$$T|_{\Omega_\pm - \Omega_\pm \cap \Sigma} = \tilde{T}_\pm|_{\Omega_\pm - \Omega_\pm \cap \Sigma}\,.$$

Then the jump $[T]$ of T across Σ is defined by

$$[T] = \tilde{T}_+|_\Sigma - \tilde{T}_-|_\Sigma\,.$$

$[T]$ can be extended to a tensor field smooth on a neighbourhood of Σ. The *inner covariant derivative* $\tilde{\nabla}$ is defined on the smooth tensor fields of Ω defined on Σ (Maugin, 1976), in local coordinates, by

$$\tilde{\nabla}_\mu T{:}{:}{:} = h_\mu{}^\nu \nabla_\nu T{:}{:}{:}$$

where $h_\mu^\nu = \delta_\mu^\nu - n_\mu n^\nu$ is the projection tensor onto Σ. A global, coordinate-free definition can also be given (Anile, 1983a).

Then the compatibility relations of first order are

$$[\nabla T] = \tilde{\nabla}[T] + [\nabla_{\underline{n}} T] \otimes n. \tag{14}$$

These relations will be used in order to study the evolution of an acoustic shock wave in relativistic fluid dynamics. In order to do this, first of all it is necessary to have explicit expressions for the jumps $[j^\mu]$ and $[T^{\mu\nu}]$, for an acoustic shock.

Let $\underset{+}{u}^\mu$ be the 4-velocity of the fluid ahead of the shock front and V_Σ the normal speed of propagation of Σ with respect to the family of observers defined by $\underset{+}{u}^\mu$ (Lichnerowicz, 1971). It is convenient to construct the 4-vector k^μ,

$$k^\mu = \frac{1}{\Gamma_\Sigma}(\underset{+}{u}^\mu - \Gamma_\Sigma V_\Sigma n^\mu)$$

where $\Gamma_\Sigma = (1-V_\Sigma^2)^{-\frac{1}{2}}$ is the Lorentz factor of the shock.

Then $k_\mu k^\mu = -1$ and k_μ is tangent to $\Sigma^{(1)}$.

It can easily be seen that (Lichnerowicz, 1975)

$$[j^\mu] = -\Gamma_\Sigma f_+ [1/\tau] k^\mu \tag{15}$$

$$[T^{\mu\nu}] = [1/\tau] f_+^2 \Gamma_\Sigma^2 k^\mu k^\nu + [p] h^{\mu\nu} \tag{16}$$

where $f=(\rho+p)/n$ is the index of the fluid and $\tau=f/n$ the dynamical volume.

The way to proceed is now the following. One starts with the conservation equations

$$\nabla_\mu T^{\mu\nu}=0$$

$$\nabla_\mu j^\mu=0$$

and takes the jumps of these equations across Σ, obtaining

$$[\nabla_\mu T^{\mu\nu}]=0$$

$$[\nabla_\mu j^\mu]=0$$

and by using the compatibility relations (14),

$$\tilde{\nabla}_\mu [T^{\mu\nu}] + n_\mu [n^\alpha \nabla_\alpha T^{\mu\nu}] = 0 \tag{17}$$

$$\tilde{\nabla}_\mu [j^\mu] + n_\mu [n^\alpha \nabla_\alpha j^\mu] = 0 . \tag{18}$$

By inserting the explicit expressions (15), (16) for $[T^{\mu\nu}]$, $[j^\mu]$, into (17), (18), after laborious manipulations, one obtains a set of relations involving the Thomas displacement derivative of $[1/\tau]$, $[p]$, V_Σ , the values of these quantities and of the normal gradients of the fluid variables (Anile, 1983b). In general this set of relations is rather complicated. However, in some circumstances, one can infer from it some useful results, for instance on the evolution of the shock front geometry or on the tendency to grow or decay of an initially given shock, in much the same way as is customary in classical continuum mechanics (Chen, 1976).

The most interesting result that can be obtained by this method is one which is typical of relativistic shocks. For the sake of simplicity consider the case of a relativistic shock propagating into a radiation fluid in special relativity (the treatment in fact can be carried over to the general case). Let the shock wave be extremely relativistic, $\Gamma_\Sigma \to 1$. Then, from the set of compatibility relations one obtains to the order Γ_Σ^{-1}, in the global inertial frame,

$$\frac{16}{3} n_+ f_+ \delta V_\Sigma / \delta S = 3(8\Gamma_\Sigma^6)^{-1} (\partial_{\underline{\nu}} w)_- - \frac{5}{3}(n_+ f_+ / \Gamma_\Sigma^2) \underline{\nu} \cdot (\partial_{\underline{\nu}} \underline{V})_-$$

$$2n_+ f_+ \delta V_\Sigma / \delta S = (8\Gamma_\Sigma^6)^{-1} (\partial_{\underline{\nu}} w)_- + (n_+ f_+ / \Gamma_\Sigma^2) \underline{\nu} \cdot (\partial_{\underline{\nu}} \underline{V})_-$$

where $w=\rho+p$ is the enthalpy, $\underline{V}$ is the fluid's 3-velocity, $\underline{\nu}$ is the unit 3-vector normal to the 2-surface of the shock front. One can always set up an initial profile for which $\underline{\nu} \cdot (\partial_{\underline{\nu}} \underline{V})$ is finite in the limit $V_\Sigma \to 1$. Then it is easy to see that

$$\delta V_\Sigma / \delta S = 0(\Gamma_\Sigma^{-2}), \quad V_\Sigma \to 1. \tag{19}$$

This result, which is at striking variance with the behaviour of strong shock waves in classical fluid mechanics, had already been conjectured by Liang and Baker (1977) on the basis of a rather rough and imprecise argument. The first rigorous proof was given by Anile, Miller and Motta (1980) in the case of a plane shock wave propagating into a constant state in a relativistic barotropic fluid. This result has also been confirmed with more general state equations by numerical calculations (Anile et al., 1983).
The proof outlined above is more general in the sense that it is independent of the shock front geometry and of the state ahead of the shock.

One can draw several exact analytical results from the set of compatibility relations for various special flow profiles behind the shock. These results are very useful for checking the accuracy of numerical codes dealing with relativistic shocks. However, in general, the compatibility relations described above are not sufficient to determine the evolution of a shock wave because of the appearance in them of the jumps of the normal gradients of the field variables. Therefore, short of a direct numerical integration, one has to resort to approximation methods. For plane parallel weak shocks one of the most widely used methods is the shock-fitting technique (Landau and Lifshitz, 1959; Whitham, 1974) which has been extended to special relativistic fluid dynamics by Liang (1977a) and applied to the study of cosmological nonlinear adiabatic perturbations (Liang, 1977b). This method is based upon the equal area construction and for an isolated triangular pulse propagating into a constant state yields the following decay law

$$[v]=[v]_0/(1+\alpha t[v]_0/\lambda_0)^{\frac{1}{2}} \tag{20}$$

with $\alpha=2(1-c_s^2)$ for a barotropic fluid, λ_0 being the initial length of the pulse, v the 3-velocity. The shock-fitting method, however, covers only weak plane parallel shocks in special relativity. For many applications to relativistic astrophysics and cosmology one must consider shock waves of arbitrary geometry propagating into a general relativistic background spacetime.

Recently a new method has been devised which is capable of treating this situation, at least for weak shock waves (Anile, 1983c). This method can best be explained by analogy with the usual theory of acceleration waves. An acceleration wave is a solution of the field equations which is regular everywhere except at a hypersurface where the fields *derivatives* are regularly discontinuous (to be contrasted with shock waves where the *fields themselves* are discontinuous). For acceleration waves one can construct an exact theory of the wavefront propagation based upon the method of bicharacteristics (Whitham, 1974; Chen, 1976; Jeffrey, 1976; Boillat, 1965). One introduces the concept of *rays* as the bicharacteristics of the characteristic equation and then the problem of determining the evolution of the acceleration wave is reduced to that of solving the *transport equation* for the *amplitude* of the wave along the rays (which is an ordinary differential equation of the Bernoulli type).

In general this program is not feasible in the case of a shock wave for a genuinely nonlinear system. However it can be proved that, in the case of *weak shocks*, one can introduce the concept of *rays* and the accompanying *transport equations* in a way analogous (though not identical) to that for acceleration waves (Anile, 1983c). The resulting theory covers the case of a weak shock wave propagating into an arbitrary background space-time. In the simple case of special relativity and propagation into a uniform fluid at rest one finds the following decay law

$$[v] = [v]_0/\theta\psi \tag{21}$$

where $[v]$ is the normal velocity jump across the shock front, θ is the expansion of the rays and

$$\psi=1+(\alpha\,[v]_0/\lambda_0)\int_0^t \frac{d\tau}{\theta}$$

with $\lambda_0^{-1}=\left|\frac{1}{[v]_0}\,[\partial_{\underline{\nu}} v]_0\right|$.

For spherical wavefronts

$$\theta=r/r_0\ ,\ \psi=1+(\alpha[v]_0/c_s\lambda_0)r_0\ln(r/r_0)$$

and for cylindrical wavefronts

$$\theta=(r/r_0)^{\frac{1}{2}},\ \psi=1+(2\alpha[v]_0/c_s\lambda_0)r_0\{(r/r_0)^{\frac{1}{2}}-1\}.$$

In the newtonian limit these expressions agree with those obtained by heuristic reasonings (Landau and Lifshitz, 1959).

4. THE STRUCTURE OF A RELATIVISTIC SHOCK

To treat shock waves as discontinuities is an extreme mathematical idealization. A more realistic description is in terms of a thin layer where the fluid variables vary rapidly but smoothly. The thickness of the shock layer is determined by dissipative processes (heat conduction and viscosity) and is usually of the order of a mean free path. The flow profile within a shock layer is of great astrophysical interest in:

i) models of supernovae explosions and heavy elements production (Chapline and Weaver, 1979),

ii) theories of cosmological entropy perturbations produced by the shocking of adiabatic perturbations (Zeldovich and Novikov, 1982).

The traditional treatment of dissipative processes in relativistic fluid mechanics is that of Eckart (Taub, 1978). The energy-momentum tensor is

$$T^{\mu\nu}=(\rho+p)u^\mu u^\nu+(p+\pi)h^{\mu\nu}+\pi^{\mu\nu}+u^\mu q^\nu+u^\nu q^\mu \qquad (22)$$

with

$$\pi^{\alpha}_{\alpha}=0,\ u_{\alpha}\pi^{\alpha\beta}=0,\ u_{\alpha}q^{\alpha}=0$$

and the constitutive equations of Eckart are

$$\pi=-\zeta\theta \tag{23}$$

$$\pi^{\alpha\beta}=-2\eta\sigma^{\alpha\beta} \tag{24}$$

$$q_{\alpha}=-\kappa(h^{\beta}_{\alpha}T_{,\beta}+Ta_{\alpha}) \tag{25}$$

where θ, $\sigma_{\alpha\beta}$ and a_{α} are the expansion, shear and acceleration of the fluid flow, $\kappa\geqq 0$, $\eta\geqq 0$, $\zeta\geqq 0$ are the coefficients of heat conduction, shear and bulk viscosity respectively. These equations represent the relativistic extensions of the usual Navier-Stokes-Fourier constitutive equations of classical fluid mechanics (Serrin, 1959).

Koch (1965) has studied the shock structure problem by using the Eckart theory. By assuming one dimensional flow, in the reference frame comoving with the shock, one obtains

$$T^{10}=Q=\text{const.}$$

$$T^{11}=P=\text{const.}$$

$$nu^{1}=M=\text{const.}$$

which together with eqs. (22-25) yield a plane autonomous system for the unknowns p, u^1. The boundary conditions are

$$\lim_{x\to\pm\infty} p=p_{\pm},\quad \lim_{x\to\pm\infty} u^{1}=u^{1}_{\pm}$$

where $p_{\pm}$, $u^{1}_{\pm}$ represent the states ahead and behind the shock. The study of the singular points of the system shows that a C^1 solution to the above problem exists, and is essentially unique[(2)]. However it is well known that the Eckart approach suffers from various defects, the most serious being the fact that it allows signals propagating with infinite speed (Israel, 1976; Israel and Stewart, 1979).

In order to remedy these difficulties, starting from an earlier suggestion of Cattaneo's (1948) alternative theories have been proposed (Müller, 1967; Israel, 1976; Israel and Stewart, 1979; Dixon, 1978; Bampi and Morro, 1980; de Groot et al., 1980) which introduce relaxation terms. In the simplest form of these theories, the constitutive equations (22-25) are replaced by

$$\tau_\nu \dot{\pi} + \pi = -\zeta\theta \tag{26}$$

$$\tau_s \dot{\pi}^{\alpha\beta} + \pi^{\alpha\beta} = -2\eta\sigma^{\alpha\beta} \tag{27}$$

$$\tau_q \dot{q}_\alpha + q_\alpha = -h^\beta_\alpha (T_{,\beta} + Ta_\alpha) \tag{28}$$

where a dot denotes the convective derivative along the fluid flow lines, and τ_ν, τ_s, τ_q are three relaxation times.

It is widely thought that theories of this kind are to be preferred to the old Eckart formulation because they give rise to finite wave speeds (Israel and Stewart, 1979; Lindblom and Hiscock, 1983).

For non relativistic fluid flow eqs. (26-28) reduce to their Newtonian counterpart, which is essentially Müller's theory (1967). Lately Müller and Liu (1983) have produced a much firmer foundation for constitutive equations of the rate type on the basis of the principles of extended thermodynamics. Now the question arises whether theories with constitutive equations of the rate type give better results vis à vis of the shock structure problem.

The situation in the non relativistic limit has been analyzed by Anile and Majorana (1981) while the relativistic case is under current investigation by Majorana and Motta. The main results of the non relativistic case are the following. In the non relativistic limit the shock structure equations read

$$a_{11} d\Phi/dx + a_{12} d\Psi/dx = A_1 D_1 \tag{29}$$

$$a_{21} d\Phi/dx + a_{22} d\Psi/dx = A_2 D_2 \tag{30}$$

where

$$\Phi = (M/P)v, \quad \Psi = p/P$$

(v being the x-component of the 3-velocity),

$$D_1(\Phi,\Psi) = \Phi + \Psi - 1,$$

$$D_2(\Phi,\Psi) = \Phi\Psi/(\gamma-1) - \Phi^2/2 + \Phi - a^*$$

with

$$a^* = \gamma^2 M_0^2 (1+\gamma M_0^2)^{-2} [(\gamma-1)^{-1} + \tfrac{1}{2} M_0^2]$$

γ being the usual adiabatic index and M_0 the Mach number. The coefficients a_{ij} are some functions

$$a_{ij} = a_{ij}(\Phi, \Psi, r_1, r_2, r_3)$$

where r_1, r_2, r_3 are three constants, proportional to the relaxation times τ_q, τ_v, τ_s respectively. Kinetic theory predicts the values $r_1=1$, $r_2=1$, $r_3=2/3$ (Müller, 1967). However the main qualitative result is independent of the precise values of these constants.

The shock structure problem is to solve eqs. (29-30) with the boundary conditions

$$\lim_{x\to\pm\infty} (\Phi, \Psi) = (\Phi_\pm, \Psi_\pm)$$

where $(\Phi_\pm, \Psi_\pm)$ are the two shock states, solutions of $D_1(\Phi,\Psi) = D_2(\Phi,\Psi) = 0$. Now, at variance with the situation in the Navier-Stokes-Fourier case, the determinant

$$D = a_{11}a_{22} - a_{12}a_{21}$$

can vanish.

It can be shown that, if $r_1>0$, $r_3>0$, for a sufficiently high Mach number one of the two shock states crosses the boundary of the region $D=0$ and this makes it impossible to reduce eqs. (29-30) to normal form. Furthermore for a sufficiently high Mach number the shock state (Φ_-, Ψ_-) becomes a stable node in the (Φ, Ψ) plane and therefore it is impossible to satisfy the boundary condition

$$\lim_{x\to-\infty} (\Phi, \Psi) = (\Phi_-, \Psi_-).$$

Therefore for Mach numbers exceeding a critical Mach number M_c no smooth solution exists to the shock structure problem. For the values for r_1, r_2, r_3 given by kinetic theory one finds

$$M_c = 1.65.$$

This result shows that the theory fails badly with regard to experiments because the latter show the existence of a regular structure up to much higher Mach numbers (Schmidt, 1969). This situation is to be contrasted with the situation occurring in the Navier-Stokes-Fourier theory where one proves the existence of a regular solution to the shock structure problem (Gilbarg and Paolucci, 1953) which, as noted above, is in qualitative agreement with the experimental results (Schmidt, 1969).

Theories with relaxation times have been introduced also in order to cope with non-stationary rapidly varying phenomena such as shock waves structure (Israel, 1976; Israel and Stewart, 1979). However in this regard they fail badly and are in a worse position than the standard theories of the Navier-Stokes--Fourier type.

NOTES

(1) The interpretation of k^μ is that the operator

$$\delta/\delta S \equiv (1/\Gamma_\Sigma)\nabla_{\underline{k}}$$

is the exact relativistic analogue of the Thomas displacement derivative of classical continuum mechanics (Chen, 1976).

(2) Comparison with experimental data is very difficult in the relativistic case because in ionized hydrogen without a background magnetic field relativistic shocks tend to have a large width and therefore they are not accessible to experiment (Gross, 1971; Taussig, 1973). Mildly relativistic shocks (up to $\sim$0.01c) can be produced in a hydrogen plasma with a background magnetic field and there are measurements of their structure (Taussig, 1973). However, Koch's theory does not apply to this case because it ignores magnetic fields.

In the non relativistic limit (in which Eckart's consitutive equations reduce to the classical Navier-Stokes-Fourier ones) Koch's theory reduces to the classical analysis of Gilbarg and Paolucci (1953). Then the predictions of the theory are in *qualitative agreement* with the experimental data but there are some difficulties for a good quantitative fitting (Schmidt, 1969).

REFERENCES

Anile, A.M.: 1978, "Non-linear high-frequency waves

in relativistic cosmology", *Atti Accad. Naz. Lincei, Rend. Cl.Sci.Fis.Mat.Nat.*, 63, 375.

Anile, A.M.: 1983a, "A geometric characterization of the compatibility relations for regularly discontinuous tensor field", to be published in *Le Matematiche*.

Anile, A.M.: 1983b, "Evolution of shock waves in relativistic continuum mechanics", to be published in *Ann. Inst. Henri Poincaré*.

Anile, A.M.: 1983c, "Propagation of weak shocks in relativistic fluid dynamics", to be published in the *Proceedings of the "Journées rélativistes"*, Turin.

Anile, A.M. and Greco, A.: 1978, "Asymptotic waves and critical time in general relativistic magnetohydrodynamics", *Ann. Inst.Henri Poincaré*, 29,257.

Anile, A.M. and Greco, A.: 1979a, "Shock formation and shock structure in the cosmological radiative era", *Lett. Nuovo Cimento*, 26, 121.

Anile, A.M. and Greco, A.: 1979b, "Non-linear wave propagation in a proton-electron plasma coupled with a strong radiation field", *Nuovo Cimento*, 54, 307.

Anile, A.M. and Majorana, A.: 1981, "Shock structure for heat conducting and viscid fluids", *Meccanica*, 2, 149.

Anile, A.M., Miller, J.C. and Motta, S.: 1980, "Damping of relativistic shocks in an expanding universe", *Lett. Nuovo Cimento*, 29, 268.

Anile, A.M., Miller, J.C. and Motta, S.: 1983, "Formation and damping of relativistic strong shocks", *Phys. Fluids*, 26, 1450.

Anile, A.M. and Muscato, O.: 1983, "Simple waves in relativistic magnetofluiddynamics", to be published in *Lett. Nuovo Cimento*.

Bampi, F. and Morro, A.: 1980, "Two approaches to nonstationary relativistic thermodynamics", *J.Math. Phys.*, 21, 1201.

Barrow, J.: 1980, "Galaxy formation: the first million years", *Philos. Trans. R.Soc. London*, A296, 273.

Blandford, R.D. and McKee, C.F.: 1976, "Fluid dynamics of relativistic blast waves", *Phys. Fluids*, 19, 1130.

Bogoyavlenskii, O.I.: 1977, "General relativistic self-similar solution with a spherical shock wave", *Sov. Phys. JETP*, 46, 633.

Boillat, G.: 1965, *La propagation des ondes*, Gauthier-Villars, Paris.

Boillat, G.: 1975, "Chocs caractéristiques et ondes simple exceptionelles pour les systèmes conservatifs à intégrale d'énergie: forme explicite de la solutions", *C.R. Seances Acad. Sci. Paris*, 280A, 1325.

Boillat, G.: 1982, "Urti", in Ferrarese (ed.), *Wave propagation*, Liguori, Napoli.

Cahill, M.E. and Taub, A.H.: 1971, "Spherically symmetric similarity solutions of the Einstein field equations for a perfect fluid", *Commun. Math. Phys.*, 21, 1.

Cattaneo, C.: 1948, "Sulla conduzione del calore", *Atti Semin. Mat.Fis. Univ. Modena*, 3, 83.

Chapline, G.F. and Weaver, T.A.: 1979, "Structure of relativistic shock waves in simple gases", *Phys. Fluids*, 22, 1885.

Chen, P.J.: 1976, *Selected topics in wave propagation*, Noordhoff, Leyden.

Chevalier, R.A.: 1981, "Hydrodynamic models of supernova explosions", *Fundamentals of cosmic physics*, 7, 1.

Choquet-Bruhat, Y.: 1969, "Ondes asymptotiques et approchées pour des systèmes d'équations aux dérivées partielles non linéaires", *J.Math. Pures et Appl.*, 48, 117.

Choquet-Bruhat, Y.: 1982, "Ondes asymptotiques", in Ferrarese (ed.), *Wave propagation*, Liguori, Napoli.

De Groot, S.R., Vande Leeuwen, W.A. and Van Weert, Ch.G.: 1980, *Relativistic kinetic theory*, North Holland, Amsterdam.

Dixon, G.: 1978, *Special Relativity, the foundation of macroscopic physics*, Cambridge University Press, Cambridge.

Eckhoff, K.: 1981, "On stability for symmetric hyperbolic systems, I", *J.Differ. Equations*, 40, 94.

Friedlander, F.G.: 1975, *The wave equation on a curved space-time*, Cambridge University Press, Cambridge.

Gilbard, D. and Paolucci, D.: 1953, "The structure of shock waves in the continuum theory of fluids", *J.Rat.Mech.An.*, 2, 617.

Gross, R.A.: 1971, "The physics of strong shock waves in gases", in P.Caldirola and H.Knoepfel, (ed.), *Physics of High Energy density*, Academic Press Inc., New York.

Ikeuchi, S., Tomisaka, K. and Ostriker, J.: 1983, "The structure and expansion law of a shock wave in an expanding universe", *Astrophys. J.*, 265, 583.

Israel, W.: 1976, "Nonstationary irreversible thermodynamics: a causal relativistic theory", *Ann. Phys. (USA)*, 100, 310.

Israel, W. and Stewart, J.: 1979, "On transient relativistic thermodynamics and kinetic theory", *Proc. R. Soc., London*, A365, 43.

Jeffrey, A.: 1976, *Quasilinear hyperbolic systems and waves*, Pitman, London.

Jones, B.J.T.: 1971, "The origin and evolution of structure in the Universe", unpublished Ph. D. Thesis, Cambridge.

Jones, B.J.T.: 1976, "The origin of galaxies: A review of recent theoretical developments and their confrontation with observation", *Rev. Mod. Phys.*, 48, 107.

Kline, M. and Kay, I.: 1965, *Electromagnetic theory and Geometrical Optics*, J.Wiley and Sons, New York.

Koch, P.A.: 1965, "Relativistic shock structure", *Phys. Rev.*, A140, 1161.

Lanza, A., Miller, J.C. and Motta, S.: 1982, "Relativistic shocks in a Synge gas", *Lett. Nuovo Cimento*, 35, 309.

Landau, L.D. and Lifshitz, E.M.: 1959, *Fluid Mechanics*, Pergamon Press, London.

Lanza, A., Miller, J.C. and Motta, S.: 1983, "Relativistic shock waves in a Synge gas", preprint, University of Catania.

Liang, E.P.T.: 1977a,"Relativistic simple waves: shock damping and entropy production", *Astrophys. J.*, 211, 361.

Liang, E.P.T.: 1977b, "Galaxies and entropy from nonlinear fluctuations: a simple waves analysis", *Astrophys. J.*, 216, 206.

Liang, E.P.T. and Baker, K.: 1977, "Damping of relativistic shocks", *Phys. Rev. Lett.*, 39, 191.

Lichnerowicz, A.: 1971, "Ondes de choc, ondes infinitesimales et rayons en hydrodynamique et magnetohydrodynamique relativistes", in C.Cattaneo (ed.), *Relativistic fluid dynamics*, Cremonese, Roma.

Lichnerowicz, A.: 1975, "Ondes de choc gravitationelles et ondes de choc mangétomatérielles extrêmes", *C.R. Seances Acad. Sci. Paris*, 281, 929.

Lindblom, L. and Hiscock, A.A.: 1983, "On the stability of rotating stellar models in general relativity theory", *Astrophys. J.*, 267, 384.

Ludwig, D.: 1966, "Uniform asymptotic expansion at a caustic", *Commun. Pure Appl. Math.*, 19, 215.

MacCallum, M.A.H., and Taub, A.H.: 1973, "The averaged lagrangian and high-frequency gravitational waves", *Commun. Math. Phys.*, 30, 153.

Maugin, G.: 1976, "Conditions de compatibilité pour une hypersurface singulière en mécanique relativiste des milieux continus", *Ann. Inst. Henri Poincaré*, 24, 213.

Modugno, M. and Stefani, G.: 1979, "On the geometrical structure of shock waves in general relativity", *Ann. Inst. Henri Poincaré*, 30, 27.

Müller, I.: 1967, "Zum Paradoxon der Wärmeleitungstheorie", *Z. Phys.*, 198, 329.

Müller, I. and Liu, I.: 1983, "Extended thermodynamics of classical and degenerate ideal gases", preprint, University of Berlin.

Peebles, P.J.E.: 1970, "Non linear limit on primeval adiabatic perturbations", *Phys. Rev.*, D1, 397.

Peebles, P.J.E.: 1980, *The large-scale structure of the universe*, Princeton University Press, Princeton.

Ruggeri, T. and Strumia, A.: 1981, "Main field and convex covariant density for quasi-linear hyperbolic systems", *Ann. Inst. Henri Poincaré*,34,65.

Schmidt, B.: 1964, "Electron beam density measurement in shock waves in argon", *J.Fluid Mech.*, 39,361.

Serrin, P.: 1959, "Mathematical principles of classical fluid mechanics", in *Handbuck der Physik*, VIII/1, Springer-Verlag, New York.

Taub, A.H.: 1948, "Relativistic Rankine-Hugoniot equations", *Phys. Rev.*, 74, 328.

Taub, A.H.: 1978, "Relativistic fluid mechanics", *Annu. Rev. Fluid Mech.*, 10, 301.

Taussig, R.: 1973, "Shock wave production of relativistic plasmas", in M.J.Lighthill, I.Imai and H.Sato (eds), *Dynamics of ionized gases*, University of Tokyo Press, Tokyo.

Van Riper, K.A.: 1979, "General relativistic hydrodynamics and the adiabatic collapse of stellar cores", *Astrophys. J.*, 232, 558.

Vishniac, E.T.: 1982, "Nonlinear effects on cosmological perturbations. I. The evolution of adiabatic perturbations", *Astrophys. J.*, 253, 446.

Whitham, G.B.: 1974, *Linear and nonlinear waves*, J. Wiley and Sons, New York.

Zeldovich, Y.B. and Novikov, I.D.: 1982, *Struttura ed evoluzione dell'Universo*, (Italian edition), Editori Riuniti, Roma.

COSMOLOGICAL PARAMETERS AND COSMOLOGICAL MODELS
REPORT OF WORKSHOP B3

I.D. Novikov

Space Research Institute
Academy of Sciences
Moscow, USSR

Workshop B3 had three sessions in which 20 contributed papers (Bertotti et al., 1983) were considered.

I shall now dwell upon the results of some of the presented papers and try to outline main research directions.

To estimate fundamental parameters of the universe and to understand physical processes that occurred in it in the past, the most important vehicles are astrophysical observations of the spectrum, anisotropy and other properties of the relic radiation.

The paper by G. Smoot et al. (1983) presents the data on the relic radiation spectrum measurements at 5 wavelengths $\lambda = 0.3, 0.9, 3.0, 6, 3, 12$ cm. An essential point is that measurements at various wavelengths were made by the radiometers of the same type at a height of 3800 m; all spurious radiations were thoroughly taken into account. In consideration of their data, the authors conclude that the relic radiation temperature is

$$T = 2.78 \pm 0.10 \text{ K}$$

while it is

$$T = 2.74 \pm 0.05 \text{ K}$$

according to all ground based measurements and

$$T = 2.90 \pm 0.03 \text{ K}$$

B. Bertotti et al. (eds.), General Relativity and Gravitation, 337–341.

according to all the observations. As the data of the above authors show, there are no significant deviations from Planck spectrum in the Rayleigh-Jeans region. However, the brightness temperature at the spectrum maximum, as they measures it, is considerably lower than that reported by Woody and Richars.

In his papers D.R. Matravers (1983) considers a uniform model of the universe with an abrupt anisotropic beginning of expansion in the presence of matter flow relative to uniform space. With time the model approaches a Friedmann universe. The author employs a numerical method, he discusses the constraints imposed on the model by the observational data about the He and D abundances, about Hubble's constant and a possible degree of its anisotropy, and by the data about the anisotropy of the relic radiation. The author condludes that the presence of fluid tilt weakens the constraints of the shear and other parameters found for non tilted models.

S.A. Bonometto, F. Lucchin, F. Occhionero, N. Vittorio (1983) in their paper presented the results of $\delta T/T$ fluctuation computations under the assumption that the neutrino rest mass is much larger than 1 eV. In this case the overall mass of neutrinos exceeds the mass of ordinary matter. In the evolution of the universe there is a long period when neutrinos have already become non-relativistic and perturbations in the neutrino component are monotonically growing thus differing from oscillations with a constant amplitude of the perturbations in the relativistic matter. After hydrogen recombination the ordinary matter is condensing in gravitational potential wells associated with increased non-uniformities of the neutrino component. Thus to explain the origin of galaxies a much lower amplitude of ordinary-matter density fluctuations $\delta\rho/\rho$ is required at the moment of recombination, than would be necessary in the absence of non-relativistic neutrinos. This, in turn, leads to longer predicted fluctuations of $\delta T/T$ in the relic radiation, associated with $\delta\rho/\rho$. This paper is one of the many studies being carried out throughout the world.

A possible value of the Hubble constant is estimated in the paper by R.V. Wagoner and S. Herskowitz (1983). The authors applied the Baade method of distance determination of supernovae from their spectrum analysis, to the case of Type II supernovae in whose photosphere scattering dominates absorption. They estimate the Hubble constant as $H_o > 100$ km/s Mpc which is of the order of the upper value of the boundary as determined by other astrophysical methods.

The paper by F. Occhionero, P. Santangelo, N. Vittorio (1983) presents the data of numerical calculations for the evolution of nonlinear non-uniformities in the matter distribution under the assumption of spherical symmetry and zero pressure of the matter. Calculations provide a model of the development of the cellular large-scale structure of the universe. The excess initial energy of the expansion is responsible for the origin of the spherical voids in the authors' calculations. It is one of many such studies being carried out throughout the world. It must be remembered that, along with one-dimensional calculations to which the one given here belongs, two- and three-dimensional calculations have been conducted. Here the initial spectrum of velocity and density perturbations is selected according to certain theories of the origin of small perturbations in the expanding universe.

B.W. Tolman and R.A. Matzner (1983) mention in their paper that in uniform models of the universe a group of motions retaining its uniformity may involve clockwise or counterclock-wise rotations. In fact it may give rise to different effects. The exact cosmological solution considered in the paper of J. Ibanez and E. Verdaguer (1983) is also of interest. It deals with the motion of two pairs of gravitational solitons propagating in opposite directions from the two points against the background of the Kasner cosmological model. The solution describes the collision of two solitons that have moved toward each other. Physical properties of the solution are thoroughly analyzed. Though the given solution has never had any direct physical applications, these might appear later on.

Some fundamental problems of spacetime theory that may have important cosmological implications were discussed in the papers of V.M. Canuto and I. Goldman (1983) and of D.E. Liebscher (1983).

Modern cosmology becomes a physical cosmology to ever greater extent. Theoretical studies concentrate on physical processes that occur at different stages of universe evolution. The major problem of cosmology, not to mention the very earliest stages which are the competence of workshop B2, is the origin of the large-scale structure of the universe. To understand this process, one should have a good knowledge of the origin of the small perturbation initial spectrum in various constituents of the matter in the universe (including particles still hypothetical, such as photinos, gravitinos and other "....inos" among them), should study their evolution at various stages taking account of their interdependence, should make calculations of nonlinear evolution stages of perturbations that occur in a comparatively recent epoch.

The most significant observational test in order to check different hypotheses of the origin of the large-scale structure of the universe, is the observation of the spectrum and degree of isotropy of the relic radiation. No deviations have yet been revealed from a perfect equilibrium pattern of this radiation. Practically any version of the perturbation-evolution theory predicts that deviations from equilibrium do exist in the primordial radiation (first it is the small-scale anisotropy of intensity), of the order of accuracy of current observations. It is extremely urgent here both to develop theory and to conduct observations.

Still topical are the questions of astrophysical estimation of basic parameters of the contemporary universe, tht is, the Hubble constant H_0 (known to an accuracy of coefficient 2) and the mean density of the matter ρ_0 (known to an accuracy less than an order of magnitude).

Deviation of new cosmological solutions of gravity equations and their analysis is mainly essential with respect to the problem of singularity. However, new theoretical constructions may be important for solving the problems of the present-day universe's cosmology, such as the problem of the topology of the universe's three-dimensional space, the problem of a possible complex topology in small scales due to the presence of black (and white?) holes, etc. These are the problems of interdisciplinary interest, of cosmology and other branches of relativistic astrophysics and gravity theory.

REFERENCES

Papers presented to the GR10 Conference and published in:

Bertotti, B., de Felice, F., and Pascolini, A., (eds.): 1983, *10th International Conference on General Relativity and Gravitation. Contributed Papers*, Vol. 2, CNR, Roma.

Bonometto, S.A., Lucchin, F., Occhionero, F., and Vittorio, N.: 1983,"MW background small scale fluctuations in neutrino dominated universe models", p. 793.

Canuto, V.M., and Goldman, I.: 1983, "The strong equivalence principle; effects of its possible violation on cosmology, astrophysics and planetary dynamics", p. 802.

Ibañez, J., and Verdaguer, E.: 1983, "Propagation and collision of gravi-solitons on a Kasner background", p. 835.

Liebscher, D.-E.: 1983, "Classical spontaneous breakdown of symmetry in inertia-free mechanics and the origin of Lorentz invariance", p. 559.

Matravers, D.R.: 1983, "Helium formation and the microwave background in some Bianchi type V cosmological models", p. 850.

Occhionero, F., Santangelo, P., and Vittorio, N.: 1983, "Spherical simulations of the cellular structure of the universe", p. 857.

Smoot G., De Amici, G. Friedman, S.D., Witebsky, C., Mandolesi, N., Partridge, R.B., Sironi, G., Danese, L., and De Zotti, G.: 1983, "New multifrequency measurements of the spectrum of the cosmic background radiation", p. 865.

Tolman, B.W., and Matzner, R.A.: 1983 "Cosmological spatial curvature probed by microwave polarization" p. 872.

Wagoner, R.V., and Herskowitz, S.: 1983, "Supernova photospheres and the Hubble constant", p. 876.

ON THE STOCHASTIC PROPERTIES OF RELATIVISTIC COSMOLOGICAL MODELS NEAR THE SINGULARITY

I.M.Khalatnikov, E.M.Lifshitz, K.M.Khanin, L.N.Shchur, Ya.G.Sinai*

Institute for Physical Problems,
L.Landau Institute for Theoretical Physics,
Academy of Sciences USSR, Moscow

The oscillatory mode of approach towards the singularity was first discovered for the homogeneous vacuum cosmological model of Bianchi type IX (Belinskii et al., 1970). The character of the evolution of a model can be described by indicating three "scale functions" a(t), b(t), c(t) which determine the temporal evolution of the lengths in three different directions in space. The oscillatory mode consists of an infinite sequence of successive periods (in Belinskii et al. (1970) they were called eras) during which two of the scale functions oscillate and the third one decreases monotonically. On passing from one era to another (with decreasing time t) the monotonic decrease is transferred to another of the three scale functions. The amplitude of oscillations increases during each era but the increase is especially strong on passing from one era to another; however the product abc decreases monotonically - approximately as t. The eras become condensed with $t \to 0$; an adequate temporal variable for description of their replacements appears to be the "logarithmic time" $\Omega = -\ln t$.

We denote by κ_0, κ_1, $\kappa_2, \ldots$ the "lengths" of successive eras (measured in terms of the number of oscillations they contain), beginning from a certain initial one. It turns out that this sequence of the lengths is determined by a sequence of the numbers $x_{-1}, x_0, x_1, x_2, \ldots$ $(0<x_s<1)$ each of which arises from the preceding one by the transformation

B. Bertotti et al. (eds.), General Relativity and Gravitation, 343–349.

$$x_{s+1}=\{1/x_s\} \tag{1}$$

where the curly brackets denote the fractional part of the number. The lengths are given by

$$\kappa_s=\left[1/x_{s-1}\right]$$

the square brackets denoting the integer part of the number.

It was pointed out by Lifshitz et al. (1970) that the law of replacements of the lengths of the eras according to (1) leads to an important property; spontaneous stochastization of the behaviour of the model on approach to singularity (t=0) and the "loss of memory" of the initial conditions, prescribed at some moment of time $t=t_0>0$.

The importance of the oscillatory evolution in the homogeneous model stems from the fact that this model serves as a prototype for construction of a general inhomogeneous solution of the Einstein equations (in the neighbourhood of the singularity); the relevant work has recently been reviewed in Belinskii et al. (1982). Although the inhomogeneity and the presence of matter give rise to the appearance of certain new features (rotation of the axes to which the scale functions a, b, c refer), law (1) remains unaltered. Thus, stochasticity in the vicinity of the singularity appears to be a most general property of cosmological models based on the classical Einstein equations.

The knowledge of the source of the stochastization makes it possible to construct with a considerable completeness a statistical theory of the evolution of the cosmological model in asymptotic closeness to singularity. However for a calculation of parameters of this theory an approximate method was devised in Lifshitz et al. (1970), the degree of exactness of which is difficult to estimate beforehand. The aim of the present work is to show that these parameters can be calculated exactly.

The starting point of the theory is the formula due to Gauss

$$w(x)=1/(1+x)\ln 2 \tag{2}$$

which determines the probability distribution density of the values of $x_s \equiv x$ in the interval [0,1] after many iterations of the transformation (1) (as we shall speak - in the stationary i.e., independent of s limit[1]).Hence follows the formula

$$W(\kappa)=\frac{1}{\ln 2}\ln\frac{(\kappa+1)^2}{\kappa(\kappa+2)} \tag{3}$$

for the probability distribution of the integer values of the era lengths. This function decreases with $\kappa\to\infty$ merely as κ^{-2}; such slowness makes it necessary to use logarithmic physical quantities in order to obtain for them stable statistical distributions and mean values.

The basis of the following analysis constitute the recurrence formulas (obtained in Lifshitz et al., 1970) for successive eras:

$$\frac{\Omega_{s+1}}{\Omega_s}=1+\delta_s\kappa_s\left(\kappa_s+x_s+\frac{1}{x_s}\right)\equiv\exp\xi_s, \tag{4}$$

$$\delta_{s+1}=1-\frac{\delta_s(\kappa_s/x_s+1)}{1+\delta_s\kappa_s(\kappa_s+x_s+1/x_s)}\ . \tag{5}$$

They are valid in asymptotic limit when $\ln\Omega/\Omega\to 0$ (in Lifshitz et al., 1970 formula (5) was given with a misprint in the denominator). Here Ω_s is the moment of the beginning of the s-th era; quantity δ_s is the measure (in units of Ω_s) of the initial (in the same era) amplitude α_s of the oscillations of the logarithms of the scale functions (lna, lnb, lnc) $\alpha_s=\delta_s\Omega_s$ $(0\leq\delta_s\leq 1)$. Quantity δ_s has a stable stationary statistical distribution $P(\delta)$ and a stable (small relative fluctuations) mean value. For their determination in Lifshitz et al. (1970) was used (with due reservations) an approximate method based on the assumption of statistical independence of the random quantity δ_s of the random quantities κ_s, x_s. Now an exact solution of this problem is given.

Since we are interested in statistical properties in the stationary limit, it is reasonable to introduce the so-called natural expansion of transformation (1) by continuing it indefinitely to negative indices. Such a "doubly-infinite" sequence $X=(\ldots, x_{-1}, x_0, x_1, x_2,\ldots)$ is uniform in its statistical properties over its entire length (and x_0 loses its meaning of an "initial" condition). Sequence X is equivalent to a sequence of integers $K=(\ldots, \kappa_{-1}, \kappa_0, \kappa_1, \kappa_2, \ldots)$, constructed by the rule $\kappa_s=[1/x_{s-1}]$. Inversely, every number of X is determined by the integers of K as an infinite continuous fraction

$$x_s = \cfrac{1}{\kappa_{s+1} + \cfrac{1}{\kappa_{s+2} + \cfrac{1}{\kappa_{s+3} + \dots}}} \equiv x^+_{s+1} \; . \tag{6}$$

We also introduce quantities which are defined by a continuous fraction with a retrograde sequence of the denominators

$$x^-_s = \cfrac{1}{\kappa_{s-1} + \cfrac{1}{\kappa_{s-2} + \cfrac{1}{\kappa_{s-3} + \dots}}} \; . \tag{7}$$

By means of some rearrangements eq. (5) can be brought into the form

$$x_s \frac{1+\delta_{s+1}}{\delta_{s+1}} = \left[\kappa_s + x_{s-1} \frac{1-\delta_s}{\delta_s}\right]^{-1} .$$

Hence by iterations: $x_s(1-\delta_{s+1})/\delta_{s+1} = x^-_{s+1}$ and finally

$$\delta_s = x^+_s / (x^+_s + x^-_s) . \tag{8}$$

The quantities x^+_s and x^-_s have a joint stationary distribution $P(x^+, x^-)$ which can be found starting from the joint transformation

$$x^+_{s+1} = \left\{\frac{1}{x^+_s}\right\} , \quad x^-_{s+1} = \frac{1}{\left[1/x^+_s\right] + x^-_s} \; . \tag{9}$$

In contrast to (1) this is a one-to-one mapping[2] (in the unit square of variation of x^+ and x^-). Therefore the condition for the distribution to be stationary is expressed simply by the equation

$$P(x^+_{s+1}, x^-_{s+1}) = P(x^+_s, x^-_s) J(x^+_s, x^-_s) \tag{10}$$

where J is the Jacobian of transformation (9).

The normalized solution of this equation is

$$P(x^+, x^-)=1/(1+x^+x^-)^2\ln 2 \tag{11}$$

(its integration over x^+ or x^- yields (2)). Since by (8) δ_s is expressed in terms of x_s^+ and x_s^-, the knowledge of (11) makes it possible to find the distribution $P(\delta)$:

$$P(\delta)=1/(|1-2\delta|+1)\ln 2. \tag{12}$$

The mean value is $<\delta>=\frac{1}{2}$, already as a result of the symmetry of this function.

According to Lifshitz et al. (1970) the "doubly--logarithmic" time interval for a succession of a given number s of eras is

$$\tau_s\equiv\ln(\Omega_s/\Omega_0)=\sum_{p=1}^{s}\xi_p \ . \tag{13}$$

The mean value is $<\tau_s>=s<\xi>$. The expression for ξ_s from (4) can be reduced to the form

$$\xi_s=\ln(\delta_s/(1-\delta_{s+1})x_{s-1}x_s).$$

Since $<\ln\delta_s>=<\ln(1-\delta_{s+1})>$ and $<\ln x_{s-1}>=<\ln x_s>$, we obtain

$$<\xi>=-2<\ln x>=\pi^2/6\ln 2=2.37. \tag{14}$$

For large s the values of τ_s are distributed around $<\tau_s>$ according to Gauss' law with the density

$$\rho(\tau_s)=(2\pi D)^{-\frac{1}{2}}\exp\{-(\tau_s-<\tau_s>)^2/2D\} \tag{15}$$

(cf. Lifshitz et al., 1970 §4). The calculation of variance D is more complicated since it demands not only the knowledge of $<\xi^2>$ but also the mean values $<\xi_{p_1}\xi_{p_2}>$ (which actually depend only on the difference $p=|p_1-p_2|$). It appears to be useful to rearrange the terms in the sum (13) and omit the terms which do not increase with s. Thus one can obtain

$$\sum\xi_p=\sum\ln(1/x_p^+x_p^-)\equiv\sum\eta_p \ .$$

The variance

$$D=s\{\langle\eta^2\rangle-\langle\eta\rangle^2+2\sum_{p=1}^{\infty}(\langle\eta_0\eta_p\rangle-\langle\eta\rangle^2)\}.$$

The mean value $\langle\eta\rangle=\langle\xi\rangle$, and for the mean square one can obtain $\langle\eta^2\rangle=9\zeta(3)/2\ln 2=7.80$. Without taking into account correlations we would obtain D=2.17s. By taking into account correlations with p=1,2,3,4 (calculated with the aid of an electronic computer) we arrive at the value D=(3.5±0,1)s.

ACKNOWLEDGEMENT. Reproduced here by kind permission of Zhurnal Eksperimentalnoi i Teoricheskoi Fiziki (JETP Letters).

NOTES

* The lecture was delivered at the Conference by E.M.Lifshitz.

(1) The regular evolution of the model according to rule (1) can be interrupted by the appearance of "anomalous" eras (which were called in Belinskii et al. (1970) the case of small oscillations). However it is important that in the asymptotic vicinity of the singularity (as $t\to 0$) the probability of occurrence of such "dangerous" cases tends to zero as was proved in §4 of Lifshitz et al. (1970).

(2) The reduction of the transformation to the one-to--one mapping was used before by Chernoff and Barrow (1983) - for other variables and without applications to the problems which are considered here. As to the preceding papers by Barrow (1981, 1982a, b), they contain nothing beyond the main idea (taken from Lifshitz et al., 1970) about the connection of stochasticity in cosmological models with the transformation (1) and the distributions (2) and (3) (and the repetition of some well known statements of the general ergodic theory).

REFERENCES

Barrow, J.D.: 1981, "Chaos of the Einstein Equation" *Phys. Rev. Lett.*, 46, 963.

Barrow, J.D.: 1982a, "General Relativistic Chaos and Nonlinear Dynamics", *Gen. Rel. Grav.*, 14, 523.

Barrow, J.D.: 1982b, "Chaotic behavior in General Relativity", *Phys. Rep.*, 85C, 1.

Belinskii, V.A., Khalatnikov, I.M., and Lifshitz, E.M.: 1970, "Oscillatory Approach to a Singular Point in the Relativistic Cosmology", *Adv. Phys.*, 19, 525.

Belinskii, V.A., Khalatnikov, I.M., and Lifshitz, E.M.: 1982, "A general solution of the Einstein Equations with a time singularity", *Adv. Phys.*, 31, 639.

Chernoff, D.F. and Barrow, J.D.: 1983, "Chaos in the Mixmaster Universe", *Phys. Rev. Lett.*, 50, 134.

Lifshitz, E.M., Lifshitz, I.M., and Khalatnikov, I.M.: 1970, "Asymptotic Analysis of oscillatory mode of approach to a singularity in Homogeneous cosmological models", *Zh. Eksp. Teor. Fiz.*, 59, 322. (*Sov. Phys. JETP*, 32, 173 (1971)).

THE EARLY UNIVERSE
REPORT OF WORKSHOP B2

J. B. Hartle

Department of Physics
University of California
Santa Barbara, CA 93106, USA

In the search for explanations of the large scale structure and contents of our universe, questions of origin must inevitably be addressed. What is the origin of the present large scale homogeneity and isotropy of the matter distributions (both of radiation and particles)? What is the origin of near flatness of the large scale spatial geometry? What is the origin of the deviations from exact homogeneity, isotropy and spatial flatness represented by matter condensations such as galaxies and their clusters? What is the origin of the matter content itself - the baryons, the background electromagnetic radiation and the putative backgrounds of neutrinos and gravitons? Following J.A. Wheeler (e.g., 1973) one might even ask for the origin of the laws of physics themselves.

Questions such as these can only be answered by a study of the physics of the very early universe. This is a large subject. There is almost no branch of modern physics which does not bear on some interesting issue associated with this area. One needs, for example, quantum gravity for the very beginning. One needs unified theories of the particle interactions for the origin of the baryons and dynamical effects, such as inflation, associated with symmetry breaking. One needs statistical physics and nuclear physics for the epoch of nucleosynthesis, classical gravity and hydrodynamics for the era of galaxy formation and so on and so on.

The workshop on the early universe at GR10 was intended to span this large spectrum of topics. Circumstances prevented any organization of the workshop

B. Bertotti et al. (eds.), General Relativity and Gravitation, 351–354.

before the time of the meeting so that the effective organization was provided by the contributors able to speak. Five brief reports were presented at two sessions led by the present substitute chairman[1].
It would, of course, be impossible that so limited a discussion could cover the field in all its breadth or even encompass all the major contributions made in recent time[2]. Nonetheless the contributions made do indicate pretty fairly the wide range of physical problems raised by the early universe. It seems appropriate, therefore, to summarize this workshop by briefly summarizing some of these problems and indicating where the individual contributions fit in. For details the reader is referred to the contributors themselves and their abstracts published in the volume devoted to that purpose (Bertotti et al., 1983). There also, will be found the abstracts of the contributed papers which were not presented.

Viewed as a problem in time evolution, the problem of constructing a cosmological model is equivalent to specifying boundary conditions at the big bang. Quantum mechanically this means specifying the initial state of the universe. Without some principle to specify the initial conditions there can be no complete understanding of the structure of the present universe. The phenomenon of inflation, for example, can act to dramatically ease the constraints of the initial conditions required by the present homogeneity, isotropy and age but it will not suffice to remove these altogether. The exploration of principles to specify initial conditions is therefore important. The paper presented by A. Vilenkin (1983) is an example of such a proposal in which the initial universe is "spontaneously created by quantum tunneling from nothing". Related ideas are discussed by S.W. Hawking elsewhere in these proceedings.

The boundary conditions for the universe must specify the initial state for the matter degrees of freedom as well as those characterizing the geometry. One must then ask how this state evolves into the matter backgrounds observed today. Physical processes which act to create or transmute matter in the early universe are therefore of interest. The most striking example is perhaps the creation of the present baryon number asymmetric state from an initial thermal state with the baryon number of the vacuum by particle interactions described by grand unified theories with CP violation [2]. Another example, is the creation of particle pairs in an expanding universe from a matter vacuum [2]. Yet another type of example is provided by the contribution of D. Boccaletti and A. Agostini (1983) to this workshop. They discuss the classical conversion of gravitational to electromagnetic radiation in an anisotropic universe with magnetic field.

The details of the interaction between the matter degrees of freedom can significantly affect the dynamics of the universe when these degrees of freedom are coupled to gravity. An important example is inflation in which a non-zero vacuum energy associated with an unbroken symmetry in the early universe leads to an exponential expansion on the time scale associated with grand unification energies[(2)]. This expansion can go far toward explaining the observed large scale homogeneity and isotropy by smoothing initial irregularities. If it ends in a properly controlled way it can even explain the origin of the fluctuations that produced the galaxies. Some of the possibilities and constraints for an inflation driven by supersymmetric GUT's were discussed by H. Minn (1983) in his contribution to the workshop.

The calculation of the effect of symmetry breaking on the evolution of the universe is not always simply a matter of plugging the results of flat space field theory into the right hand side of Einstein's equations. Matter is as coupled to gravity as the other way around and as illustrated in the talk of B.L. Hu (1983) spacetime curvature can influence both the formalism and results of finite temperature field theory in the early universe.

The end of an inflationary era does not signal the end of the influence of particle interactions for the dynamics of the early universe. Phase transitions or transitions which change the energy density can continue to produce significant dynamical effects. An example is contained in the work of Bonometto, Sakellariadou, Matarrese and Pantano (1983) on the quark-hadron phase presented at this workshop. While the detailed behaviour of this phase transition has not been conclusively established (e.g. whether it is first or second order), quarks and hadrons do not behave as an ideal gas in the temperature interval from about 500 MeV to the pion mass in which the transition takes place. The authors show how the thermal history of the universe in this regime will deviate from the behaviour which would be associated with pure radiation.

Taken as a whole perhaps nothing was as impressive about the workshop as the variety of physical problems in the papers of the contributors. It serves to emphasize the variety of challenges which a search for understanding of the origin of the present universe presents.

NOTES

(1) Dr. S.A. Bonometto from Padova and Dr. D. Boccaletti from Roma acted as scientific secretaries of the workshop.

(2) The literature on the early universe is too vast to be cited here. For current reviews of the issues, ideas and references to the literature the reader might consult Gibbons, Hawking and Siklos (1983).

REFERENCES

Papers presented to the GR10 conference and published in:

Bertotti, B., de Felice, F., and Pascolini, A. (eds.): 1983, *10th International Conference on General Relativity and Gravitation. Contributed Papers*, Vol. 2., CNR, Roma.

Boccaletti, D., and Agostini, A.: 1983, "Conversion of gravitational waves into electromagnetic waves in a Bianchi type I universe with a uniform magnetic field", p. 757.

Bonometto S.A., and Matarrese, S.: 1983, "Cosmological relevance of the quark-hadron phase transition", p. 761.

Hu, B.L.: 1983, "Spacetime dynamics and finite temperature effect on symmetry restoration", p. 1086.

Minn, H.: 1983, "Fluctuations in a supersymmetric inflation during the hadron era", p. 768.

Vilenkin, A.: 1983, "Creation of universes from nothing", p. 774.

Other references

Gibbons, G.W., Hawking, S.W., and Siklos, S. (eds): 1983, *The Very Early Universe*, Cambridge University Press, Cambridge.

Wheeler, J.A.: 1973, "From relativity to mutability", in J. Mehra (ed.), *The Physicists Conception of Nature*, Reidel, Dordrecht.

RECENT DEVELOPMENTS IN THE PHYSICS AND ASTROPHYSICS OF BLACK HOLES
REPORT OF WORKSHOP B1

Robert M. Wald

Enrico Fermi Institute
University of Chicago, Chicago, IL 60637, USA

The rate of development of the theory of black holes has slowed considerably since the late 1960's and early 1970's when a number of major breakthroughs occurred. Nevertheless some important progress has been made during the past few years in our understanding of both the theoretical properties of black holes and their role in astrophysics. Much of this progress was reported at the workshop on "Physics and astrophysics of black holes" and abstracts of these contributions appear among "Contributions to Workshop B1" of the first part of the conference proceedings (Bertotti et al., 1983). In this report, I shall mention a number of other significant results which have been obtained recently and I also will discuss some important unresolved issues. My report will be biased in the direction of "theoretical" topics as opposed to "astrophysical" topics since I have followed developments of the former topics much more closely than those of the latter.

One important result obtained recently concerns the black hole uniqueness theorems. Among the most significant advances made in the late 1960's and early 1970's was the proof that the Kerr family of metrics is the only family of solutions of Einstein's equation in vacuum which describes stationary black holes. It has long been conjectured that the charged Kerr solutions (and their generalizations permitting magnetic monopoles) are the unique stationary black hole solutions in the electrovac case, i.e. when the possible presence of electromagnetic fields is taken

B. Bertotti et al. (eds.), General Relativity and Gravitation, 355–361.

into account. Israel's theorem (which states that a static vacuum black hole must be Schwarzschild) was generalized to the electrovac case (Israel, 1968) and Hawking's theorem (which states that a stationary black hole must either be static or axisymmetric) also holds in the electrovac case (Carter, 1973). However, a generalization to the electrovac case of the Carter--Robinson theorem (which states that a stationary, axisymmetric vacuum black hole must be Kerr) has eluded researchers for many years. Recently, however, exactly such a generalization was obtained independently by Mazur (1982) and Bunting (unpublished). Thus, the charged Kerr metrics are indeed the unique stationary electrovac black hole solutions. The proofs of Mazur and Bunting also provide insight into how Robinson's identity (used in the Kerr uniqueness proof in the vacuum case) may be derived.

While the Mazur-Bunting theorem puts to rest a problem that had received much attention ten years ago, some very recent work of Detweiler and Ove (1983) has resurrected a problem that was generally considered to have been settled ten years ago: the stability of the Kerr metric. The stability of static black holes can be proven analytically but no analytic proof of stability of the Kerr black hole has ever been given. However, Press and Teukolsky (1973) and others investigated the stability of the Kerr metric numerically and concluded that it is stable, i.e., that there are no scalar, electromagnetic, or gravitational perturbations which grow exponentially with time. However, in their analysis they examined only a few modes of low angular momentum. Detweiler and Ove used analytic techniques to examine scalar perturbations of nearly extreme (i.e. $a^2+e^2 \cong m^2$) charged Kerr black holes. They considered modes of high angular momentum which are very near the superradiant limit, and concluded that - at least for certain values of the black hole parameters - some of these modes are unstable. It will be very important to determine whether this instability occurs for all non-static black holes and also applies to electromagnetic and gravitational perturbations. It also will be of great importance to determine if the instability can occur on an astrophysically relevant timescale. (The instabilities found by Detweiler and Ove have much too long a growth time to be astrophysically relevant but only a small class of possibly unstable modes are susceptible to their analytic analysis). Thus, it seems clear that we need to learn a great deal more about the stability of the Kerr metric.

Some progress was made in the past few years toward putting the generalized second law of thermodynamics on a firmer foundation. As first conjectured by Bekenstein, this law states that the total generalized entropy - equal to entropy of ordinary matter outside a black hole plus $\frac{1}{4}$ of the area, A, of the black hole - never decreases. However, as pointed out long ago by Bardeen, Carter, and Hawking (1973), this conjectured law apparently can be violated by putting matter containing entropy into a box and lowering it quasi-statically to the horizon of the black hole before dropping it in: in this way, the entropy of matter outside the black hole is decreased, but the area increase of the black hole can be kept arbitrarily small by lowering the box sufficiently close to the horizon. It would seem that quantum effects like those of Hawking radiation would be of no help here since this *Gedankenexperiment* works as well for macroscopically large black holes as for small black holes. Bekenstein (1974, 1981) proposed a resolution of this difficulty by conjecturing the existence of a fundamental limit on the entropy, S, of a box of size R containing energy E, namely $S/E \leqq 2\pi R$. Such a limit, it was argued, would prevent one from lowering a box sufficiently close to the horizon to violate the generalized second law. Thus, according to this hypothesis, the validity of the generalized second law would rest upon the detailed properties of ordinary matter which give rise to the Bekenstein limit. Recently, however, Unruh and I found an alternate resolution, involving a quantum effect which is large near the horizon of a macroscopic black hole. There is an effective buoyancy force felt by a box that is quasi-statically lowered toward a black hole and this force affects the energy transfer in such a way as to produce an increase in the area of the black hole which always is just sufficient to ensure that the generalized second law holds (Unruh and Wald, 1982). Thus, the validity of the generalized second law in this *Gedankenexperiment* has been demonstrated without the need to invoke any additional assumptions.

In my opinion, the major outstanding issues in the theory of black holes remain (i) a formulation and proof (or disproof) of the cosmic censor hypothesis and (ii) a deeper understanding of the nature of black hole thermodynamics. The basic idea of the cosmic censor hypothesis is that all singularities of gravitational collapse are hidden within black holes. However, it has long been known that "shell crossing" and other somewhat more serious types of "naked singularities" (such as those discussed by Christodoulou in this volume) can occur in the collapse of dust and other fluids. Since these naked singularities are known to occur only under very special circumstances (e.g., special initial conditions or equations of state), most researchers do not consider these known violations to be true counterexamples. However, it remains a task for the future to prove that no "true counterexamples" exist. If "true counterexamples" are found, the physical relevance of black holes would become very much in doubt.

With regard to black hole thermodynamics, the mathematical analogy between laws of black hole physics and laws of thermodynamics has long been known, as has the fact that black holes Hawking radiate exactly like black bodies. Furthermore, as indicated above, there is theoretical evidence for the validity of the generalized second law in that *Gedankenexperimente* have failed to produce a counterexample. Most researchers feel that the explanation of this situation is that the laws of black hole thermodynamics simply *are* the laws of ordinary thermodynamic applied to black holes; that, in particular $\frac{1}{4}A$ represents the physical entropy of a black hole, so that the generalized second law is really the ordinary second law applied to a system containing a black hole. However, it remains a mystery why the entropy of a black hole should be related in such a direct and simple manner to its surface area. Nor it is at all understood how the laws of thermodynamics would arise in this context from the underlying microscopic physics. Unfortunately, it is quite possible that we will need to have a satisfactory quantum theory of gravitation before further progress can be made on these issues.

With regard to the astrophysics of black holes, an important observational discovery was made recently. Although many binary X-ray sources consisting of an ordinary star in close orbit with a compact object are known, in many cases the X-ray emission has periodic structure, indicating that the compact object is a neutron star. In essentially all other cases, the possibility that the compact object is a neutron star cannot be ruled out. For the past ten years, only for the X-ray source Cyg X-1 have clear cut arguments been made that the mass of the compact object is well above the neutron star upper mass limit and hence that it must be a black hole. However, recently an optical identification has been made of the source LMC X-3 in the Large Magellanic Cloud (Cowley et al.,

1983) and an analysis of this system (Paczynski, 1983) reveals that in this case also the compact object has a mass well above the upper limit for neutron stars. Thus, Cyg X-1 no longer is the only example for which a clear cut case for the existence of a black hole can be made.

Accretion of matter onto black holes of stellar mass is believed to be responsible for the production of X-rays from sources such as Cyg X-1 and LMC X-3. Accretion of matter onto large black holes (of mass ranging up to $\sim 10^{10} M_\odot$) at the centers of active galactic nuclei and quasars is believed to be responsible for the enormous output of energy observed from these objects. In both cases, research has continued during the past few years on issues dealing with the structure and stability of accretion disks around black holes produced by the infalling matter and the mechanisms by which energy is given off. In the case of large black holes, research also is being actively pursued on how accretion onto such black holes can result in the production of the jets which are observed in many active galaxies and quasars. Since I have not followed developments in this area closely, I shall not attempt here to summarize recent progress in research in these areas. However, I will mention one noteworthy development. In essentially all models of the accretion process, the black hole plays the basically passive role of providing a sufficiently deep "potential well" for the matter to flow into. In these models, the physical processes responsible for the production of X-rays or jets result from the plasma and hydrodynamical properties of the matter, not the general relativistic properties of the black hole. However, recently, Rees et al. (1982) have proposed a model for active galactic nuclei and quasars applicable in the case of low accretion rates onto the black hole. In this model, the accreting ionized gas forms a torus structure which is supported by gas pressure rather than radiation. The magnetic field lines carried by the gas into the black hole couple the black hole to the outside matter. Rees et al. propose that in this manner, the black hole can be made to spin down, and its rotational energy may thereby be extracted. Thus, in this model, the black hole plays an active role in powering the quasars, and general relativity plays an essential role in the physics of the process.

Further progress in our understanding of X-ray sources and quasars will depend in large measure on our ability to analyze the detailed hydrodynamics of the accreting matter. In order to do so, one usually must resort to numerical computations. Tremendous advances in the technology of computers have been made in recent years, so that many numerical calculations which could not have been done in the past are now feasible or soon may become so. Thus, I believe that prospects are bright for further progress in the near future on some aspects of black hole astrophysics.

NOTE ADDED IN PROOF

The claimed proof of the instability of some black holes by Detweiler and Ove (1983) has now been retracted.

REFERENCES

Bardeen, J.M., Carter, B., and Hawking, S.W.: 1973, "The Four Laws of Black Hole Mechanics", *Commun. Math. Phys.*, 31, 161.

Bekenstein, J.D.: 1974, "Generalized Second Law of Thermodynamics in Black-Hole Physics", *Phys. Rev.*, D9, 3292.

Bekenstein, J.D.: 1981, "Universal upper Bound on the Entropy-to-Energy Ratio for Bounded Systems", *Phys. Rev.*, D23, 287.

Bertotti, B., de Felice, F., and Pascolini, A. (eds): 1983, *10th International Conference on General Relativity and Gravitation. Contributed Papers*, Vols. 1-2, CNR, Roma.

Carter, B.: 1973, in C. DeWitt and B.S. DeWitt (eds), "Black Hole Equilibrium States", Gordon and Breach, New York.

Cowley, A.P., Crampton, D., Hutchings, J.B., Remillard, R., and Penfold, J.: 1983, "Discovery of a Massive Unseen Star in LMC X-3", *Astrophys. J.*, 272, 118.

Detweiler, S., and Ove, R.: 1983, "Instability of Some Black Holes", *Phys. Rev. Lett.*, 51, 67.

Israel, W.: 1968, "Event Horizons in Static Electrovac Space-Times", *Commun. Math. Phys.*, 8, 245.

Mazur, P.O.: 1982, "Proof of Uniqueness of the Kerr-Newman Black Hole Solution", *J.Phys.*, A15, 3173.

Paczynski, B.: 1983,"Mass of LMC X-3",*Astrophys.J.Lett.*, 273, L81.

Press, W.H., and Teukolsky, S.A.: 1973,"Perturbations of a Rotating Black Hole.II. Dynamicas Stability of the Kerr Metric", *Astrophys. J.*, 185, 649.

Rees, M.J., Begelman, M.C., Blanford, R.D., and Phinney, E.S.: 1982,"Ion-Supported Tori and the Origin of Radio Jets", *Nature*, 295, 17.

Unruh, W.G., and Wald, R.M.: 1982,"Acceleration Radiation and the Generalized Second Law of Thermodynamics", *Phys. Rev.*, D25, 942.

Experimental Gravitation

TESTING RELATIVITY WITH SOLAR SYSTEM DYNAMICS

Ronald W. Hellings

Jet Propulsion Laboratory
California Institute of Technology
Pasadena, California 91109, USA

1. INTRODUCTION

On November 20, 1982, the Mutch Memorial Station (Viking Lander 1) on the surface of Mars interrupted its transmission to Earth due to a loss of the information in its antenna pointing table. It was not heard from again. However, in over six years of operation since its landing in July, 1976, it and Viking Lander 2 have provided a set of ranging data which are sufficiently accurate to enable the Earth-Mars distance to be modeled to ±10m during that period. This unprecedented accuracy in a Solar System data set has been responsible for a major breakthrough in the accuracy of Solar System dynamical tests of relativistic gravity. It is the purpose of this talk to explain the data which are used in the solutions for the relativistic effects (Section 2), to explain the Solar System model which is used to fit the data and the process by which such fits are performed (Section 3), and to discuss the results (Section 4, see also Hellings et al., 1983) of our tests for the effects of the following:

1. The Large Numbers Hypothesis and a time-varying gravitational constant,
2. PPN parameters β,γ, and α_1 along with the value of the solar quadrupole moment J_2,
3. Moffat's Nonsymmetric Gravitational Theory (which does not fall into the formalism of the PPN).

B. Bertotti et al. (eds.), General Relativity and Gravitation, 365–385.

2. SOLAR SYSTEM ASTROMETRIC DATA

Certainly the greatest volume of data comes from the optical observations. Over 44.000 arc-second-accuracy observations of the positions of the sun, moon, and planets relative to stars of the FK4 catalog have been taken using the 6 and 9 inch transit circles at the U.S. Naval observatory in Washington, D.C. The instrument consists of an optical telescope (see Figure 1) which is free to move only in the plane of the local meridian. The elevation measurement is read off the large circle and the azimuth measurement comes from the determination of the time of local meridian transit. Since 1911, these measurements have been made using the "impersonal micrometer", in which the observer keeps the center-of-light or the limb of the planet centered in the crosshairs while a timer electronically defines the moment of meridian passage. Several corrections for known systematic effects must be made to the data, the values of the corrections being determined in the process of the least square fits. First, there is a known systematic trend in the measured position of a planet as a function of the planet's phase. Second, there is a correction made for a drift of the FK4 catalog relative to local inertial space.

The data set possessing the greatest absolute accuracy is the lunar laser ranging data (see Figure 2), with sensitivity of about ±10cm. In a typical measurement, 10^{18} photons are sent in a short pulse from a ruby laser at McDonald Observatory at the University of Texas and bounced off the corner reflector left on the surface of the moon by the Apollo astronauts. A typical return rate is one photon detected in the photomultiplier tube per burst. Any photon received in the tube will of course be counted, and so the light must be filtered in order to assure that it is the proper photon which is being seen. The light entering the telescope is filtered in three ways - spatially, in frequency, and in time. A window is opened around the expected time of arrival of the pulse, and the number of photons counted in each time bin is compared with the number in adjacent bins. A statistically significant increase in the number of photons counted in a bin gives the estimated return time of the pulse. Corrections must be made for a number of known systematic effects, named in the table in the lower left-hand section of Figure 2. Those without stars following them are known a priori while the starred effects require corrections which must be determined in the fits.

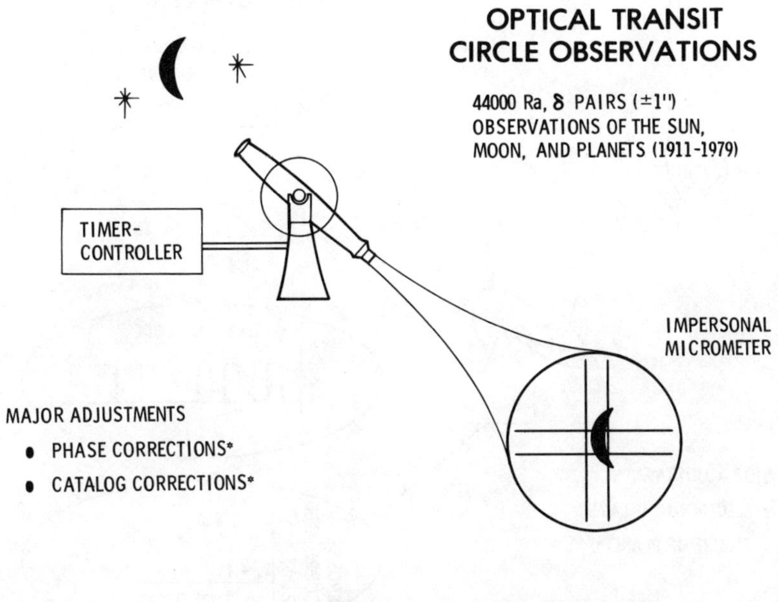
OPTICAL TRANSIT
CIRCLE OBSERVATIONS
44000 Ra, δ PAIRS (±1")
OBSERVATIONS OF THE SUN,
MOON, AND PLANETS (1911-1979)
TIMER-
CONTROLLER
IMPERSONAL
MICROMETER
MAJOR ADJUSTMENTS
• PHASE CORRECTIONS*
• CATALOG CORRECTIONS*

Fig. 1.

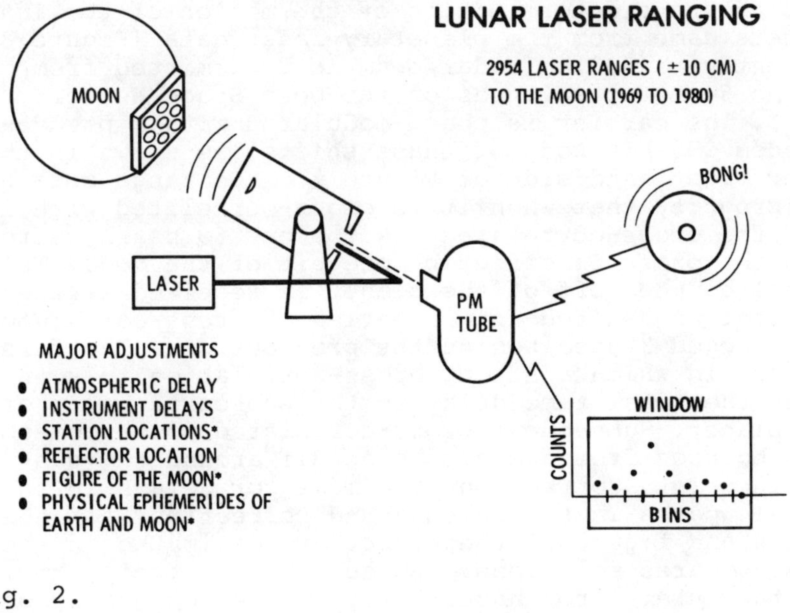
LUNAR LASER RANGING
2954 LASER RANGES (±10 CM)
TO THE MOON (1969 TO 1980)
MOON
BONG!
LASER
PM
TUBE
MAJOR ADJUSTMENTS
• ATMOSPHERIC DELAY
• INSTRUMENT DELAYS
• STATION LOCATIONS*
• REFLECTOR LOCATION
• FIGURE OF THE MOON*
• PHYSICAL EPHEMERIDES OF
EARTH AND MOON*
WINDOW
COUNTS
BINS

Fig. 2.

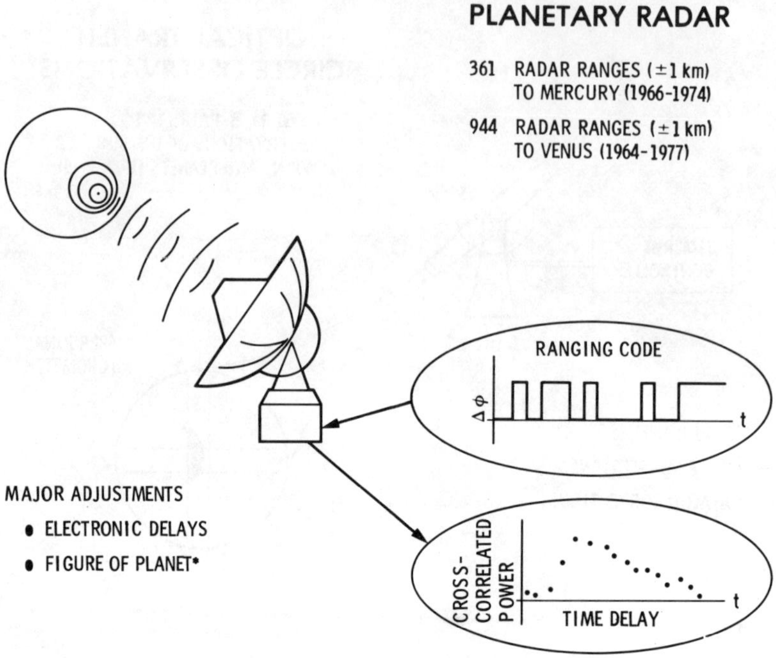

Fig. 3.

Prior to the acquisition of spacecraft data, the most accurate determination of the motion of the inner planets came from the planetary radar data (Figure 3). Currently, a 300 km radar beam is transmitted from one of the 64 meter antennas of the Deep Space Network (DSN). The carrier is phase-modulated with a pseudo--random 255-bit code of phase shifts, as shown in the lower right-hand side of Figure 3. This range code has the property that when it is cross-correlated with itself, the cross-correlated power drops to nearly zero when the phase is offset by one bit of the code. The signal bounced off of the planet is received at the DSN station and fed into a series of cross-correlators, each lagged 3 μsec behind the previous one. The first channel in which a strong cross-correlation is seen gives the total time delay to the sub-earth point of the planet. Subsequent cross-correlated power comes from bounces from successive annuli around the sub--earth point. Corrections are made for delays in the signal paths in the antenna and corrections for the radius and figure of the planet are estimated in the least squares solutions. The current limit of ±1km in accuracy stems from uncertainty in the topography of the planet.

But the star of the show is the spacecraft data. When a spacecraft is in orbit around a planet, or, better, sitting on the surface of a planet, then ranging to the spacecraft gives a very accurate determination of the planet's position. We currently have data from flybys of Jupiter by Voyager and Mercury by Mariner 10, data from the Mariner 9 and Viking orbiters around Mars, and data from the Viking landers on the surface of Mars. The most accurate data come from the landers. The S-band tracking signal from the DSN station is phase-modulated in a regular 1 MHz pattern, as shown in the lower right-hand side of Figure 4. The signal is coherently transponded in a phase-lock loop on board the spacecraft and retransmitted at a translated frequency. The received signal is cross-correlated with current uplink signal and, in a separate channel, with the current uplink,offset in range-code phase by 90 degrees. The theoretical cross-correlation functions are shown as solid lines in the graphs in the lower right-hand side of Figure 4. By comparing the relative

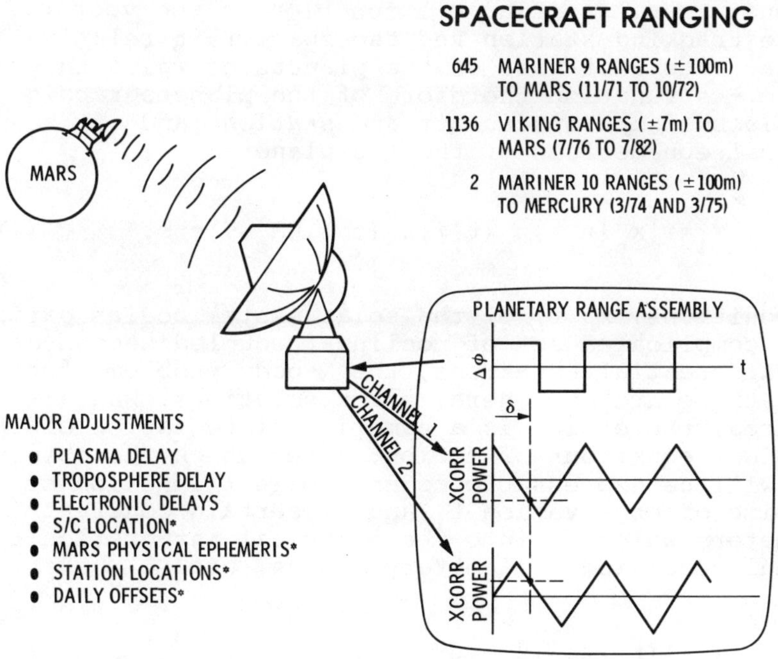

Fig. 4.

cross-correlated power in the two channels, the time delay can be unambigously determined, as shown. A ranging measurement actually consists of a series of these delay measurements, each one using a wider range code, up to a maximum period of 0.25 secs. Corrections are made for electronic delays and for delay caused by the interplanetary plasma content. The residual uncertainty of about ±10m is thought to be due to unmodelable fluctuations in the interplanetary plasma. The locations of the DSN stations and the Viking landers and the parameters of the physical ephemeris of Mars are also determined in the least squares solutions.

3. THE SOLAR SYSTEM MODEL

The ith observation of some measured quantity, such as the time-of-flight τ of an electromagnetic tracking signal from a tracking station on Earth to a spacecraft on the surface of Mars, is a function of the position of Mars $x_m(t_i)$, and of Earth, $x_e(t_i')$ at the appropriate times, and of the parameters such as the position and mass of the sun which determine the path through curved spacetime which the tracking signal must take. It is also a function of the vectors of the tracking station and the spacecraft relative to the centers-of-mass of the planets on which they reside - a function therefore of the planetographic locations of the spacecraft and stations and of the physical ephemerides of the two planets:

$$\tau_i=\tau\left(x_e(t_i),x_m(t_i'),x_s(t_i),GM_s\ldots\text{etc.}\right) \qquad (1)$$

The positions, $x_n(t)$ of the solar system bodies satisfy a complicated set of nonlinear coupled second-order differential equations, the N-body problem, for which there are no general exact solutions. What is required, therefore, is a numerical integration of the dynamical equations of motion, after which the observable will be expressable as an explicit function of the time of observation t_i and of certain constant parameters which go into the numerical integration - initial conditions, planetary masses, etc.

$$\tau_i=\tau(t_i,x_{n0},\dot{x}_{n0},GM_n,J_2,\ldots\text{etc.}) . \qquad (2)$$

The parameters currently used in our solar system

model include:

1. $(x_0, y_0, z_0, \dot{x}_0, \dot{y}_0, \dot{z}_0)$ - the initial heliocentric positions and velocities of nine planets and the moon.
2. GM_n - the masses of the planets and of Ceres, Pallas, and Vesta.
3. MADC, MADS - the mean densities of 200 more of the largest asteroids whose radii are fairly well known, divided into two classes (C or S) based on their compositional type.
4. AU - The length of the astronomical unit in meters (equivalent to determining the GM of the sun).
5. $(J_2, \beta, \gamma,$ etc.) - parameters that specify the relativistic gravitational field of the sun.
6. Other parameters unique to individual data sets, such as the physical ephemerides (rigid body motion) of the planets, the locations of tracking stations, corrections for known systematic errors in the data, etc.

The values of these parameters along with the computer code used to numerically integrate the equations of motion constitute the solar system model. Using this model, one may compute the predicted value of an observable like τ, using Equation (2), and compare the computed value with the actual value observed to form what is called the "residual", y_i:

$$y_i = \tau_{i,\text{observed}} - \tau_{i,\text{computed}} \ . \qquad (3)$$

An interesting question then arises. Why are the values of y_i *not* all zero? There are two answers to this question. First, the observed values may be wrong - i.e., there may be some random, uncorrectable error ν_i in each measurement τ_i. Second, the computed values may be wrong due to incorrect values of the solar system parameters. The purpose of the data analysis is to eliminate, as far as possible, the second kind of error - to adjust the parameters of the solar system model so that $y_i = \nu_i$ only.

The method of finding the necessary adjustments in the parameters so that they will better fit the data is the method of linear least squares estimation. In this process, it is assumed that the required changes in the parameters $(x_1, x_2, x_3, \ldots)$ will be small, so that one may write

$$y_i = \frac{\partial \tau_i}{\partial x_1} \Delta x_1 + \frac{\partial \tau_i}{\partial x_2} \Delta x_2 + \frac{\partial \tau_i}{\partial x_3} \Delta x_3 + \ldots + \nu_i \qquad (4)$$

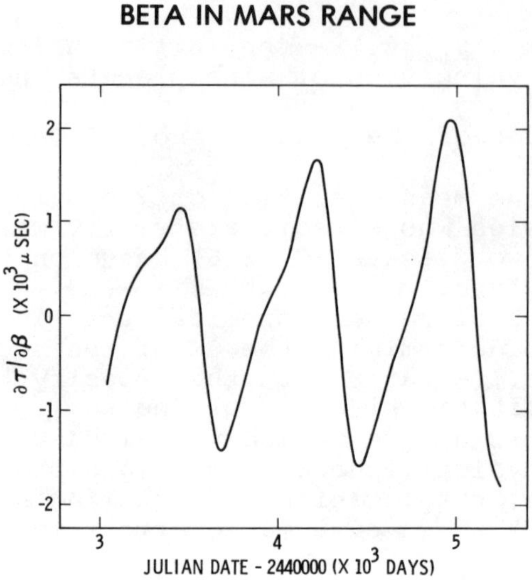

Fig. 5. Partial derivative of range to Mars wrt PPN parameter beta.

The partial derivatives in Equation (4) give the effect produced in each range measurement τ_i due to a change in the value of each parameter x_j. Each partial derivative gives the signature of its parameter in the data - a signature which can be searched for in order to determine the size of the correction Δx_j which must be made in the parameter x_j in order to soak up observed trends in the data. For example, the partial derivative of τ with respect to PPN parameter β during the time span of the Viking data is shown in Figure 5. If the data exhibit a systematic trend with just this shape, then the size of the trend observed gives the size of the correction which should be made to the nominal value of β in order to absorb it and leave a smaller set of residuals. The partial derivative of τ with respect to J_2, the solar quadrupole moment, is shown in Figure 6.

It is instructive to overlay the two graphs, as has been done in Figure 7. It should be noted that the two signatures are very similar, but that they are not the same. The similarity means that it will be very difficult to tell the difference between fitting the data by making a small increase in the value of β or by a small decrease in the value of J_2. As a result, the uncertainty in either parameter in a combined

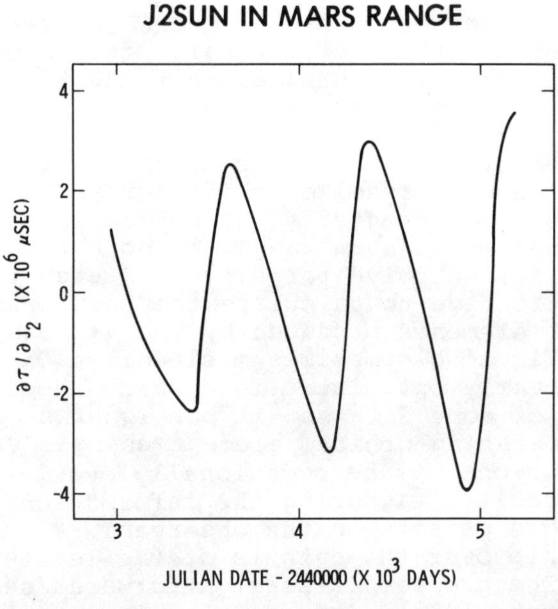

Fig. 6. Partial derivative of range to Mars wrt Solar quadrupole moment, J_2.

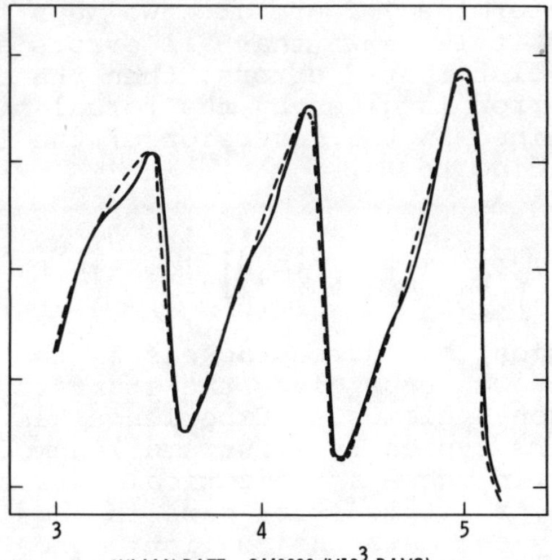

Fig. 7. Partial derivatives of range to Mars wrt β (solid) and J_2 (dashed).

solution will be larger than what one would get if one were to solve for one of the parameters alone. On the other hand, the signatures are not identical, so at a reduced level of sensitivity it should be possible to tell the difference between β and J_2.

It may be noted that in none of this has there been any mention of perihelion shifts or any other changes in orbital elements. We do not measure perihelion position; we measure range. We do not solve for perihelion shift; we solve for β, γ, J_2, etc. It is sometimes instructive to calculate the secular changes in the orbital elements produced by the various parameters, but it is also sometimes misleading. One may be led to be overly optimistic in occasion about the possibilities of some data set if one ignores difficulties in separating orbital element changes via range alone, or one may be occasionally overly pessimistic as a result of ignoring the information available in periodic effects in the observable.

I will skip over the details of the least squares mathematics. The problem is straightforward (see, for example, Bierman, 1977). One needs a single solution - a single inversion of an "information matrix" - in order to determine the best values for the solar system parameters, including the values of the gravitational field parameters for which we are searching.

Once the values of Δx_j have been determined, another critical question arises. How accurately are the parameters determined? There are two ways to answer this question. If it is known that all errors are random, Gaussian-distributed errors, then the answer is simple - the error is given by the formal standard deviation, determined in the inversion of the information matrix according to

$$\sigma_j^2 = \left(\sum_i \nu_i^2\right)\left[\sum_k \left(\partial \tau_k/\partial x_m\right)\left(\partial \tau_k/\partial x_m\right)\right]^{-1}_{jj} \quad (5)$$

where the expression in square brackets is the information matrix whose inverse gives the standard deviations as its diagonal elements. Experience has shown, however, that there typically exist small unmodelled systematic errors in these astrometrical data sets and, in this case, a little extra care is required in order to correctly estimate the realistic uncertainty which should be assigned to the values of the parameters. All uncertainties quoted for the results given in Section 4 are realistic uncertainties. These are typically several times the size of the formal standard deviations given by Equation (5).

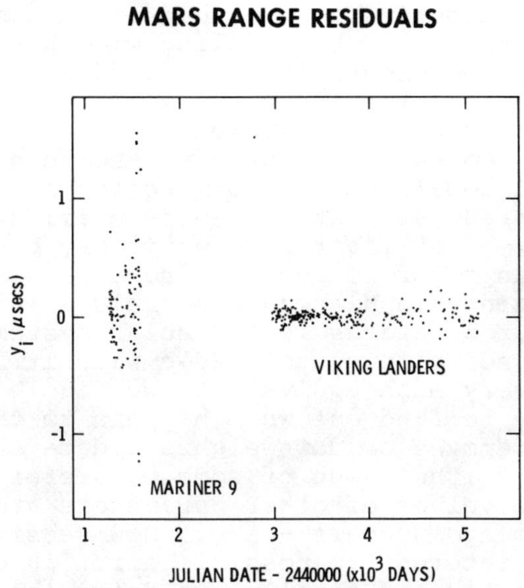

Fig. 8. Mars range residuals.

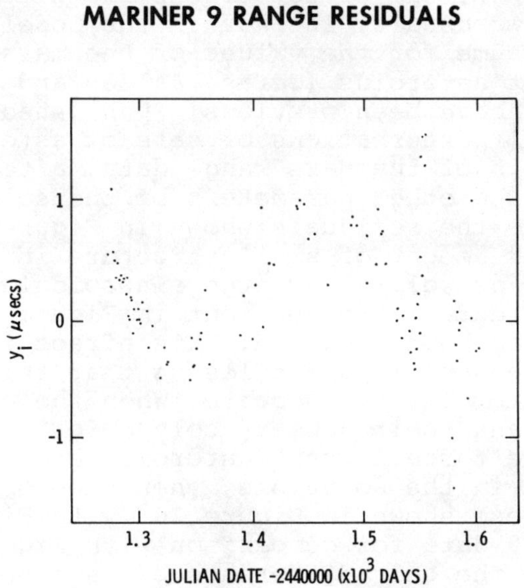

Fig. 9. Mariner 9 range residuals.

A laboratory experimenter responds to the possibility of systematic error by fiddling with his equipment. He will try the experiment at a different time of day or at a different voltage. He will purposely degrade some element which he suspects to be a source of error in order to see how sensitive the measurement is to it. He will do all he can to make the systematic error become apparent so that it can be eliminated, or corrected for, or at least so that it may be taken into account as an error source and increase his final published uncertainty in his measured value. In the dynamics tests our apparatus is the solar system itself, and we are not allowed to fiddle with it. Instead, we do something very much equivalent. We fiddle with the model. We try to find out what happens to the solution when one removes or downweights a data set. We ask what happens to the value of some parameter of interest when the values of other parameters are fixed to some predetermined numbers - i.e., how sensitive is the parameter of interest to possible error in other parametrized effects? And finally, we determine how well the data are fit when such values are forced into the solution. Does the chosen particular value produce visible trends in the data which are not reasonably due to errors in the measurements?

Let me give an example of this last type of test. If we were to ignore the effect of the 200 little asteroids which we usually include in the model, and if we were to assume for the values of the masses of the three largest asteroids (Ceres, Pallas and Vesta) the values which have been previously published, derived from mutual perturbations of sets of asteroids, then the residuals of the Mars range data, after the best fit to all the other parameters of the solar system model, are the residuals shown in Figure 8. One may notice a fair amount of small structure in the Viking residuals as well as a large sinusoidal trend in the Mariner 9 data, more apparent in Figure 9 where the time scale has been expanded. This effect is far larger than is reasonable and tells us that this model of the solar system is not adequate. When the asteroids are put back in and their density solved for and when the masses of the three largest asteroids are allowed to be determined in the solutions, then the post-fit residuals are those shown in Figure 10 (with Figure 11 for the Mariner 9 data for comparison with Figure 9). I reiterate that the only difference between these two sets of plots is that in the second case we have allowed there to be small changes in the masses of three asteroids and we have included the combined effect of

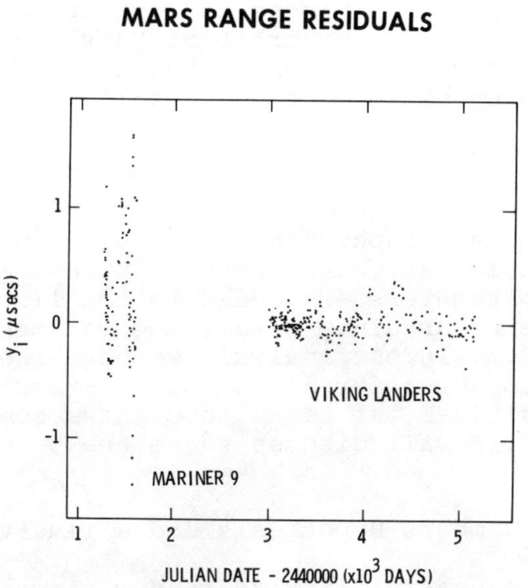

Fig. 10. Mars range residuals with asteroids in the solution.

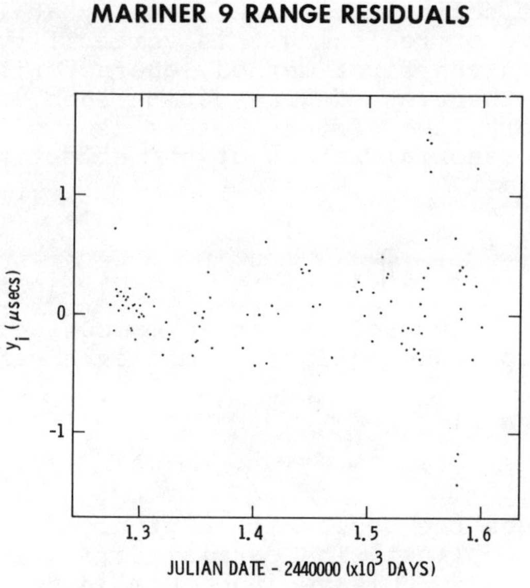

Fig. 11. Mariner 9 range residuals with asteroids in the solution.

200 more of these little pieces of rock at an average distance of 1.5 AU from the orbit of Mars. The message is that you can't do very much to the solar system without messing up the Mars range data.

4. RESULTS

Figure 7 makes graphic the difficulty involved in simultaneously estimating several parameters. As more and more parameters are added to the list to be determined, the uncertainty in all the parameters invariably degrades. In our analysis we have therefore not attempted to solve for all possible effects at once but have divided our tests into three somewhat separable areas. I will discuss these one by one.

4.1. The Large Numbers Hypothesis and a Time-Varying G.

It has been nearly fifty years now since Dirac (1937) first proposed that the coincidence of certain dimensionless large numbers relating fundamental cosmic and atomic quantities is not such a coincidence after all, but that the constants of local physics, notably the Newtonian gravitational constant, may be changing on a cosmic time scale. There are actually at least two separate ways of realizing this variability (Adams et al., 1983). In the first method, characteristic of auxiliary-field theories of gravitation such as the Brans-Dicke theory, the effect appears as a simple time-dependent renormalization of the effective gravitational constant:

$$G_{eff}=G\left[1+\frac{\dot{G}}{G}(t-t_0)\right] . \qquad (6)$$

This time variability produces an anomalous acceleration in the Newtonian equations of motion given by

$$\delta a^i=-\frac{\dot{G}}{G}\left[\frac{GM}{r^2}\,\frac{x^i}{r}(t-t_0)\right] . \qquad (7)$$

Equation (7) gives the acceleration produced by $\dot{G}/G$ in the orbit of a planet. The partial derivative of δa^i with respect to $\dot{G}/G$ leaves the term in square brackets, which can be numerically integrated to give the effect of the parameter in the position of each planet. The two position effects are then combined to deduce the effect in the range.

A second way to realize the effects suggested in the Large Numbers Hypothesis is to assume that cosmic effects couple to local atomic physics, causing atomic clocks to drift relative to the implicit clock of gravitational dynamics. In this approach, one writes the relation between the two proper times, atomic time τ_A and dynamical Einstein time τ_E, as

$$\frac{d\tau_E}{d\tau_A} = 1+\dot{\phi}(t-t_0). \tag{8}$$

When the gravitational geodesic equation is transformed to reflect this new atomic proper time as the affine parameter, there arises an anomalous term on the right-hand side of the equation:

$$\frac{d^2x^\mu}{d\tau_A{}^2}+\Gamma^\mu_{\alpha\beta}\frac{dx^\alpha}{d\tau_A}\frac{dx^\beta}{d\tau_A}=\dot{\phi}\left(g^{\mu 0}-\frac{dx^\mu}{d\tau_A}\frac{dx^0}{d\tau_A}\right). \tag{9}$$

This new term produces an anomalous acceleration given by

$$\delta a^i=\dot{\phi}\left[\frac{GM}{r^2}\frac{x^i}{r}(t-t_0)-\frac{dx^i}{dt}\right]. \tag{10}$$

Equation (10) is numerically integrated as was equation (7), and both $\dot{G}/G$ and $\dot{\phi}$ were looked for in the solutions.

Before presenting the results of our analysis for these two effects, I would remind the reader that the basis of the Large Numbers Hypothesis is that the changes in the physical constant are tied to cosmic evolution, so that the expected rate of change in G should be the inverse Hubble time:

$$\dot{G}/G \sim H_0 \cong 5\times 10^{-11}\,\mathrm{yr}^{-1}.$$

Previous determinations of $\dot{G}/G$ have been made. One method (Van Flandern, 1981) has been to exploit a difference in the rate of decay of the lunar orbit measured by lunar laser ranging data (atomic time) and measured by ancient eclipse data (gravitational dynamic time). The result in this case was

$$\dot{G}/G=(-6.9\pm 2.4)\times 10^{-11}\,\mathrm{yr}^{-1}.$$

A second method (Reasonberg and Shapiro, 1978) uses planetary radar ranging to find the limit

$$\dot{G}/G < 24\times10^{-11}\text{yr}^{-1}.$$

Our latest results, for which the improvement in sensitivity comes largely from the Viking lander data, are:

$$\dot{G}/G = (0.2\pm0.4)\times10^{-11}\text{yr}^{-1}, \tag{11a}$$

$$\dot{\phi} = (0.1\pm0.8)\times10^{-11}\text{yr}^{-1}. \tag{11b}$$

The uncertainties quoted are realistic error estimates, several times the formal standard deviations, and come from uncertainties in the masses of the asteroids. The conclusion from Equation (11) is that we see no evidence for the existence of these effects at a level below that expected from simple cosmic scale arguments. This puts severe limits on the size of cosmic effects in local gravitational ($\dot{G}/G$) and nongravitational ($\dot{\phi}$) physics.

4.2. PPN Metric Parameters and J_2

I will assume that the reader is generally familiar with the Parametrized Post-Newtonian (PPN) metric formalism, referring him to the literature (e.g. Will, 1981, pp. 86ff) for additional detail. Very briefly, a large number of metric theories of gravity have post-Newtonian metrics which can be written in a general form, with theory-dependent values for certain dimensionless parameters. The form we use in our solar system model is the following:

$$g_{00} = 1 - 2\frac{GM}{r}\left[1 - \frac{J_2 R_s^2}{r^2}\, P_2(\hat{k}\cdot\hat{r})\right] + 2\beta\frac{G^2M^2}{c^2r^2} + \alpha_1\frac{GM}{r}\left(\frac{w}{c}\right)^2 \tag{12a}$$

$$g_{0k} = \alpha_1\frac{GM}{r}\frac{w^k}{c} \tag{12b}$$

$$g_{ij} = -\left(1 + 2\gamma\frac{GM}{r}\right)\delta_{ij}, \tag{12c}$$

where β, γ, and α_1 are PPN Parameters, R_s is the radius of the sun, and J_2 is the value of the solar quadrupole moment. P_2 is the usual Legendre polynomial and the vector $\hat{k}$ is a unit vector normal to the sun's equator, taken to be

$$\hat{k}=(\sin I_s \sin\Omega_s, -\sin I_s \cos\Omega_s, \cos I_s), \tag{13a}$$

with I_S=7°15' and Ω_S=750°84'. The preferred frame velocity w^k is taken to be the average of the determinations of the solar system velocity relative to the 3K background (Corey and Wilkinson, 1976; Smoot et al., 1977),

$$w^k=(-353,29,34)\text{km/sec} \tag{14}$$

in mean Earth equator and equinox of 1950 coordinates. There are generally two additional PPN parameters mentioned in the semi-conservative form of the PPN metric, α_2 and ζ_w, but these parameters are known to be small from other kinds of experiments, so they have been ignored in our solar system dynamics tests (see Will, 1981, p.204). Geodesic equations using the metric of Equation (12) are derived, leading to coordinate acceleration equations which are parametrized with β, γ, α_1, and J_2. The accelerations are numerically integrated, in the same way as discussed for Equation (7), and partial derivatives of range with respect to each of the parameters are calculated.

As was shown in Figure 7, it is difficult to separate β and J_2, so previous dynamical PPN metric tests have simply assumed a value for one of the parameters while solving for the value of the other. It has generally been assumed that the solar quadrupole moment had a value consistent with uniform rotation of the sun, $J_2<10^{-7}$. For purposes of comparison, we have also made this assumption and then solved for the three PPN parameters only. When this is done, the results are:

$$\beta-1=(-0.2\pm1.0)\times10^{-3} \tag{15a}$$

$$\gamma-1=(-1.2\pm1.6)\propto10^{-3} \tag{15b}$$

$$\alpha_1 =(2.2\pm1.8)\times10^{-4}. \tag{15c}$$

These J_2=0 results are consistent with previously published results for γ (Reasenberg et al., 1979), based on some of the same Viking lander data we have used, and represent an improvement by a factor of 5 over previous results for β (Reasenberg and Shapiro,

1977). The α_1 result is better than previous determinations based on Earth tide data by about three orders of magnitude. The values of the parameters required for the validity of General Relativity are $\beta=1$, $\gamma=1$, and $\alpha_1=0$. These results are thus seen as a confirmation of General Relativity to about a tenth of a percent, a little worse in γ and a little better in α_1.

About a year ago, Henry Hill at the University of Arizona published a value for the quadrupole moment, derived from his measurements of normal mode oscillations of the sun (Hill, 1982). He pointed out that, given his value of $J_2=6\times10^{-6}$, General Relativity would predict motion for the planet Mercury which was inconsistent with observation by about two standard deviations. We have tested the implications for gravitational theory of his J_2 value by using all available Solar System data, forcing his calculated value into the equations of motion, and solving for β and γ. With $J_2=6\times10^{-6}$, the derived values for these PPN parameters are

$$\beta-1=(12.1\pm0.8)\times10^{-3} \tag{16a}$$

$$\gamma-1=(-3.5\pm1.5)\times10^{-3}, \tag{16b}$$

a 14σ effect in β! The conclusion here should be clear. If Hill's value for the solar quadrupole moment is correct, then General Relativity is seriously in disagreement with observation. Hill's value, however, is determined only indirectly and depends for its calculation on having an adequate model of the solar interior. The assumptions which Hill makes in order to do his calculations certainly seem to be reasonable ones, but they are not compulsory, and it would clearly be desirable in testing a fundamental theory like gravitation to have an independent, direct dynamical determination of J_2.

This is what the acquisition of the Mars data has given. With Viking data from Mars and radar data from Mercury, it is possible to separate β and J_2 and solve for them simultaneously. One usually thinks of the effects of Relativity as being important for Mercury only, since they fall off so quickly as one goes away from the sun, but the Mars data are so accurate that β could be determined over an order of magnitude more accurately using the Mars range data alone than with Mercury radar only. Even J_2 can be seen four times more clearly in the Mars range than in our available Mercury radar.

The preferred values of the PPN parameters and J_2

estimated in least squares fits to the solar system astrometric data are:

$$J_2=(-1.4\pm1.5)\times10^{-6}, \tag{17a}$$

a value consistent with $J_2=0$ to within one sigma, and

$$\beta-1=(-2.9\pm3.1)\times10^{-3}, \tag{17b}$$

$$\gamma-1=(-0.7\pm1.7)\times10^{-3}, \tag{17c}$$

$$\alpha_1=(2.1\pm1.9)\times10^{-4}. \tag{17d}$$

The conclusions then of the dynamical PPN tests are that all PPN parameters have values consistent with the values predicted by General Relativity and that the value for the solar quadrupole moment determined from solar oscillations seems to be in error.

4.3. The Nonsymmetric Gravitational Theory

There is, however, an important possible exception to the above statement. Shortly after the publication of Hill's results, J.W.Moffat of the University of Toronto pointed out that his previously published Nonsymmetric Gravitational Theory (Moffat, 1979) would make predictions for the motion of Mercury which would be consistent with the observed motion and with Hill's value of J_2. Since his theory falls outside the formalism of the PPN metric, this possibility is not tested in the cases shown above. In Moffat's theory, there is an additional parameter ℓ^4 which appears as an integration constant in the Schwarzschild-like solution, giving as the line element

$$ds^2=\left(1+\frac{\ell^4}{r^4}\right)\left(1-2\frac{GM}{r}\right)c^2dt^2+\left(1-2\frac{GM}{r}\right)^{-1}dr^2+r^2d\Omega^2 \tag{18}$$

in Schwarzschild coordinates. In isotropic, PPN coordinates the g_{00} component of the solar system metric is

$$g_{00}=1-2\frac{GM}{r}\left[1-\frac{J_2R_s^2}{r^2}P_2(\hat{k}\cdot\hat{r})\right]+2\frac{G^2M^2}{c^2r^2}+\frac{\ell^4}{r^4}, \tag{19}$$

giving an anomalous acceleration of

$$\delta a^i = 2 \frac{c^2 \ell^4}{r^6} x^i . \tag{20}$$

We have solved simultaneously for the values of J_2 and ℓ^4, again using data from the two planets. The results are

$$J_2 = (1.7 \pm 2.4) \times 10^{-7} \tag{21a}$$

$$\ell^4 = (-0.1 \pm 1.0) \times 10^{13} \text{ km}^4 ,$$

or $\ell < 1800$ km. We conclude once again that the value of the solar quadrupole moment is much less than the value found by Hill and that there is no evidence in the solar system astrometric data for the necessity of the Nonsymmetric Gravitational Theory.

5. CONCLUDING REMARK

On the first day of this conference, I discussed the results presented here with one of the participants. His response was, "Well, I'm not surprised," as if to say that General Relativity is so obviously correct that it barely even needs to be tested. Let me conclude this presentation then with the reminder that General Relativity is Physics, and that Physics is an experimental science. General Relativity has had many opportunities to fail in the past, and it will have many more opportunities to fail in the future. When, as a result of any experimental test, it is found that General Relativity is still consistent with what is actually observed in the universe, then you may be pleasantly surprised if you wish... but you must always be surprised.

REFERENCES

Adams, P.J., Canuto, V.M., Goldman, I., and Hellings, R.W.: 1983, "Testing for a Cosmological Influence on Local Physics using Atomic Antigravitational cloks", *Phys. Rev.*, D28, 1822.

Bierman, G.J.: 1977, *Factorization Methods for Discrete Sequential Estimation*, Academic Press, New York.

Corey, B.E., and Wilkinson, D.T.: 1976, "A Measurement of the Cosmic Microwave Background Anysotropy at

19 Ghz", *Bull. Am. Astron. Soc.*, 8, 351.

Dirac, P.A.M.: 1937, "The Cosmological Constants", *Nature*, 139, 323.

Hellings, R.A., Adams, P.J., Anderson, J.D., Keesey, M.S., Lau, E.L., Standish, E.M., Canuto, V.M., and Goldman, I.: 1983, "An Experimental Test of the Variability of G, using Viking Lander Ranging Data", submitted to *Phys. Rev.*, D.

Hill, H.A., Bos, R.J., and Goode, P.R.: 1982, "Preliminary Determination of the Sun's Gravitational Quadrupole Moment from Rotational Splitting of Global Oscillations and its Relevance to Tests of General Relativity", *Phys. Rev.*, 49, 1794.

Moffat, J.W.: 1979, "Static Spherically Symmetric Solution for the Field of a Charged Particle in a Theory of Gravity", *Phys. Rev.*, D19, 3562.

Reasenberg, R.D., and Shapiro, I.I.: 1977, "Solar System Test of General Relativity", in B.Bertotti (ed.), *Proceedings of the International Meeting on Experimental Gravitation*, Accademia Nazionale dei Lincei, Roma, pp. 143-168.

Reasenberg, R.D., and Shapiro, I.I.: 1978, "Viking Relativity Experiment: Vanification of Signal Retardation by Solar Gravity", in L.Halpern (ed.), *On the Measurement of Cosmological Variations of the Gravitational Constant*, University Presses of Florida, Gainesville, pp. 71-86.

Reasenberg, R.D., Shapiro, I.I., Mac Neil, P.E., Goldstein, R.B., Briedenthal, J.C., Brenkle, J.P., Cain, D.L., Kaufmann, T.M., Komarek, T.A., and Zygielbaum, A.I.: 1979, "Viking Relativity Experiment: Verification of Signal Retardation by Solar Gravity", *Astrophys. J.*, 234, L219.

Smoot, G.F., Gorenstein, M.V., and Müller, R.A.: 1977, "Detection of Anisotropy in the Cosmic Blackbody Radiation", *Phys. Rev. Lett.*, 39, 898.

Van Flandern, T.C.: 1981, "Is the Gravitational Constant Changing?", *Astrophys.J.*, 248, 813.

Will, C.M.: 1981, *Theory and Experiment in Gravitational Physics*, Cambridge University Press, Cambridge.

SPACE EXPERIMENTS
REPORT OF WORKSHOP C2

Peter L. Bender *

Joint Institute for Laboratory Astrophysics
National Bureau of Standards and
University of Colorado
Boulder, CO 80309, USA

1. INTRODUCTION

The keynote report at GR10 on new results from space experiments was the invited paper "Testing relativity with solar system dynamics", which was presented by R.W. Hellings earlier in the Conference. In that paper, the major improvements in solar system tests of general relativity made possible by adding the Viking Lander and Viking Orbiter tracking data to other existing information were discussed (see also Hellings et al., 1983, and Reasenberg, 1984). It is striking that the accuracy of some of the very interesting results is limited more by uncertainties in modeling the effects of many different asteroids on the orbit of Mars (Williams, 1984) than by any other factor. Further studies of the effects of the asteroids appear to be very important.

The first workshop session concerned a wide variety of space experiments using either earth satellites or suggested new interplanetary missions. The one with the greatest potential impact on gravitational physics over the next decade is the widely know Stanford Gyro Relativity Experiment. The prospects for flight of this experiment by about 1990 fortunately appear to be quite good. This would be the first test of the "gravitomagnetic" effects predicted by general relativity, and would determine whether the fundamental rotational effects of massive bodies are built into the theory correctly (Thorne, 1984). There also are important applications in astronomy, since frame-dragg-

B. Bertotti et al. (eds.), General Relativity and Gravitation, 387–395.

ing effects of super-massive rotating black holes may well be responsible for the alignment of jets emitted by quasar and active galactic nuclei.

The second session of the workshop was devoted to searches for low frequency gravitational waves using microwave tracking of distant spacecraft and accurate timing of pulsar signals. Although the combined Galileo and International Solar Polar Mission gravitational wave experiments are not scheduled until 1987, a continuing effort is needed in order to be ready for this opportunity. Since the largest source of noise is expected to be plasma density fluctuations, success in achieving some approximation to dual frequency uplink capability would be particularly important for achieving the highest possible scientific benefits from the experiments. The pulsar timing measurements now limit the possible isotropic gravitational wave background to about four orders of magnitude below the cosmic closure level for frequencies near 10^{-8} Hz.

2. GRAVITATIONAL PHYSICS WITH EARTH SATELLITES AND SUGGESTED NEW INTERPLANETARY MISSIONS

The Standford Gyro Relativity Experiment is a very strong candidate for flight about 1990. It currently is being studied intensively by NASA, and approval of construction of an engineering model of the complete experimental package appears likely. Fairbank et al., (1983) described a thorough redesign which has been carried out recently to reduce the total size and cost while retaining all of the main scientific requirements for the mission. A major improvement is a reduction in the size of the apparatus so that a Dewar similar to the one flown on the Infrared Astronomical Satellite or the one being built for the Cosmic Background Explorer can be used. A new approach to testing the complete experimental package is being considered. It involves flying the engineering package in 1988 on the shuttle in order to test the gyro operation in a much lower acceleration environment than on the earth. The experimental package would then be refurbished and the 14 months duration experiment would be carried out on a free-flier in a polar orbit. The size of the geodetic precession and the Lense-Thirring precession due to frame-dragging are about 6.9 arcsec/yr and 44 milliarsec/yr respectively. The experimental accuracy is expected to be 1 milliarsec/yr, so a thorough and reliable test of the frame-dragging effects should be obtained. This experiment has been rated by the Space Science Board of the U. S. National Research Council as being of first priority scientifically.

In another paper, Ashby and Bertotti (1983) discussed detailed calculations of the relativistic corrections to the equations of motion for a satellite in the neighborhood of the earth (see also Ashby and Bertotti, 1984). One satellite of interest is a completely passive and dense geodetic satellite called LAGEOS, which is in a very stable orbit with 5900 km altitude and small eccentricity. The tracking accuracy currently is about 3 cm for a number of laser ranging systems, and may be improved to somewhat better than 1 cm in the next few yers. The frame of reference used by Ashby and Bertotti is the generalization, in the "slow-motion, weak-field" approximation, of the local inertial frame which can be constructed in the neighborhood of a geodesic in the field of the sun. The generalization involved including the field of the earth and allowing for nonlinear interactions. This work was an extension of an earlier study by Ashby and Allan (1979) on methods for extablishing a practical coordinate time system in the neighborhood of the earth's surface which includes the known relativistic effects. In addition to the advance of perigee and change in mean motion for the satellite due to the second-order term in the earth's potential, plus the expected geodetic precession due to the sun, the following new relativistic accelerations were found: corrections to the tidal field of the sun which are quadratic in its mass; "magnetic" terms arising because the sun is in motion in this frame; and cross terms between the solar tidal potential and the earth's Newtonian potential. These additional accelerations are very small for artificial satellites with accurately determined orbits such as LAGEOS. However, the first one may be barely observable in the future for the moon.

Next, Cowling (1983) described relativistic effects which may be observed by the HIPPARCOS satellite, which is planned for launch in a few years. This space astrometry satellite is a major component of the scientific program of the European Space Agency. It will make repeated determinations of the positions of about 100,000 stars with an accuracy of about 0.005 arcsec for each observation. This means that corrections to the positions due to light bending will be significant over the whole sky. The expected precision for checking on the PPN parameter γ is 0.16%. Other possible observations of interest to gravitational physics are transitory displacements or brightenings of images due to black holes or compact objects near the line of sight.

Paik (1983) described a proposed null test of the inverse square law in earth orbit. A three-axis gravity gradiometer is being developed for orbital mapping of the earth's gravity field. If the inverse square law holds, the sum of the gradients in three perpendicular directions must be zero. For a suitable elliptical orbit, a check to better than one part in 10^9 is expected over the distance range of roughly 10^2 to 10^4 km. A single axis superconducting gradiometer has been developed and used in a preliminary earth-based test (Chan et al., 1982). The sensitivity achieved by using the gradiometer in three perpendicular orientations and a large test mass was about 3% for deviations from the inverse square law at one meter distance. Improvements to 10^{-4} sensitivity with a three-axis gradiometer and extension of the ground-based results to distances of up to 10 km are expected.

The other possible future missions discussed at the workshop are at a much earlier stage of investigation. One is laser heterodyne gravitational wave experiment involving three spacecraft separated by 10^6 km or more, which was described by Faller et al. (1983). The difference in length of two nearly equal but roughly perpendicular paths between carefully shielded test masses in the three spacecraft would be measured to search for gravitational waves with periods of up to hours. The precision expected with 1 mW of laser power at 628 nm and with 50 cm diameter optics is about 10^{-21} for a 10^4 sec averaging time. Efforts recently have been focused on studies of possible effects of short period laser noise on the results. Optical methods look promising for sensing the shield positions with respect to the test masses so that they can be controlled to 10 μm precision under up to 10^{-11} g variable accelerations due to the solar wind and radiation pressure variations.

Ashby et al. (1983) reported on investigations of what could be learned about gravitational and solar physics from accurate microwave measurements of the Earth-Mercury distance. They used a modified form of "worst case" analysis because the dominant errors are expected to be systematic errors at frequencies similar to those occurring in the signatures of the parameters of interest. A small relativity satellite in a suitable orbit with accurate dual-frequency ranging and Doppler systems was assumed. The expected improvements in accuracy for determining the solar quadrupole moment, possible changes with time of the gravitational constant, and the various PPN parameters were discussed.

3. GRAVITATIONAL WAVE EXPERIMENTS USING MICROWAVE TRACKING AND PULSAR TIMING

A report on the prospects for detection of low frequency gravitational waves by microwave tracking of distant spacecraft was give by Bertotti and Estabrook (1983), An important coordinated effort is now under way for the NASA Galileo mission to Jupiter and the ESA out-of-the-ecliptic International Solar Polar Mission (ISPM). In October, 1987 the two spacecraft will be at solar opposition almost at the same time and it is planned to have a continuous, joint observational period of several weeks. The Galileo spacecraft will be able to receive either S-band or X-band uplink signals, the ISPM spacecraft will be able to receive at S-band, and both spacecraft will transmit dual frequency downlinks. Four of the main noise sources are electron density fluctuations along the path, clock frequency fluctuations, Deep Space Network (DSN) system noise, and tropospheric delay fluctuations. S-band experiments with Viking (Armstrong et al., 1979) and Voyager (Hellings et al., 1981) indicated that the electron density fluctuations gave about 3×10^{-14} Doppler noise for 1000 sec averaging time near solar opposition. This means that about 3×10^{-15} Doppler noise is expected for the X-band uplink on Galileo. Still better results might be achieved if it is possible to switch between the X-band and S-band uplinks rapidly enough or to use two different X-band frequencies. Clock noise hopefully will be reduced below 10^{-15} by 1987 by using a combination of improved hydrogen masers and oscillators stabilized to superconducting cavities. Tests at DSN stations in 1983 indicated system noise of 3×10^{-15}, and efforts to reduce the system noise to 10^{-17} over the next 10 years are under way. Measurements by Armstrong and Sramek (1982) using the Very Large Array in New Mexico indicate that about 10^{-15} Doppler noise would be contributed by tropospheric delay fluctuations at that site. Thus the basic overall Doppler noise level in the 1987 Galileo experiments probably will be about 3×10^{-15} or lower.

A.J. Anderson (1983) continued the discussion of the expected accuracy and the various noise sources for the combined Galileo and ISPM experiments. The precise Voyager gravitational wave experiments were described. In some of them both Goldstone and the Owens Valley Radio Observatory tracked with independent hydrogen masers, so that the clock, troposphere and ground systems noise contributions were decorrelated. The use of the autocorrelation of the signal at the round-trip delay time in the analysis to reduce the effects of noise sources also was discussed. The importance of plasma and tropospheric noise was emphasized, and the possibility of separately monitoring the wet component of the tropospheric delay was mentioned. Independently equipped Mark III VLBI observatories in Europe, North America and Japan are expected to take part in the Galileo and ISPM experiments, as well as the DSN stations.

The combined analysis of signals from the two spacecraft was discussed by Wahlquist (1983) and by Bertotti and Iess (1983). Wahlquist emphasized the importance of having a coincidence experiment in order to increase confidence in a positive result. In addition, if the two spacecraft were tracked independently from different sites, most of the serious noise sources would be independent, while the gravitational wave signals would be highly correlated. Some information on the source direction also might be obtainable. Bertotti and Iess showed that, for reasonable models of the plasma turbulence, the plasma noise for the two spacecraft would indeed be uncorrelated for reasonable separations.

Gravitational wave experiments in 1981 an 1982 using Pioneer 10 near solar opposition were described by J.D. Anderson and Lau (1983). The round trip light-time in 1982 was 7 hr 42 min and is increasing by about 40 min/yr. The overall strain sensitivity was about 10^{-13} for 1000 sec integration time. Discrete lines with amplitudes of 2×10^{-14} would be nearly detectable.

The remaining two contributions were on the possible effects of gravitational waves on pulsar timing experiments. Hellings and Downs (1983) described the effects of very low frequency gravitational waves which could cause motion of the earth with respect to different pulsars. The motions of the various pulsars would be different, but the earth's motion should affect the timing for the various pulsars in a consistent way. For frequencies between 4×10^{-9} and 10^{-7} Hz, the resulting upper limit to the spectrum of the isotropic gravitational radiation background was below the critical density level. Below 10^{-8} Hz the level was 1.4×10^{-4} of the critical density level. Pulsar experiments thus provide valuable gravitational wave limits in the very low frequency region.

The possibility that irregularities in the radiation period observed in many pulsars may be a consequence of the absorption of gravitational waves was discussed by Cerdonio and Vitale (1983). The mechanism proposed for coupling absorbed gravitational wave energy into a change in neutron star rotation rate is an annihilation or pinning-unpinning process for superfluid vortices in the star. No direct evidence for the proposed process is claimed, but a statistical analysis of data from about 30 pulsars within a few kiloparsec of the solar system indicates a possible location for a hypothetical source of disturbances.

NOTE

* Staff Member, Quantum Physics Division, National Bureau of Standars.

REFERENCES

Papers presented to the GR10 conference and published in:

Bertotti, B., de Felice, F., and Pascolini, A. (eds): 1983, *10th International Conference on General Relativity and Gravitation. Contributed Papers.* Vol. 2, CNR, Roma.

Anderson, A.J.: 1983, "Precise spacecraft tracking using VLBI digital tone extraction for the detection of a cosmic background of gravitational wave radiation", p. 945.

Anderson, J.D. and Lau, E.K.: 1983, "Doppler detection of gravitational radiation with Pioneer 10", p. 946.

Ashby, N., and Bertotti, B.: 1983, "Relativistic perturbations of an earth satellite", p. 948.

Ashby, N., Bender, P.L., and Wahr, J.M.: 1983, "Accuracy of gravitational physics tests and of the solar quadrupole moment from ranging to a Mercury orbiter", p. 947.

Bertotti, B., and Estabrook, F.B.: 1983, "The prospects for low frequency gravitational wave detection",

p. 951.

Bertotti, B., and Iess, L.: 1983, "Correlation between two spacecraft in the Doppler detection of a cosmic gravitational background", p. 952.

Cerdonio, M., and Vitale, S.: 1983, "Noisy pulsars, superfluid vorticity and gravitational waves", p. 953.

Cowling, S.A.: 1983, "Relativistic effects in observations by the HIPPARCOS satellite", p. 956.

Fairbank, W.M., Breakwell, J.V., Cabrera, B., Chen, J. H., Cheung, W.S., DeBra, D.B., Duhamel, T., Everitt, C.W.F., Farnsworth, R.A., Gill, J., Keiser, G.M., Levine, P., Lipa, J.A., Lockhart, J.M., Pollock, S., Turneaure, J.P., and Van Patten, R.A.: 1983, "The Stanford gyro relativity experiment".

Faller, J.E., Bender, P.L., Chan, Y.M., Hall, J.L., Hils, D., and Hough, J.: 1983, "Laser gravitational wave experiment in space", p. 960.

Hellings, R.W., and Downs, G.S.: 1983, "Upper limits on the isotropic gravitational radiation background from pulsar timing analysis", p. 963.

Paik, H.J.: 1983, "Null test of the inverse square law in the earth's orbit", p. 966.

Wahlquist, H.D.: 1983, "Correlation analysis for dual spacecraft Doppler gravity wave experiments", p. 967.

Other references

Armstrong, J.W., and Sramek, R.A.: 1982, "Observation of tropospheric phase scintillations at 5GHz on vertical paths", *Radio Sci.*, 17, 1579.

Armstrong, J.W., Woo, R., and Estabrook, F.B.: 1979, "Interplanetary Phase Gravitational Radiation", *Astrophys. J.*, 230, 570.

Ashby, N., and Allan, D.W.: 1979, "Practical implications of relativity for a global coordinate time scale", *Radio Sci.*, 14, 649.

Ashby, N., and Bertotti, B.: 1984, "Relativistic perturbations of an Earth satellite", *Phys. Rev. Lett.*, 52, in press.

Chan, H.A., Moody, M.V., and Paik, H.J.: 1982, "Null test of the gravitational inverse square law", *Phys. Rev. Lett.*, 49, 1745.

Hellings, R.W., Adams, P.J., Anderson, J.D., Keesey, M.S., Lau, E.L., Standish, E.M., Canuto, V.M., and Goldman, I.: 1983, "Experimental test of the variability of G using Viking lander ranging data", *Phys. Rev. Lett.*, 51, 1609.

Hellings, R.W., Callahan, P.S., Anderson, J.D., and Moffet, A.T.: 1981, "Spacecraft-Doppler gravity-wave detection. II. Results.", *Phys. Rev.*, D 23, 844.

Reasenberg, R.D.: 1984, "The constancy of G and other gravitational experiments", *Phyl. Trans. Roy. Soc. Lond. A*, in press.

Thorne, K.S.: 1984, "Gravitomagnetism, test in Quasar and the Stanford gyroscope experiment", in C.W.F. Everitt, B.S. Deaver, and P. Michelson (eds), *Near Zero: New Frontiers in Physics, Essays in Honour of William M. Fairbank*, W.H. Freeman, San Francisco, in press.

Williams, J.G.: 1984, "Determining asteroid masses from perturbations on Mars", *Icarus*, in press.

LASER INTERFEROMETER GRAVITATIONAL WAVE DETECTORS

R.W.P. Drever

California Institute of Technology, USA
and
University of Glasgow, UK

1. INTRODUCTION. WHY DETECT GRAVITATIONAL WAVES?

In the last few years an increasing effort has gone into the development of gravitational radiation detectors based on laser interferometers and to introduce the subject it might be useful first to consider some of the reasons for attempting to detect gravitational radiation at all. One objective is clearly just to observe the radiation and confirm its existence, and thus provide a check of one of the more interesting predictions of general relativity. The beautiful observations of changes in orbital period of the binary pulsar are of course giving good evidence for loss of energy corresponding to that expected by emission of gravitational radiation from the system. However, it would be very significant to detect the radiation directly. And if more than mere detection is achievable, many of the predicted properties of the radiation might be experimentally checked. For example, simultaneous observation of a single gravitational wave burst by a number of wide-band gravity wave detectors at widely separated locations on the earth could give information on the velocity of the radiation, and possibly on polarization properties. If a gravitational wave pulse could be associated with another observed astronomical event, such as a supernova outburst, more precise information on the velocity could be obtained. Other possibilities arise if continuous gravitational waves, such as from a pulsar or a fast binary system, are detected and observed for an extended period. Measurements of per-

B. Bertotti et al. (eds.), General Relativity and Gravitation, 397–412.

iodic changes in amplitude of signal recorded by a detector rotating with the earth, and of changes in its phase due to the earth's orbital motion, could give fairly precise data on polarization and on velocity of propagation. Such experiments might be regarded as analogous to the discovery and experimental study of a new predicted particle in high energy particle physics. And although one may think one knows what the results will be, the experiments are still extremely important.

A quite separate, and possibly even stronger, reason for wishing to detect gravitational radiation is its use as a tool for astrophysics. Detection of gravity wave pulses from collapse events, and particularly studies of waveforms as well as pulse amplitudes and durations, would clearly give detailed information on collapse processes which could not be found in any other way; for the core of the collapse is often screened for other radiation by sorrounding material. Studies of the continuous signal from a pulsar and its harmonic content, could give information on pulsar structure. And by suitable analysis of the output from a broadband gravitational wave detector it may be possible to discover signals from pulsars not found by their radio emission, due, for example, to an unfortunate direction of beaming at the source. Detection of a stochastic background of gravitational radiation would give very interesting information of quite another type, relating possibly to very early stages of the universe, or to epochs of black hole formation. For any, or all, of these reasons, observation of gravitational radiation would almost certainly lead to new advances in astrophysics.

There is thus ample justification for the serious efforts which have been made in the past, and are now continuing and expanding, to detect gravitational radiation.

2. TARGETS FOR EXPERIMENTS

In thinking about experimental techniques for detection of gravitational waves, it is useful to consider the possible and expected wave amplitudes which might be present. There have been several reviews and summaries of the large amount of theoretical work done on gravitational radiation production in various astrophysical processes. (A few examples are Thorne, 1978; Tyson and Giffard, 1978; Douglass and Braginsky, 1979; Epstein and Clark, 1979; Smarr, 1979; Thorne, 1980.) Here it is appropriate just to summarize some conclusions. First there are large uncertainties in expected fluxes or event rates for impulsive sources, and for the strongest periodic sources, and the maximum limits set by general arguments do not rule out detection at experimental levels of sensitivity already achieved. However expected fluxes from known phenomena are considerably smaller. For example in the case of pulses in the kilohertz region of the spectrum from such processes as stellar collapse to form neutron stars or black holes, or coalescence of black holes, detectors for signals with strain amplitude of order 10^{-21} might be required to bring a sufficient number of possible sources within range to give an event rate of order of one per month. This is one interesting target for future experiments. Expected amplitudes for periodic signals are generally smaller, and for example estimates of amplitudes of the periodic signals from known pulsars such as the Crab and Vela pulsar are of order 10^{-24} to 10^{-27}, at frequencies between 10 and 100 hz. Although these amplitudes are smaller than for pulses, the possibility of using integration over an extended time to narrow the experimental bandwidth can enhance signal to noise ratio for continuous signals, and the experimental difficulties of reaching the pulse and periodic sensitivities just quoted may be roughly comparable. At lower frequencies somewhat larger gravitational wave amplitudes may be expected, but ground-based experiments below about 10 hz are likely to be limited by local gravitational gradient noise. The signals from known binary stars, for example, are likely only to be detectable in space experiments. The possibilities of laser interferometer experiments in space are very interesting, but we shall not consider them here.

3. EXPERIMENTAL APPROACHES

The pioneering experimental technique in gravitational wave detection has been the resonant bar detector, developed originally by J. Weber, and subsequently by many others. This technique is a good one, and the most sensitive gravitational wave detectors existing at present are of this type. However when one considers trying to achieve much higher sensitivity along with wide bandwidth, detectors based on use of laser interferometers to sense relative motion of widely separated test masses appear to have some advantages. The long baseline over which the gravitational wave strain may be sensed with a laser interferometer-from tens of meters to possibly several kilometers - makes the displacements to be measured relatively larger than with a bar detector, whose length is determined by its required mechanical resonant frequancy. Thus high sensitivity may in principle be achieved without serious problems arising from thermal motions in the test masses, and cryogenic operation is not necessary; and the standard quantum limits set by the uncertainty principle are not likely to be so important. Also, as there is no resonant storage of the signal in most modes of operation wide bandwidth is automatically available, and a wide range of frequencies may be covered in one experiment. There are, of course, disadvantages with the laser interferometer technique too; such as the requirement for long evacuated pipes for the laser beams, and the technical difficulties of achieving adequate sensitivity in the optical measurement. However the expected eventual performance makes it seem well worth putting some effort into overcoming these difficulties.

3.1 Basic Interferometer Detection Systems

Most suggested arrangements for interferometric gravity wave detectors involve making a comparison between changes in two baselines arranged perpendicular to one another, so that there is no need for an exceptionally stable standard of length or frequency with which to monitor a single baseline. A simple arrangement might consist of three suspended test masses arranged to give the required two perpendicular baseline, with mirrors and a beamsplitter attached to the masses in such a way that a Michelson interferometer could be used to compare relative changes in the baselines. Early experiments with an interferometric gravitational wave detector similar to this were performed by R.L. Forward (Moss et al., 1971; Forward, 1978). The difficulty with this scheme is that the sensitivity of the Michelson interferometer for small changes in optical path is limited by the statistical fluctuations in the photon counting rate registered by any photodetector used to sense small changes in the fringe pattern. The resulting "photon shot noise" limit to sensitivity depends on the light power in the arms of the interferometer - and to make this limit approach our gravitational wave target sensitivities would require a laser source giving hundreds of kilowatts of lights. The various experimental techniques now being developed arise from different methods proposed for avoiding this requirement for excessive light power.

3.2 Delay Line Michelson Interferometers

One approach, suggested by R. Weiss (1972), uses an optical delay line to reflect the light in each arm of a Michelson interferometer many times back and forth along each baseline between mirrors attached to the test masses, so that the optical phase shift arising from a given displacement of the masses is increased. A schematic diagram of such a system is given in Figure 1. In this figure, the three squares represent

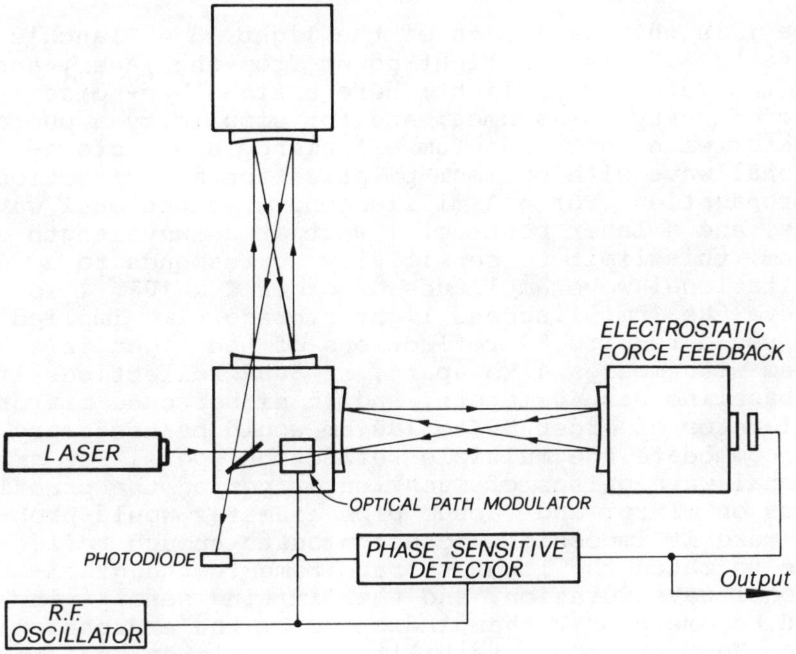

Fig. 1.

the test masses, with attached concave mirrors. A radio-frequency phase modulator is also indicated in an arrangement designed to feed back a small electrostatic force to one of the test masses so that it takes up a position giving a dark fringe at the photodetector, at which point maximum sensitivity is obtained. Differential strains in the two arms may be recorded by monitoring changes in the voltage applied to the feedback plates, or error signals from other points in the feedback loop.

The limit to the sensitivity of this type of system determined by photon shot noise is a function of the number of times the light in each arm traverses the space between the test masses, and if losses of light at the mirrors are negligible, it improves with number of traversals until the time spent by the light within each arm becomes comparable with the period of the gravitational wave. In this case, the sensitivity for a pulse of duration τ corresponds to a minimum gravitational wave amplitude given approximately by

$$h_{min} = (\lambda\hbar/8\pi I c \tau^3)^{1/2}$$

where τ is the wavelength of the light, $\hbar$ = Planck's constant/2π, I is the light power from the laser, and c is the velocity of light; here a signal-to-noise ratio of unity is assumed; and for simplicity a photodetector with unity quantum efficiency and a gravitational wave with optimum polarization and direction of propagation. For a 1 millisecond gravitational wave pulse, and a laser power of 1 watt at a wavelength of 500 nm, this limit to sensitivity corresponds to a gravitational wave amplitude of order 2×10^{-21}. To achieve the 1 millisecond light storage time implied here would require 80 reflections of the light in a system with masses 4 km apart, or 8000 reflections if the baseline was 40 meters, and in either case mirrors of diameter of order 50 to 100 cm would be necessary to accommodate the multiple reflection spots. For gravitational wave pulses of much longer period the practical limits on mirror and vacuum pipe diameter would probably make it impossible to accommodate enough reflections to match the light storage time to the gravitational wave duration, and the limiting sensitivity would become poorer than indicated by the expression above. However, the possibility that this general technique might give sensitivities for short pulses approaching that indicated above makes the method seem interesting, and experimental work to investigate practical problems started in several laboratories in the early 1970's. Experiments at the Max Planck Institute in Munich (Billing et al., 1979), and at the University of Glasgow, showed that scattering of light at the multireflection mirrors could be serious, for even a small amplitude of scattered light reaching the photodetector by a different route from that of the main beam could give a phase error which would be modulated by any changes in laser frequency and thus become a source of noise. The Munich group devised an effective way of reducing this effect, in which they modulate the laser frequency in such a way that phase errors arising from any particular path difference are averaged out (Schilling et al., 1981). A related method is being developed at MIT, in which random phase modulation is applied to the laser output to give the light a relatively short coherence length, making scattered light contributions corresponding to large path differences unimportant. At Glasgow, the scattering problem led to investigation and development of quite a different optical technique - the use of Fabry-Perot cavities (Drever et al., 1983a). This approach seems to have some general advantages, unrelated to scattering, over delay-line Michelson interferometers; and it is now being developed at the California Institute of Technology as well as at Glasgow.

3.3 Fabry-Perot Interferometers

In this technique concave mirrors attached to the test masses form long Fabry-Perot optical cavities spanning each arm of the gravitational wave detector. Both of these cavities are arranged to be in resonance with the light from the laser, and small changes in cavity length are sensed by measuring changes in phase of the light within the two cavities. One way of doing this is indicated schematically in Figure 2. Here light from a laser passes via a beamsplitter and two phase modulators P1 and P2 to the optical cavities. The first mirror of each cavity is arranged to have lower reflectivity and higher transmission than the second, highly-reflecting mirror, and light penetrating the first mirror is reflected many times back and forth within the cavity. At resonance the intensity of the light within the cavity may be many times that from the laser itself, and light emerging from the first mirror may have an amplitude almost equal to that of the input light.

The phase of the light from each cavity may be monitored separately using slightly-reflecting beamsplitters, shown dotted in the Figure 2, to divert some of the light to photodiodes D2 and D3. If the phase modulators are driven at a suitable radio frequency, the light reaching each photodiode may be regarded as having a phase modulated component due to light directly reflected by the first cavity mirror, and an unmodulated component arising from stored light emerging from the cavity with its modulation sidebands

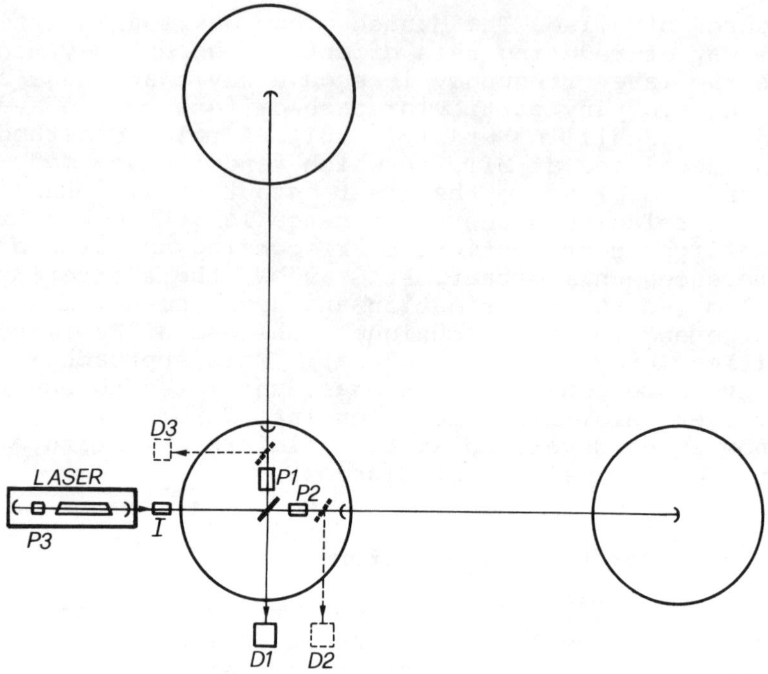

Fig. 2.

removed. Interference between these components results in a radio frequency output from the photodiode whose amplitude, measured with a phase sensitive detector , is proportional to the phase difference between the light in the cavity and the light from the laser, for small phase differences. In one mode of operation of the system, the demodulated signal from one photodetector, D2 for example, is amplified and fed back to a phase modulator P3 within the laser itself so that it adjusts the wavelength of the light and holds it closely in resonance with the cavity. The signal from the second photodetector, D3, may be applied to a piezoelectric transducer supporting one of the mirrors of the second cavity, or to capacitor plates applying a force to one of the test masses, so that the length of that cavity is adjusted to maintain resonance with the laser light. The gravitational wave signal may then in principle be found from the fed back displacement or force applied, if the gains in both feedback loops are high. In an alternative, in principle more sensitive, mode of operation the light from the two cavities is optically combined at the main beamsplitter. Phase modulators P1 and P2 are driven in antiphase with one another, and with optical path lengths suitably adjusted, the radio frequency output from the third photodiode D1 can give a direct measure of the difference in phase in the two cavities, and may be used to control the second cavity, or give an output signal. If the error signals are not negligible, or if the two cavities are not matched in length of finess to one another, correction signals may be obtained from a suitable combination of the signals from D2 and D3. An optical isolator prevents feedback of light from the interferometer to the laser.

The photon shot noise limit to the sensitivity of this type of interferometer depends on the storage time of the light within the cavities. In an interferometer with short arms this may be determined by the maximum available mirror reflectivity, and in this case it may be shown that the sensitivity is in principle better by a factor of about 2 than the sensitivity of a delay line Michelson interferometer in which the useful number of reflections is also limited by equivalent mirror reflectivity, if it is assumed that scattering and absorption losses in the input mirrors are small compared with their transmission. In practice these losses could make the two systems approximately equally sensitive. In a large interferometer used for millisecond pulse searches the available mirror reflectivity is not likely to be the limiting factor; and for equal storage times the systems would again be expected to have comparable sensitivity. For low frequency signals, however, there may be a significant difference. In this case limits to practicable mirror diameter may set a limit to the number of reflections and thus to storage time for a delay line; while with a Fabry-Perot system only the mirror losses limit the effective number of reflections, and much longer storage times may be achievable. In this region the Fabry-Perot interferometer has a potential for significantly higher sensitivity.

The cavity mirrors in a Fabry-Perot system can be fairly small - a mirror diameter of 17 cm is sufficient for even a cavity 5 km long - and this gives several practical advantages. Vacuum pipes can be smaller, or several separate interferometers may be accommodated within one vacuum system; it is likely that higher reflectivity and optical quality may be obtainable in small mirrors than in large ones; and design of test masses for low internal thermal noise may be simplified. The Fabry-Perot interferometer does, however, require that the frequency of the light from the laser be very well controlled to match the narrow resonances of the long cavities. A considerable amount of experimental work has gone into development of the the laser stabilization method outlined above, and the laser frequency stabilization achieved is already adequate for the present prototype interferometers and close to being good enough for the larger instruments planned. Techniques developed in this work (Drever et al. 1983b; Hough et al. 1983) are proving useful in other fields.

4. CURRENT STAGES OF THE EXPERIMENTAL WORK

Experimental development of laser interferometer prototypes is currently being carried out in several laboratories. Delay-line Michelson interferometers are being developed at MIT, with a 1-metre prototype, and at the Max Planck Institute, now at Garching, with a 3-metre and a new 30-metre instrument. Fabry-Perot interferometers are being developed at Glasgow, with a 10-metre system, and at Caltech with a 40-metre interferometer. All of these instruments are fairly complex, with several auxiliary servosystems required to control the orientation and position of the test masses and the associated mirrors. Currently, most of these instruments have been operated at relatively low power levels, of order 10 mW, but have achieved performance not far from the photon shot noise levels for the powers and mirrors used. Steady development is continuing, and no major problems have become apparent.

5. SOME POSSIBILITIES FOR ENHANCING SENSITIVITY

The estimates of sensitivity limits for the interferometer systems given above are interesting ones, but even higher sensitivity would be desirable to give better overlap with expected source strengths. Two possible methods for achieving better photon shot noise to sensitivity in large interferometers have been suggested recently (Drever, 1983; Drever et al., 1983c), and may be worth mentioning here. Both are mainly applicable to the large interferometers which are now being planned, and exploit the fact that with low-loss mirrors it should be possible to achieve light storage times long compared with the periods of the gravitational waves, at least in the kilohertz part of the spectrum.

In both the delay line Michelson interferometer and the Fabry-Perot interferometer described earlier the optical paths are normally adjusted to that the main output photodetector is at a dark fringe in the interference pattern. This implies that most of the light leaves the system from the other side of the beamsplitter, and raises the possibility that this escaping light be fed back into the interferometer input with its phase adjusted to reinforce the original light from the laser. Ways of doing this in principle, are indicated in Figure 3 for both a Michelson

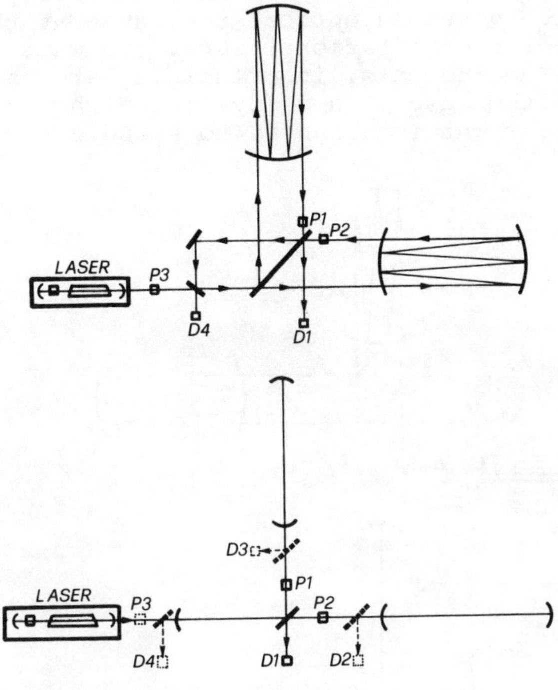

Fig. 3.

interferometer and Fabry-Perot system. In either case a mirror of suitable reflectivity returns the escaping light to the interferometer, using an auxiliary high frequency phase modulator P3 and photodetector D4 to adjust P1 and P2 or the wavelength of the light for correct phasing. The whole interferometer becomes effectively contained within a large optical cavity giving an increase in effective light flux. If the system is arranged so that the dominant losses are due to the finite reflectivity of the delay line mirrors, for the Michelson, or for the cavity mirrors in the Fabry-Perot, then the photon shot noise limit to the sensitivity is improved by a factor of order the ratio of (overall light storage time/storage time in each arm)$^{1/2}$. For a large system, with arms 5 km long, mirrors of reflectivity $(1-10^{-4})$, and a laser power of 10 watts, this gives in principle a photon shot noise limit to sensitivity for a 1 millisecond pulse of order 2×10^{-22}. These parameters are not impossible ones for future interferometers.

In experiments designed to search for periodic gravitational waves of known frequency it may be possible to obtain a further improvement in sensitivity by using a system which gives an optical resonance at the expected period of the gravitational wave. Two ways we have suggested for doing this, in principle, are shown in Figure 4. In the case of a delay line Michelson interferometer the light is arranged to spend a time equal

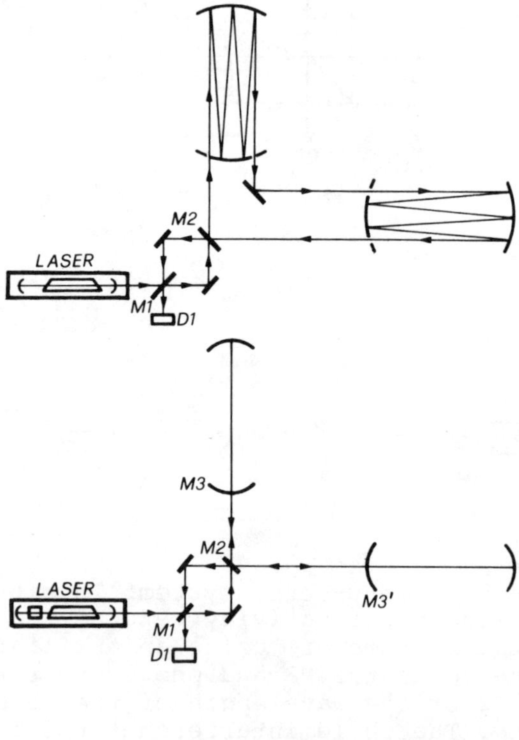

Fig. 4.

to half of the gravity wave period in one arm of the system, and then pass to the other arm for an equal time, subsequently being returned to the first by a mirror of high and suitably chosen reflectivity M2. Phase shifts in the light accumulate steadily as the light cycles round the system in one direction, while light circulating in the opposite direction (against the arrows), builds up a phase shift of opposite sign. The resultant phase shift is registered by photodetector D1. In the case of the Fabry-Perot interferometer, light passes from one cavity to the other via mirror M2, with an effective periodicity dependent on the reflectivity of mirrors M3 and M3', and again there is a build up of the phase signal(1). The photon shot noise limit to sensitivity of these systems is expected to be better than that of a non-resonant system by a factor of order the ratio of the overall storage time to the gravitational wave period. This arrangement might in principle give such a good photon shot noise limit that detection of expected fluxes from the Crab or Vela pulsars might seem possible in a relatively short integration time, but it should be noted that this may be misleading since other noise sources, such as thermal noise, and even the quantum limit, are likely to be important problems in this case. The technique may be more useful for signals at higher frequencies, such as from fast pulsars or following some collapse or coalescence processes.

These suggested arrangements for enhancing the photon shot noise limit to gravitational wave interferometers are only intended to indicate some of the possibilities in principle. However they are probably sufficient to show that laser interferometer gravitational wave detectors may be capable of making useful searches for signals from a number of quite different types of sources, with potential for very interesting sensitivities.

A considerable amount of practical development of the techniques remains to be done, but no serious problems have been encountered so far, and overall the possibilities for these, and other, types of gravitational wave detectors look very promising indeed.

ACKNOWLEDGEMENTS

The author would like to acknowledge the essential contributions of his colleagues at the California Institute of Technology and the University of Glasgow both in relation to discussions of principles and ideas and to underlying experimental work. The research in the USA is supported by the National Science Foundation (Grant NSF PHY82 04056) and in the UK by the Science and Engineering Research Council.

NOTE

(1) An alternative way of looking at this system is to regard the two main cavities as being coupled via the small cavity formed between mirrors M3 and M3', with the parameters chosen to give one resonance of the coupled cavities at the frequency of the laser light and a second resonance at the sum of the laser frequency and the frequency of the expected gravitational wave.

REFERENCES

Billing, H., Maischberger, K., Rudiger, A., Schilling, S., Schnupp, L., and Winkler, W.; 1979, "An argon laser interferometer for the detection of gravitational radiation", *J. Phys.*, E12, 1043.

Douglass, D.H. and Braginsky, V.B.: 1979, "Gravitational-radiation experiments", in S.W. Hawking and W. Israel *General Relativity: an Einstein Centenary Survey*, Cambridge University Press, Cambridge, p. 90.

Drever R.W.P.: 1983,"Interferometric detectors for gravitational radiation", in N. Deruelle and T. Piran (eds) *Proceedings of the Advanced Physics Institute, Les Houches, France, in Gravitational Radiation*, North Holland, Amsterdam, p. 321.

Drever, R.W.P., Ford, G.M., Hough, J., Kerr, I., Munley, A.J., Pugh, J.R., Robertson, N.A., and Ward, H.: 1983a, "A gravity wave detector using optical cavity sensing", in E. Schmutzer (ed.) *Proceedings of the Ninth International Conference on General Relativity and Gravitation, GR-9, Jena, 1980*, Veb Deutscher Verlag der Wissenschaften, DDR-1080, Berlin, p. 265.

Drever, R.W.P., Hall, J.L., and Kowalski, F.V.: 1983b, "Laser phase and frequency stabilization using an optical resonator", *Appl. Phys.*, B31, 97.

Drever, R.W.P., Hough, J., Munley, A.J., Lee, S.-A., Spero, R., Whitcomb, S.E., Ward, H., Ford, G.M., Hereld, M., Robertson, N.A., Kerr, I., Pugh, J.R., Newton, G.P., Meers, B., Books III, E.D., and Gursel, Y.: 1983c, "Gravitational wave detectors using laser interferometers and optical cavities: ideas, principles and prospects", in P. Meystere and M.O. Scully (eds), *Proceedings of the NATO Advanced Study Institute, Bad Windsheim, West Germany, in Quantum Optics, Experimental Gravity and Measurement Theory*, Plenum, New York, p. 503.

Epstein, R., and Clark, J.P.A.: 1979, "Sources of gravitational radiation", in L. Smarr (ed.) *Sources of Gravitational Radiation*, Cambridge University Press, Cambridge, p. 477.

Forward, R.L.: 1978, "Wideband laser-interferometer gravitational-radiation experiment", *Phys. Rev.*, D17, 379.

Hough, J., Drever, R.W.P., Munley, A.J., Lee, S.-A., Spero, R., Whitcomb, S.E., Ward, H., Ford, G.M., Hereld, M., Robertson, N.A., Kerr, I., Pugh, J.R., Newton, G.P., Meers, B., Brooks III, E.D., and Gursel, Y.: 1983, "Gravitational wave detectors using laser interferometers and optical cavities: some practical aspects and results", in P. Meystere and M.O. Scully (eds), *Proceedings of the NATO Advanced Study Institute, Bad Windsheim, West Germany, in Quantum Optics, Experimental Gravity and Measurement Theory*, Plenum, New York, p. 515.

Moss, G.E., Miller, L.R., and Forward, R.L.: 1971, "Photon-noise limited laser transducers for gravitational antenna", *Appl. Opt.*, 10, 2495.

Schilling, R., Schnupp, L., Winkler, W., Billing, H., Maischberger, K., and Rudiger, A.: 1981, "A method to blot out scattered light effects, and its application to a gravitational wave detector", *J. Phys. E. Sci. Instrum.*, 14, 65.

Smarr, L.: 1979, *Sources of Gravitational Radiation*, Cambridge University Press, Cambridge.

Thorne, K.S.: 1978, "General-relativistic astrophysics", in N.R. Leibovitz, W.H. Reid, and P.O. Vandervoort (eds), *Theoretical Principles in Astrophysics and Relativity*, University of Chicago Press, Chicago, p. 149.

Thorne, K.S.: 1980, "Gravitational wave research: current status and future prospects", *Rev. Mod. Phys.* 52, 285.

Tyson, J.A., and Giffard, R.P.: 1978, "Gravitational-wave astronomy", *Ann Rev. of Astro. Astrophysics*, 16, 521.

HIGH FREQUENCY GRAVITATIONAL WAVES: ATTEMPTS TO LABORATORY DETECTION
REPORT OF WORKSHOP C1

Edoardo Amaldi

Dipartimento di Fisica, Università di Roma
"La Sapienza"
Roma, Italy

Franco Fuligni

Istituto di Fisica dello Spazio Interplanetario
CNR
Frascati, Italy

Eighteen papers were actually presented to Workshop C1, of which 8 concerned resonant antennas, 5 laser interferometric detectors and the remaining 5 various problems ranging from gravitational radiation sources to methods of detection.

1. RESONANT DETECTORS

P.F. Michelson reported on the work in progress at Stanford (Fairbank et al., 1983a). At the moment they are preparing their 4800 kg detector for cooling down, making the following improvements: 1) He^4 refrigerator, to cool the antenna to 1K and test vibration isolation concepts; 2) improved transducer and D.C. SQUID (Superconductive Quantum Interference Device). With these improvements they expect to achieve a pulse noise temperature of $T_N \lesssim 1$mK.

They are also constructing a second detector with a cryostat designed to cool the antenna to $\sim$ 3mK. With this system they expect to achieve a pulse noise temperature near the quantum limit and a sensitivity to a continuous source that will be within the detector bandwidth of $(\Delta\ell/\ell) \sim 10^{-26}/\text{day}^{1/2}$. This is of interest

B. Bertotti et al. (eds.), General Relativity and Gravitation, 413–422.

because of the recent discovery of a rapidly rotating neutron star.

Finally, it should be noted that a resonant mass detector can achieve a significant bandwidth. They envisage that a bandwidth approaching 50-100Hz is possible with present technology. Significantly more bandwidth and sensitivity could be obtained with an array of bar detectors.

Concerning the problem of detecting monochromatic gravitational radiation M. Bassan (Fairbank et al., 1983b), also from Stanford, showed that the disadvantage of operating at frequencies different from the antenna resonance can be partly compensated by using a resonant transducer tuned to the source frequency. The results of the analysis of this problem were presented with special emphasis on the tuning conditions and bandwidth limitations which are much less stringent than the Q width of the antenna.

A.G. Mann reported on the work of the Louisiana State University group (Mann et al., 1983) which is involved in the problem of minimizing the Q degradation from attachments to their high Q 5056 Al bar. They had previously conducted some room temperature tests on an Al 6061 bar of identical size (2400kg). They found no Q degradation (at the 10^7 level) due to welding. Bolting on 1-4 kg loads (transducers) caused Q degradation at the 5×10^5 level but this can be extended to about 3×10^6 if solid Stanford type flanges (and clamps) are employed. Q degradation roughly proportional to cross sectional area was seen for wires attached to the transducers: at the 2×10^6 level for several 0.17mm copper wires and 2×10^5 for two 0.8mm coaxial cables. The vibration introduced by these wires did not heat the bar, at least not by more than the measurements uncertainty (20K), provided that at least one Stanford type wire vibration isolator was used.

G. Pizzella reported on the work carried out by the Rome group (Amaldi et al., 1983a)

In April 1983 the 5056 Al bar, M=2320kg realized by the Rome group and installed at CERN was cooled for the first time. The bar, suspended by a cable, was equipped with the first prototype of a capacitive resonant transducer of new design and with a FET amplifier.

The overall cryogenic behaviour of the system was good except for a vacuum leak which only made it possible to maintain the antenna to 4.2K for a few hours. It was possible however to make measurements at temperature above 16K while the system was warming up at a rate of about 1K/day with a thermal input of the order of 0.5W. The resonance frequency of the transducer was just 1Hz below the resonant frequency of the bar allowing an energy transfer of 99%. At 16K, they measured $Q\sim 2.5\times 10^6$ for the system bar+transducer and observed a steep decrease for higher temperatures. The extrapolated value of Q to 4.2K was of the order of 10^7. The brownian noise was measured at the expected value indicating that the mechanical filters are satisfactory.

During next year they plan to cool again with the final version of the capacitive transducer and with a D.C.SQUID coupled to the transducer by means of a superconducting transformer.

The program at Maryland (Davis et al., 1983) was reported by H.J. Paik. The main research effort consists of three parts: 1) milli-Kelvin cryostat and vibration isolation; 2) three mode superconducting transducer, and 3) point contact D.C. SQUID. Three cryostats are under construction or operation. The one in the test stage has a 1200kg, Al 5056 antenna with a 3-mode superconducting transducer and an SHE D.C. SQUID. This system should be capable of giving a pulse detection sensitivity of $T_p \lesssim 10^{-3}$K. Another cryostat with a 1400kg Al 6061 antenna has been operating for over a year. It was instrumented with a capacitor transducer and exhibited $T_p = 1.8$K.

Progress has been made in the Q-measurements of the three mode transducer and the operation of the point contact D.C. SQUID.

This was reported in a second contribution also by Paik (Folkner et al., 1983). The mass ratio of 400 was chosen between neighbouring masses in the three-mass-system. This permits a resolution time of 7msec. After annealing, the niobium transducer exhibited a Q of 3×10^6 for the intermediate mass and 5×10^6 for the final conic mass. High Q's are expected when the transducer is mounted on the antenna. Electrical Q is being studied.

Noise measurements have been carried out for point-contact D.C. SQUID's. In addition to the Johnson noise, an input flux noise of $10^{-4}\sim 10^{-5}\ \Phi_0 \mathrm{Hz}^{-1/2}$ was observed. This excess noise is believed to be due to rf noise picked up by the tightly coupled input coil. If this noise could be eliminated, the Johnson noise observed would agree with the value predicted by the theoretical model and would correspond to an input energy resolution of 20 $\hbar$.

The work at Rochester (Bocko et al., 1983) was reported by L. Narici. The detector under development in Rochester features a superconducting, non resonant transducer which inductively couples the motion of the ends of an aluminium 5056 cylinder to an ultra low noise SQUID. The transducer is not in physical contact with the cylinder. This, in conjuction with the newly designed four point suspension of the bar, has given the highest mechanical Q's yet achieved in an antenna, $Q \gtrsim 20 \times 10^6$. The transducer consists of two planar inductors which are held very close ($\sim 20 \mu m$) to the ends of the aluminium cylinder. These inductors are connected together to form a superconducting loop into which a large persistent current is injected. The SQUID is connected across the planar inductors. In order to improve mechanical reliability, the group is developing a new kind of planar inductor, two prototypes of which have already been successfully tested. Also the isolation from terrestrial noise has recently been improved by the newly designed four point suspension, giving by itself an attenuation of 130dB at 2kHz. The total attenuation of the system thus results over 200dB.

The activity of the University of Western Australia was reported by D.G. Blair who described the main features of this experiment (Blair et al., 1983): 1) 1.5 tonne superconducting niobium antenna with a $Q \gtrsim 50 \times 10^6$; 2) non-contacting transducer (microwave parametric up converter); 3) two-mode vibration isolation suspension.

The main results are: a) $Q > 5 \times 10^6$ at 300K; b) Q increasing rapidly as T falls below 1K; c) long term cryogenic performance demonstrated (1 May 1983 to present).

The following advantages of the parametric transducer were underligned: 1) high coupling without Q degradation; 2) no problem of noise introduced by leads; 3) no problem of Q-degradation by screws.

The sensitivity expected in 1983: $h = 9 \times 10^{-19}$. Immediate improvements should bring it to $h = 3 \times 10^{-19}$.

2. LASER INTERFEROMETRIC DEVICES

R. Schilling presented a status report of the Munich group (Schilling et al., 1983). The work on the (old) laser interferometer (with an arm length of 3m, and with 138 passes in the delay lines) is being continued. In particular a considerable improvement was achieved at frequencies below 3kHz, mainly by better stabilization of the laser frequency and by a further reduction of mechanical noise. At 3kHz they are very close to their aim of reaching the shot noise limit. Towards lower frequencies, they observe a rise with (approximately) 1/f, which can be quite clearly distin-

The construction of a larger prototype, with 30m arms, has been completed. To start with, they have chosen N=50 bounces, leading to a total light path of 1.5km. Initial interferometric tests, begun in mid-June 1983, showed a satisfactory quality of the delay line mirrors, which will certainly allow higher numbers of bounces. In the first week of tests, they observed an interferometer noise greater than a factor of ten above shot noise. They are convinced that the main contribution comes from the strong frequency noise of the laser used (which is not yet as well stabilized as the one of the 3m-experiment).

A report by L. Schnupp followed on the possibility of enhancing the sensitivity of the GW interferometer by employment of antennas with very long light paths (Schnupp et al., 1983). The main idea is to receive GW events with duration times less than travel time of light in the antenna. It is shown that the output signal of long antennas may contain a sequence of repetitions of the received GW pulse. When all repetitions are added up, the sum attains a better discrimination against shot noise. They also showed that the gain in signal/noise ratio which is achieved by signal repetition is of the same magnitude as the square root of the intensity gain achieved by light storage in the antenna.

D.Z. Anderson reported for the Caltech group (Spero et al., 1983). Their interferometer consists of two 40m Fabry-Perot cavities at right angles to one another. The two cavities were successfully locked to a common resonance frequency in June 1982. Long term locking was obtained by September 1982. A search for the radiation emitted by the discovered 1.5msec pulsar, performed in January and February 1983, served as a rehearsal for future, more sensitive searches and as an information test of the interferometer. The noise is white above 10kHz but the sensitivity is considerably degraded around the frequencies of interest (~1kHz).

They are currently investigating the noise sources and inserting a mode selector into the optical path. Further improvement to the interferometer will include the use of new cavity mirrors that should provide a storage time of about 1msec, a factor of about 30 larger than presently exists.

The work by the Glasgow group (Ward et al., 1983) was reported by H. Ward. The 10m baseline detector under development at Glasgow uses two optical cavities arranged at right angles. The initial configuration, in which a separate detection of the signal from the triangular cavities was used, allowed the development of the high precision laser frequency stabilization techniques required, together with a study of the sources of the excess noise observed in the detector output.

The detector has now been rebuilt incorporating a number of significant improvements. Two mirror cavities are now being used to provide a simpler layout and alignment of the optics, and a polarization scheme is included in order to help prevent feedback of light to the laser. With this technique the output of the two cavities can now be optically combined, providing for the possibility of compensating for residual fluctuations in laser frequency and also giving improved output fringe contrast. Active and passive systems for reducing fluctuations in laser beam geometry have been installed and additional seismic isolation included. With these improvements it is likely that the sensitivity will be limited at low frequencies by the thermal noise of the interferometer components on the central test mass. Mounting the cavity mirrors on separate high Q masses should then allow the detector sensitivity to approach the limit due to photon shot noise.

P.S. Linsay presented a progress report from the MIT group (Benford et al., 1983). The innovations incorporated in the prototype antenna (of 1.5m arm length) are: 1) a fiber optic coupler between the laser and the antenna to clean the spatial modes of the laser beam; 2) the addition of wide band phase noise to the laser light in order to reduce the ability of scattered light in the delay lines to produce spurious signals. Concurrent with the construction of the laboratory antenna they are designing a large (5-10km arm) laser antenna. The design goal for this antenna is a strain sensitivity of 10^{-23} Hz^{-1} at 30Hz. The probable construction is a pair of L's at continental separation.

3. OTHER PROBLEMS

R.V. Wagoner (1983) from Stanford reported on the gravitational radiation from neutron stars with relatively weak magnetic fields which are accreting matter from a disk. They may acquire enough angular momentum to become unstable against the emission of gravitational radiation. When a steady state is reached (viscous damping time equals instability growth time), the metric perturbation h is related to the observed time-average X-ray flux ℓ_γ by

$$h \sim 3\times10^{-27}(1\ \mathrm{kHz/mf})^{1/2}(\ell_\gamma/10^{-8}\ \mathrm{erg\ cm^{-2}s^{-1}})^{1/2}$$

where m ~4 is the mode number and f is the frequency of the monochromatic gravitational waves. The small nonaxisymmetric distortion of the neutron star should produce some modulation of the X-ray emission at the gravitational-wave frequency. This frequency should lie in the range $200 \leqq f \leqq 300$ Hz, below the rotational frequency of the neutron star.

S. Frasca of the Rome group (Amaldi et al., 1983b) presented the results of the analysis of the data taken with two gravitational antennas, one in Frascati, the other in Geneva operated in successive periods of time for a total of 2503 hours from June 1978 to April 1982. The Fourier analysis of the data above a few thousand Kelvin gives a prominent peak at $\tau = 718.03 \pm 0.05$min in agreement with the sidereal semi-period within an experimental error of $\pm$ 3seconds. Arguments are reviewed which seem to give evidence against an interpretation in terms of gravitational radiation.

A.C. Melissinos (Reece et al., 1983) of the University of Rochester reported on a detector for high frequency gravitational effects based on a parametric converter using superconducting cavities operating at 10GHz. The performance of the detector was tested by setting the end-wall of the cavity in motion at the electrical resonant frequency of ~ 1MHz. With an electrical quality factor $Q_o = 5\times10^8$ they detected harmonic displacements of amplitude $\Delta x = 3.4\times10^{-17}$cm. Such a detector will respond directly to gravitational fields at the resonant frequency provided the dimentionless potential $h \geqq 10^{-23}$. A possible application is to high energy particle storage rings.

T. Damour reported for the group of Institut Poincaré-Observatoire de Meudon-Laboratoire de Physique des lasers (Tourrenc et al., 1983) on ultrahigh resolution laser spectroscopy of atoms as a possible probe of gravitational fields including gravitational radiation.

The possible relevance of a spectroscopic sensor of gravitational fields was put forward four years ago (Rautian and Smirnov, 1978; Nesterikhin et al., 1978).

A controversy recently arose over the question as different points of view seemed to lead to different results. The authors presented a general covariant method which gives the gravitational correction to non-linear optical spectra of atoms. The equivalence between the different points of view is explicitly proved, putting an end to the above mentioned controversy. Several experimental schemes were discussed.

Finally G. Marföldi (1983) from Budapest discussed

the derivation of the gravitational wave-parameters from general relativity introducing limit and initial values.

REFERENCES

Papers presented to the GR10 conference and published in:

Bertotti, B., de Felice, F., and Pascolini, A. (eds): 1983, *10th International Conference on General Relativity and Gravitation. Contributed Papers*, CNR, Roma, vol. 2.

Amaldi, E., Bonifazi, P., Castellano, G., Coccia, E., Cosmelli, C., Frasca, S., Modena, I., Ogawa, Y., Pallottino, G.V., Pizzella, G., Rapagnani, P., and Ricci, F.: 1983a, "Progress report on the gravitational wave experiment of the Rome group".

Amaldi, E., Bonifazi, P., Cavallari, G., Frasca, S., Pallottino, G.V., Pizzella, G., Rapagnani, P, and Ricci, F.: 1983b, "Analysis of the Frascati and Geneva gravitational antenna background exhibiting to one half sideral period".

Benford, R., Dewey, D., Linsay, P.S., Livas, J., Saulson, P., Shoemaker, D., and Weiss, R.: 1983, "Program report on the MIT gravity wave interferometer".

Blair, D.G., Bryant, J., Buckingam, M.J., Edwards, C., James, R.N., van Kann, F.J., Mann, L.D., Ferreirinho, J., Suzuki, T., and Veitch, P.: 1983, "Progress in the development of a niobium gravitational radiation antenna".

Bocko, M.F., Douglass, D.H., Gram, R.Q., Johnson, W.W., Karim, M., Muhlfelder, B., Narici, L., and Zucher, M.: 1983, "The Rochester gravitational wave detector: a progress report".

Davis, A.S., Desrosier, F., Folkner, W., Gretz, D., Krack, K., Moody, M.V., Paik, H.J., Richard, J.-P., Weber, J., and Wilmot, G.: 1983, "The gravity wave antenna program at Maryland".

Fairbank, W.M., Bassan, M., Mapoles, E., McAshan, M.S., Michelson, P.F., and Moskovitz, R.: 1983a, "Progress report on the Stanford gravitational radiation experiment".

Fairbank, W.M., Bassan, M., Mapoles, E., Mc Ashan, M.S., Michelson, P.F., and Moskovitz, R.: 1983b, "Detection of Monochromatic G.W. radiation with resonant detectors".

Folkner, W., Moody, M.V., and Richard, J.P.: 1983, "Three mode transducer and D.C. SQUID development".

Mann, A.G., Hamilton, W.O., Campisi, I.E., DeWitt, D., Oelfke, W.C., Pipes, P.B., Sibley, A., Spetz, G.W., Wang, G., and Xu, B.: 1983, "The gravitational radiation experiment at Louisiana State University".

Marföldi, G.: 1983, "Gravitational wave-parameters, derived from GR introducing limit and initial values".

Reece, C.E., Reiner, P.J., Rogers, J., and Melissinos, A.C.: 1983, "A detector for high frequency gravitational effects based on parametric convertion at 10 GHz".

Schilling, R., Schnupp, L., Winkler, W., Billing, B., Maischberger, K., and Rüdiger, A.: 1983, "Laser interferometry for gravitational wave detection: status and sensitivities for the Garching prototypes".

Schnupp, L., Winkler, W., Maischberger, K., Rüdiger, R., and Schilling, R.: 1983, "A possibility of enhancing the sensitivity of the G.W. interferometer by employment of antennas with very long Garching prototypes".

Spero, R., Anderson, D.Z., Drever, R.W.P., Gursel, Y., Gutt, G., Hereld, M., Kaufman, J., and Whitcomb, S.E.: 1983, "The Caltech gravitational wave detector".

Tourrenc, P., Damour, T., Sharma, J., and Bordé, C.J.: 1983, "Ultra-high resolution laser spectroscopy of atoms as a probe of gravitational fields including gravitational radiation".

Wagoner, R.V.: 1983, "Monochromatic gravitational radiation from accreting neutron stars".

Ward, H., Drever, R.W.P., Hough, J., Munley, A.J., Newton, G.P., Meers, B.J., Hoggan, S., and Kerr, G.A.: 1983, "Development of a long baseline laser interferometric gravitational radiation detector using resonant optical cavities".

other references

Nesterikhin, Y.U.E., Rautian, S.G., and Smirnov, G.I.: 1978, "Laser detectors of gravitational waves", *Sov. Phys. JETR*, 48, 1

Rautian, S.G., and Smirnov, G.I.: 1978, "Non linear resonances of accelerated atoms and molecules", *Sov. Phys. JETR*, 47, 678.

LABORATORY EXPERIMENTS
REPORT OF WORKSHOP C3

H. Hirakawa

Department of Physics,
The University of Tokyo
Bunkyo, Tokyo 113, Japan

The test of the laws of gravitational force and the test of the equivalence principle were the main topics of the Workshop on Laboratory Experiments.

1. GRAVITATIONAL FORCE

Fujii reviewed some of the theoretical ideas (Fujii, 1974, 1981, 1982, 1983) from which one expects a possible deviation from the inverse-square law of the static gravitational force in the laboratory distance scale (Long, 1981). The Einstein gravity, like the electromagnetism, yields the long-range force because of the underlying gauge symmetry, the translational invariance. However, a number of alternatives to the classical and macroscopic theory of Einstein have been proposed from the point of view of microscopic, and hence quantum, physics. As an example, the scalar-tensor theory reinterpreted in this context gives a force from the scalar field which is more likely massive than being massless. There is an interesting theoretical possibility predicting a crude order-of-magnitude estimate for the force-range, $\sim$ (Newton's constant)$^{-1/2}$ $\times$ (typical particle mass $\sim$ 1 GeV)$^{-2}$ $\sim$ 10^5cm (in the unit system with $c = \hbar = 1$). It was emphasized that this additional force can occur in a manner fully consistent with general covariance and the equivalence principle. Other theories resulting in the similar effect were also reviewed (Wagoner, 1970; O'Hanlon, 1972; Matsuyama et al., 1979).

B. Bertotti et al. (eds.), General Relativity and Gravitation, 423–429.

Hoskins and Newman (1983) presented the result of a new experiment on the inverse-square law which compared the gravitational forces generated by a mass of 44 g at distance r = 5 cm and by two masses of each 7 kg at 105 cm, using a test body, a 60-cm long bar of mass 523 g, suspended by a torsion fiber with oscillation period 30 min. The result is given in terms of Newton's gravitational constant with possible dependence on r,

$$G(r) = G_0(1 + \varepsilon \ln r), \quad \varepsilon = (-1\pm3) \times 10^{-4}.$$

The experimental limits are due to seismic noises. This is compared with the results obtained earlier by Long (1976), $\varepsilon = (20\pm4) \times 10^{-4}$, at r = 4.5 cm and 30 cm. Ogawa, Tsubono and Hirakawa (1983) reported the result of a dynamic gravity experiment (Ogawa et al., 1982), in which a 400-kg steel bar is rotated at 30 Hz and the resulting dynamic field is measured by a resonant 1400-kg square-plate antenna at 60 Hz, as a function of the distance r between the rotor and the antenna. The range of r extended from 2.6 m to 10.7 m. The result, after correcting for the finite size of the bodies, is presented in a form of gravitational potential, $1/r^{1+\delta}$, with $\delta = (2.1 \pm 6.2) \times 10^{-3}$, the accuracy being limited by the Brownian motion of the antenna and the metrical uncertainties. A future extension of the range of experiment up to r = 60 m, with a 4000-kg rotor and a cryogenic antenna was suggested. Experiments of this type measure the gravitational force directly and are able to observe the possible vacuum polarization process, which might evade null-experiments of certain kind (Long, 1980).

In the χ-g framework (Ni, 1977), the gravitational coupling to the electromagnetic field is given by χ which has 21 independent components in general. Ni (1983) argued that only one component of χ is not constrained by the present experiments and observations. If only this component is different from the metric theory, it is the counter-example proposed earlier to the Schiff's conjecture (Ni, 1977). This theory can be formulated as a torsion theory, which gives non-Einstein spin-dependent forces. To test the theory he proposed experiments using a toroidal test-body made of two kinds of magnetic materials wrapped with μ-metals so that the magnetic field is confined and the test-body has a net spin. Graham and Newman (1983) reported on the progress of their experiment on a similar line, testing the role of spin in gravitation. They use a cryogenic torsion balance to search for anomalous long-range interactions between samples with polarized electron spins. The anomalous interactions would normally be masked by the magnetic forces associated with aligned spins. To discuss the sensitivity of this experiment they defined α to be the ratio of an anomalous torque to the unshielded magnetic torque between the two spin samples. An existing limit, $\alpha < 10^{-8}$, may be inferred from the agreement of the measured electron g-factor with the QED prediction. They plan to improve this limit by three to six orders of magnitude, reducing the magnetic coupling by a factor of the order of 10^{15}. This will be accomplished with a combination of techniques such as encasing the sample in superconducting shields, and using ring-shaped magnets consisting of $GdNi_5$ and NdNi semirings, so that when the ring is magnetized the flux lines of B circulate continuously around the ring yet the two halves of the ring have spin alignments in the same direction. An attenuation of magnetic interaction of the order of 10^{13} is achieved so far.

2. EQUIVALENCE PRINCIPLE

The Einstein equivalence principle consists of three parts: 1) weak equivalence principle (WEP) which has been verified to high accuracy by Eötvös-type experiments, 2) local Lorentz invariance (LLI) and 3) local position invariance (LPI). The usual gravitational red-shift for two identical clocks of frequency f is given by $z = \Delta f/f = \Delta U/c^2$ where ΔU is the difference of gravitational potential for two locations. If LPI were violated, assuming that this violation is due to non-universal coupling to an external gravitational potential U, we would expect the locally measured frequency f_A of clock A to be

$$f_A = (f_A)_0 [1 - \alpha_A \, U/c^2 + O((U/c^2)^2)] ,$$

where $(f_A)_0$ is the frequency of A relative to the standard clock (of a different type) at some fiducial spacetime location from which U is measured and α_A is a dimensionless parameter that measures the extent of LPI violation for clock A. If LPI is valid $\alpha_A = 0$ for any clock type. For a comparison of two clocks of type A and B, the relative frequency would vary according to

$$f_A/f_B = (f_A/f_B)_0 [1 - (\alpha_A - \alpha_B) U/c^2].$$

Such a comparison, a "null" gravitational red-shift experiment, is therefore a powerful test of LPI and of the equivalence principle. Turneaure, Will, Farrell, Mattison and Vessot performed an experiment (Turneaure et al. 1983a,b) by comparing the frequencies of two hydrogen masers with those of three superconducting-cavity stabilized oscillators (SCSO's) as a function of solar gravitational potential. The experiment was carried out at Stanford University in April 1978. During this period the solar gravitational potential at the laboratory varied diurnally with an amplitude of 3 parts in 10^{13} because of rotation of the earth and also varied approximately linearly with time at 3 parts in 10^{12} per day due to the earth's eccentric orbital motion. An upper limit on the relative frequency variation of 1.7 parts in 10^2 of the external potential was set. The accuracy was limited by the frequency stability of the clocks and by unmodeled environment effects. The result is consistent with LPI and the Einstein equivalence principle at the 2% level.

Keyser, Hoskins and Faller (1983) reported on the progress of a precise Eötvös-type experiment testing the equivalence principle (WEP). They arranged 24 test masses, 12 of copper and 12 of lead, around the perimeter of a cylinder. The total mass of each 12-piece test mass is 480 kg. The cylinder, 1.27 m in diameter, floats on water at temperature 3.98°C. There are balls on the lid of the float and electrically insulated balls under the lid of the float container. AC potentials across them serve to center the float and to give the torque. A breakdown of equivalence principle would result in a difference in the acceleration of the two sides of the float toward the sun, and this would evidence itself as a 24-hour-period variation in the position of the float. The sensitivity of this type of apparatus can be expected to improve with size D, scaling the fundamental thermal-noise limited sensitivity as $D^{5/2}$. For their apparatus the calculated one-standard-deviation Brownian-motion limit on the equivalence of gravitational and inertial masses is $\eta \sim 7 \times 10^{-14}$ day $^{-1/2}$. In both the Princeton ($|\eta| < 3 \times 10^{-11}$) and Moscow ($9 \times 10^{-13}$) experiments (Roll et al., 1964; Braginskii et al., 1971), where masses were suspended by a quartz fiber, seismic disturbances turned out to be a primary noise limit for the system. To this type of noise the fluid approach seems to be remarkably insensitive. Improved results on the equivalence principle is expected in due time.

3. CONCLUSION

The last decade has seen the explosive growth of the field of gravity experiments. Developments in experimental technique and theoretical analysis have stimulated renewed interests in laboratory experiments. Hopefully, before the next GR11 conference the experiments now in progress as well as new experiments will establish interesting results, particularly on the inverse-square law, the spin-dependent forces and the equivalence principle.

REFERENCES

Papers presented to the GR10 conference and published in:

Bertotti, B., de Felice, F., and Pascolini, A. (eds.): 1983, *10th International Conference on General Relativity and Gravitation. Contributed Papers*, Vol. 2., CNR, Roma.

Anandan, J.S.: 1983, "Relativistic gravitational effects on the magnetic field produced by an electric current", p. 1019.

Chiba, J.: 1983, "Generation and detection of dynamic Newtonian gravitational fields and Morse-code communication in very-near-zone", p. 972.

Fujii, Y.: 1983, "On the theoretical aspects of the possible finite-range gravitational force".

Graham, D., and Newman, R.D.: 1983, "Experimental test of the role of intrinsic spin in gravitation", p. 981.

Hoskins, J.K., and Newman, R.D.: 1983, "Test of the inverse square law of gravity at distances of 5 cm and 1 m", p. 984.

Keyser, P.T., Hoskins, J.K., and Faller, J.E.: 1983, "New laboratory test of the equivalence principle", p. 993.

Ni, W.-T.: 1983, "Equivalence principles, torsion and polarized test-body experiments", p. 997;

Ogawa, Y., Tsubono, K., and Hirakawa, H.: 1983, "Experimental test of the law of gravitation", p. 999.

Pappas, P.T.: 1983, "A simple laboratory experiment limits relativistic theories", p. 1002.

Turneaure, J.P., Will, C.M., Farrell, B.F., Mattison, E.M., and Vessot, R.F.C.: "1983a, "Test of the principle of equivalence by a null gravitational red-shift experiment", p. 1009.

Unruh, W.G.: 1983, "Squezeed state technique in precision measurement of oscillation".

Other references

Braginskii, V.B., and Panov, V.I.: 1972, "Verification of the equivalence of inertial and gravitational mass", *Sov. Phys. - JETP*, 34, 463.

Fujii, Y.: 1974, "Scalar-tensor theory of gravitation and spontaneous breakdown of scale invariance", *Phys. Rev.*, D9, 874.

Fujii, Y.: 1981, "Composition indipendence of the possible finite-range gravitational force", *Gen. Rel. Grav.*, 13, 1147.

Fujii, Y.: 1982, "Origin of the gravitational constant and particle masses in a scalar-invariant scalar-tensor theory", *Phys. Rev.*, D26, 2580.

Long, D.R.: 1976, "Experimental examination of the gravitational inverse square law", *Nature*, 260, 417.

Long, D.R.: 1980, "Vacuum polarization and non-Newtonian gravitation", *Nuovo Cimento*, 55 B, 252.

Long, D.R.: 1981, "Current measurements of the gravitational 'Constant' as a function of the mass separation", *Nuovo Cimento*, 62 B, 130.

Matsuyama, S., and Miyazawa, H.: 1979, "Long range forces between hadrons", *Prog. Theor. Phys.*, 61 942.

Ni, W.-T.: 1977, "Equivalence principles and electromagnetism", *Phys. Rev. Lett.*, 38, 301.

Ogawa, Y., Tsubono, K., and Hirakawa, H.: 1982, "Experimental test of the law of gravitation", *Phys. Rev.*, D26, 729.

O'Hanlon, J.: 1972, "Intermediate-range gravity: a generally covariant model", *Phys. Rev. Lett.*, 29, 137.

Roll, P.G., Krotkov, R., and Dicke, R.H.: 1964, "The equivalence of inertial and passive gravitational mass", *Ann. Phys. (New York)*, 26, 442.

Turneaure, J.P., Will, C.M., Farrell, B.F., Mattison, E.M., and Vessot, R.F.C.: 1983b, "Test of the principle of equivalence by a null gravitational red-shift experiment", *Phys. Rev.*, D27, 1705.

Wagoner, R.V.: 1970, "Scalar-tensor theory and gravitational waves", *Phys. Rev.*, D1, 3209.

Quantum Gravity

FREE DIFFERENTIAL ALGEBRAS AND RELATIVISTIC THEORIES [+]

Tullio Regge*

CERN
Geneva 23
Switzerland

Local lagrangian field theories can be conveniently expressed in terms of differential algebras. It has been known since the work by Cartan and subsequently by Sciama, Utiyama and Kibble, that general relativity can be cast into the language of differential forms by introducing the concept of tetrad or vierbein. In this formalism instead of the metric tensor one introduces the forms

$$V^a = V^a_\mu dx^\mu \qquad V^a_\mu V^b_\nu \eta_{ab} = g_{\mu\nu} \tag{1}$$

$$\omega^{ab} = \omega^{ab}_\mu dx^\mu$$

satisfying Cartan's structural equation

$$\begin{aligned} R^a &= dV^a - \omega^{ab} \wedge V_b = 0 \\ R^{ab} &= d\omega^{ab} - \omega^{at} \wedge \omega^b_t \end{aligned} \tag{2}$$

and having the Einstein-Cartan action:

$$S = \text{const.} \int R^{ab} \wedge V^c \wedge V^d \varepsilon_{abcd} \,. \tag{3}$$

The task of obtaining a local lagrangian theory

B. Bertotti et al. (eds.), General Relativity and Gravitation, 433–437.

from a given Lie group is usually called the "gauging of groups". The example of general relativity corresponds to the gauging of the Poincaré group. One can indeed consider the forms V^a and ω^{ab} as Yang-Mills potentials for the Poincaré group.

There are other theories which have similar formal properties. The best known of these is the celebrated N=1 supergravity in which one gauges the Poincaré supergroup. In this theory besides the standard vierbein V^a and spin connection one also has a spinor potential Ψ for the gravitino modes. The structural equations are then:

$$R^a = dV^a - \omega^{ab} \wedge V_b - \frac{i}{2}\bar{\Psi}\Gamma^a\Psi$$

$$R^{ab} = d\omega^{ab} - \omega^{at} \wedge \omega_t^{\ b}$$

$$\rho = D\Psi = d\Psi + \frac{1}{4}\Gamma_{ab}\omega^{ab} \wedge \Psi$$

and besides defining the standard curvature and torsion R^a, R^{ab} also define a spinor curvature ρ. The action is then given by

$$S = \int (R^{ab} \wedge V^c \wedge V^d \varepsilon_{abcd} + 4\bar{\Psi}\gamma_5\gamma_a\rho V^a). \qquad (5)$$

I should like to remind you that in its original form these theories were written on a 4-dimensional differentiable manifold, locally Minkowsky. There is however another approach, the so called "group manifold approach", originally proposed by the author and Ne'eman, in which equations are written directly on a differentiable manifold which is locally diffeomorphic to the group manifold. In the conventional approach one has to introduce a number of constraints on the curvature by hand. If we work on the group manifold these constraints appear naturally from the variational principle. A distinct disadvantage of the group manifold approach is that no quantum theory has been developed so far; in fact the theory has the largest possible set of fields and of corresponding constraints. The theory is therefore useful whenever we need a geometrical interpretation of the fields.

The main idea is that fields are all represented by suitable forms. In the original setting these were all forms of degree 1 which could be interpreted as Yang-Mills potentials of a suitable Lie group. If the set of forms is then ω^A we have curvatures defined as:

$$R^A = d\omega^A - \frac{1}{2} C^A{}_{BF}\, \omega^B \wedge \omega^F \tag{6}$$

where the C^A_{BF} are the structure constants of the Lie group. These formulae include as a particular case Cartan's structural equations. Left invariant forms on the group manifold will have vanishing curvatures. The corresponding equations $R^A=0$ are then known as the Maurer-Cartan equations. One states then general rules for the construction of an action. These rules work in a finite number of steps, the algorithm stops after a finite number of operations and checks and either produces an action or the program stops with no result. If the action exists, one has a general rule for the construction of the transformation of the fields under supersymmetry which works automatically for on shell configurations. It is not clear how these procedures will work off shell, the corresponding algorithm does not exist although there is a number of interesting results with hints toward a general rule.

The "Axioms" for the gauging of the groups are the following:

1. The action density is a polynomial in the forms constructed in the exterior algebra. Only the product $\wedge$ and differentiation d rules are permitted. The Hodge dual * is not allowed.
2. The degree of the action density p does not match the dimension of the group manifold. The action is obtained by integrating the density over an arbitrary submanifold M of the appropriate dimension. It is then required that the action be stationary with respect to all independent variations of all fields and to all choices of the submanifold. In this sense the imbedding of the manifold is also a field. However variation with respect to these fields does not lead to any independent equation. This result follows from the general coordinate invariance of the theory.
3. The configuration with all vanishing curvatures must be a solution of the field equations. We call it "flat space". This flatness is relative to the choice of the group G. Minkowsky space is flat with respect to the Poincaré group but not with respect to the Poincaré supergroup. Viceversa, the conventional superspace of Salam-Strathdee is flat with respect to the Poincaré supergroup.
4. The flat solution may not be the only one (non rigidity). Gravity in 3 dimensions is indeed rigid for it

admits only flat space as solution. Supergravity with extra terms (see Regge, 1983, for a discussion) is also rigid. The requirement that the theory should not be rigid is usually very restrictive and renders the action density practically unique if taken with Axiom 3.

5. Rheonomy. It states that a complete field configuration on the whole group manifold is determined, modulo inessential gauge and coordinate transformations, by the restrictions on an initial submanifold. More precisely the normal components of the curvature are expressible in terms of the tangential ones. From the algebraic point of view rheonomy seems to follow from Axiom 4.

The point of this note is to state that the scheme above can be naturally enlarged by introducing forms of arbitrary degree. In this case the Maurer-Cartan equations must be replaced with a set of differential equations forming a so called free differential algebra (FDA). From the physical point of view the expansion is natural and indeed already done in practice. The interesting 11-dimensional theory of Cremmer-Julia-Schaerk (CJS) contains, besides the usual forms, also a potential A of degree 3 which is essential in the discussion of the symmetry of the theory as well as in the dimensional reduction techniques.

A 6-dimensional supergravity, developed recently, also contains B, a 2-form potential which extends the conventional Maxwell field to higher dimensions. I shall discuss this last theory for its simplicity although the 11-dimensional one is of course far more interesting from the physical point of view. The curvatures in this theory are given by:

$$\begin{aligned} R^{ab} &= d\omega^{ab} - \omega^{at} \wedge \omega_t{}^b \qquad a,b=1\ldots6 \\ R^a &= DV^a - \frac{i}{2}\overline{\Psi} \wedge \Gamma^a \Psi \\ \rho &= D\Psi \\ R^{\otimes} &= dB - \frac{i}{2}\overline{\Psi} \wedge \Gamma^a \Psi \wedge V_a \end{aligned} \tag{7}$$

and the action is then found to be:

$$S=\int (R^{ab}\wedge\frac{1}{4}\varepsilon_{abc_1\ldots c_4}V^{c_1}\ldots V^{c_4}+$$

$$+3iR^a\bar{\Psi}\Gamma^b\Psi\wedge V_b\wedge V_a-$$

$$-3iR^{\otimes}\bar{\Psi}\Gamma^a\Psi V_a-i\bar{\rho}\Gamma_{abc}\wedge\Psi\wedge V^a\wedge V^b\wedge V^c+h.c.$$

$$\pm 2\sqrt{3}\ R^a\wedge R^{\otimes}\wedge V_a)\cdot$$

The theory can be dimensionally reduced to standard 5-dimensional supergravity and then to N=1 supergravity in d=4. All the Axioms 1-5 are then respected provided one replaces the Cartan-Maurer curvatures with the FDA ones in (7). The system (7) with all $R^A=0$ is then a FDA. It is completely integrable.

I cannot give here a complete account of FDA, the interested reader should consult van Nieuwenhuizen (1982) and Sullivan (1977). It suffices to say that they are also very natural mathematical extensions of the Maurer-Cartan algebras. Sullivan has investigated these objects extensively and given a set of general rules for their construction. It is however conceivable that they will have a role in physics similar to that of ordinary Lie groups.

NOTES

+ This is an extended abstract of the talk given at Padova. The details can be found in Regge (1983).

* On leave of absence from Istituto di Fisica Teorica dell'Università, Torino, Italy.

REFERENCES

Regge, T.: 1983, Lecture notes, Ecole d'été de Les Houches, to be published.

Sullivan, D.: 1977, "Infinitesimal Computations in Topology", IHES Publication n. 47, Bures sur Yvette, pp. 269-331.

van Nieuwenhuizen, P.: 1982, "Free graded differential superalgebras", CERN preprint TH3499.

TOPICS IN QUANTUM GRAVITY

Bryce S. DeWitt

Department of Physics
University of Texas
Austin, Texas 78712, USA

I wish to share with you today three separate developments in quantum gravity theory that have excited my interest during recent months. The first of these is a computer search, being carried out by my student Richard Rohwer, for a solution representing the Big Bang era, of the dynamical equations generated by a proposed approximate effective action for quantum gravity. The second is a surprising new development in the theory of the effective action due to Gregory Vilkovisky of Moscow. The third is a new attempt to confront and resolve the old question whether the notion that space may undergo spontaneous changes in topology can make any sense. It is being carried out in collaboration with my student Arley Anderson.

1. NUMERICAL TEST OF A PROPOSED EFFECTIVE ACTION FOR QUANTUM GRAVITY

The effective action for quantum gravity, to second order in fluctuations around flat spacetime, is uniquely determined by the requirement of gauge invariance (invariance under the diffeomorphism group) to have the functional form

$$\Gamma[g] = S[g] +$$

$$+\int d^4x \int d^4x' \Big[\tfrac{1}{2}\tilde{\Sigma}_1\big((x-x')^2\big) C_{\mu\nu\sigma\tau}(x) C^{\mu\nu\sigma\tau}(x') -$$

B. Bertotti et al. (eds.), General Relativity and Gravitation, 439–451.

$$-\frac{1}{6}\tilde{\Sigma}_2\left((x-x')^2\right)R(x)R(x')\Big] .$$

S [g] is the classical action, $C_{\mu\nu\sigma\tau}$ is the linearized Weyl tensor, R is the scalar curvature, and $\tilde{\Sigma}_1$ and $\tilde{\Sigma}_2$ are the Fourier transforms of the two independent form factors Σ_1 and Σ_2 possessed by the graviton. In momentum space the poles of the graviton propagator are located at the zeros of the functions

$$p^2+\mu_P^{-2}p^4\Sigma_1(p^2) \text{ and } p^2+\mu_P^{-2}p^4\Sigma_2(p^2)$$

where μ_P is the Planck mass.

The expressions for $\tilde{\Sigma}_1$ and $\tilde{\Sigma}_2$ obtained in one--loop perturbation theory make the above integral logarithmically divergent. This divergence can be renormalized only by introducing nonclassical counter terms and a logarithmic arbitrariness in the functions Σ_1 and Σ_2. In DeWitt (1981b) it was proposed to choose for $\tilde{\Sigma}_1$ and $\tilde{\Sigma}_2$ functions having the form

$$\text{const}\times\mu_P^2\int_0^1 \frac{d\xi}{[(x-x')^2-\xi\lambda_P^2+i\varepsilon]^2} \quad (\lambda_P=\text{Planck length}).$$

which is suggested by computations in which infinite sets of Feynman graphs are summed and which gives substance to the old idea that a quantized gravitational field "smears" the light cone and acts as its own regulator. The above ansatz yields the finite unambiguous momentum-space forms

$$\Sigma_{1,2}(p^2)=A_{1,2}\left[-\frac{\pi i}{2}\frac{H_1^{(2)}\ (\lambda_P^2p^2-i\varepsilon)^{\frac{1}{2}}}{(\lambda_P^2p^2-i\varepsilon)^{\frac{1}{2}}}-\frac{1}{\lambda_P^2\ p^2-i\varepsilon}\right]$$

which are complex-valued when p is spacelike and real-valued when p is timelike. In order that these form factors give rise to no ghosts, the constants $A_{1,2}$, which are in principle determined by all the other fields in nature, must avoid positive values lying between 0.024 and 0.054 (timelike ghost region), as well as an infinite set of discrete values lying between -0.011 and 0.009 and clustering around zero (tachyon ghost region).

Even when the constants $A_{1,2}$ assume allowed values, the functions $p^2+\mu_P^{-2}p^4\Sigma'_{1,2}(p^2)$ still have an infinity of zeros in the complex p^0 plane. These zeros, however, are harmless, for the following reason: they all lie in the first and third quadrants and correspond to the temporal behavior

$$\exp[\pm i(\omega+i\gamma)t] \ , \quad \omega,\gamma>0.$$

Such behavior *appears* to signal the existence of unstable modes in the solutions of the effective field equations

$$\delta\Gamma/\delta g_{\mu\nu}=0.$$

However, one must remember how the effective action is to be used. Its exponential $\exp(i\Gamma)$, is the transition amplitude from a coherent "in" state to a coherent "out" state. The Feynman boundary conditions for such an amplitude quench any positive frequency modes which blow up in the "out" region (as $t\to\infty$) and any negative frequency modes which blow up in the "in" region (as $t\to-\infty$), i.e., precisely the above modes. Another important consequence of the location of the zeros of the functions $p^2+\mu_P^{-2}p^4\Sigma_{1,2}(p^2)$ is that *Wick rotation is not possible*. If the above ansatz has any correspondence with reality then *the Euclidean program for quantum gravity cannot be correct*. (In Euclidean quantum gravity there is no light cone to be smeared).

A yet more important feature of the above ansatz is that $\Sigma_{1,2}$ go rapidly to zero as $|p^2|$ rises above μ_P^2. This has the consequence that the quadratic part of the effective action already determines the high energy behavior of all vertex functions, through the Ward identities. The trans-Planckian behaviour of the effective action is therefore well-described by the following generally covariant generalization of the quadratic effective action:

$$\Gamma[g]=\mu_P^2\int g^{\frac{1}{2}}R\,d^4x+\mu_P^2\int d^4x\int d^4x'\,g^{\frac{1}{2}}g'^{\frac{1}{2}}$$

$$\times\left[\frac{i}{\sigma(x,x')-\frac{1}{2}\lambda_P^2+i\varepsilon}-\frac{i}{\sigma(x,x')+i\varepsilon}\right]$$

$$\times\left[\frac{1}{4}A_1 g^{\mu\alpha'}g^{\nu\beta'}g^{\sigma\nu'}g^{\tau\delta'}C_{\mu\nu\sigma\tau}C_{\alpha'\beta'\gamma'\delta'}-\frac{1}{12}A_2RR'\right].$$

$C_{\mu\nu\sigma\tau}$ and R are the full Weyl tensor and curvature scalar respectively, $\sigma(x,x')$ is half the square of the geodesic distance between x and x' (the so-called *world-function*), $g^{\mu\alpha'}$ is the geodesic parallel displacement matrix, and $g=-\det(g_{\mu\nu})$.

An attempt has been underway for many months to solve the effective field equations generated by this effective action, on a computer. It was decided to search first for solutions having the metric form of a spatially-flat Friedman-Robertson-Walker universe. This decision was made for several reasons. Firstly, any solutions found would provide an immediate probe of the Big Bang. Secondly, such a search would constitute an acid test of the ansatz for Γ. There would be no shearing effects to help one avoid the classical singularity. The only matter term allowed in the effective field equations would be one corresponding to pure radiation. Thirdly, the Weyl tensor term would drop out of the effective field equations (FRW universes are conformally flat) and along with it the difficult-to-compute parallel displacement matrix factors $g^{\mu\alpha'}$. Fourthly, spatial flatness (which corresponds well with the real universe even during the Big Bang) makes the spatial integrals performable analytically by elementary contour integration.

The first problem encountered was the computation of the world function σ. This function satisfies a simple Hamilton-Jacobi equation and the first idea was to solve the equation (with appropriate boundary conditions) by a straightforward differencing scheme, thus avoiding the need to solve the geodesic equations geodesic by geodesic. Preliminary computer runs gave nothing but hash. At first it was suspected that caustics, which occur in the FRW setting, were to blame. But then a stability analysis was made of the difference equations that were being used. It caused considerable surprise when it was discovered that *there exists no stable differencing scheme for the Hamilton-Jacobi equation*. Present understanding of this fact runs along the following lines: The Hamilton--Jacobi equation is a delicate equation. Its solutions have elaborate and highly complicated branch structures, generating whole families of "dynamical trajectories". Consider one particular trajectory and another infinitesimally close to it. Even very simple dynamical systems can display chaotic behavior, in which two trajectories that are initially close together become widely separated with time. The very randomness that is introduced by approximating differential equations by difference equations serves to accentuate this chaotic behavior.

The following way out of this difficulty was chosen: it was observed that it is not really the light cone that is being smeared in the original ansatz for Γ, but the behavior of a Green's function *on* the light cone. In flat spacetime σ is the reciprocal of the massless scalar Green's function. Why not simply smear the scalar Green's function? In fact, why not replace σ by the reciprocal of the *conformally invariant* massless scalar Green's function? This is what was done. The advantage of choosing the conformally invariant Green's function is that it can be expressed *analytically* in an FRW universe, the latter being conformally flat.

With this choice, which greatly simplified the computerprogram, it was possible to begin rapid and efficient computer runs. According to the basic interpretation of the effective action, all runs have been and will continue to be made with end-point boundary conditions on the classical (coherent-state) values of the scale factor of the universe rather than with Cauchy data at an initial instant. (In any case it is not clear what one would use for Cauchy data since the effective field equations are nonlocal *integro*-differential equations). Once a solution is found it will be inserted into the expression for Γ and used to obtain the value $\exp(-2\mathrm{Im}\Gamma)$ giving the probability for transition from the given initial state to the given final state. The initial and final times will be chosen so as to straddle what, classically, would be the singularity at the Big Bang. $\exp(-2\mathrm{Im}\Gamma)$ will then yield the probability for tunnelling through the Big Bang.

Initial computer runs were designed as follows. All the classical terms of the effective field equation were placed on the left and all the quantum terms were placed on the right. A zeroth order time-symmetric non-singular smooth "trial" form for a solution was guessed and inserted into the right hand side. This served as a "source" for the left hand side, and a new, first order, trial solution was generated by integrating from the moment of time symmetry. The procedure was then iterated. Not surprisingly, perhaps, it failed to converge to any well defined function and gave results which, although not exhibiting obvious instabilities, departed more and more widely from the initial trial function.

When one is dealing with a wholly new equation, whose solutions have properties of which one is totally ignorant, one would like to have recourse to a relaxation scheme based on a variational principle. In the case of the effective action, although the field equations are derived from a variational principle, it is not a *minimum* principle. In order to have a minimum principle it was resolved to replace the equation $\delta\Gamma/\delta a=0$ (a=scale factor of the universe) by

$$\delta\overline{\Gamma}/\delta a=0, \qquad \overline{\Gamma}\overset{\text{def}}{=}\int|\delta\Gamma/\delta a'|^2 dt$$

$$(t=\text{selected time coordinate}).$$

This equation, which is unfortunately more complicated than $\delta\Gamma/\delta a=0$, has been programmed, and runs are now underway. A straightforward relaxation scheme has been set up which attempts to bring the value of $\overline{\Gamma}$ as close to zero as possible. There are two complications that must be dealt with: (1) Γ is a complex-valued functional, and the scale factor a (and hence the metric itself) is complex. The computer must therefore do complex arithmetic, and branch-point behavior must be taken into account. (2) Each solution of $\delta\Gamma/\delta a=0$ is isolated in a two-parameter family of solutions of $\delta\overline{\Gamma}/\delta a=0$ which, except for the solution of $\delta\Gamma/\delta a=0$ itself, do not make $\overline{\Gamma}$ vanish. To find solutions of $\delta\Gamma/\delta a=0$ one must break each stage of the relaxation process into two parts: (a) Relaxation toward a stationary solution with fixed end-point values. (b) Relaxation of the end-point values (but *not* the end point derivatives) by searching for the best constant complex number to add to the solution of $\delta\overline{\Gamma}/\delta a=0$. In this way solutions of $\delta\Gamma/\delta a=0$ have been approached, but there are still difficulties with the program and adequately small values of $\overline{\Gamma}$ have not yet been reached.

It is planned to keep on with the effort until either good solutions of the effective field equations are found or else it becomes evident that no smooth solutions exist and that the chosen ansatz for Γ does not succeed in removing singularities from gravity theory. If good solutions *are* found, attempts will be made to construct curves giving the Big Bang tunnelling probability and the associated particle production for various values of the constant A_2, as well as for various values of other adjustable parameters in the theory such as the cosmological constant.

2. THEORY OF THE EFFECTIVE ACTION

A systematic extension of the widely used background field method to multiloop process now exists (see DeWitt, 1981a). In this extension the classical background field, useful in one-loop calculations, is replaced by the quantum c-number *mean* field. New Feynman rules for non-Abelian gauge fields as well as the gravitational field are introduced which yield effective actions that are gauge invariant. Two independent calculations of the Callan-Symanzik β function to two-loop order for the Yang-Mills field have been carried out using the new rules (Hart, 1981; Abbott, 1981). The renormalization procedure was found to be greatly simplified. It was not necessary to renormalize non-gauge-invariant parts of the effective action separately, and BRS techniques were completely bypassed.

Until a few months ago a major question still remained. Although the effective action Γ obtained by the new rules is gauge invariant in the sense that it has the same local symmetries as the classical action, its functional form nevertheless depends on the choice of the gauge breaking terms that have always to be introduced. Is it possible to find a *unique* gauge invariant effective action whose form is independent of the choices of gauge breaking terms that one makes? It was believed by most experts that the answer to this question is *no*. Gregory Vilkovisky in the USSR has recently shown that the experts were wrong (Vilkovisky, 1983).

Vilkovisky's proof is an ingenious one and makes use of an idea that is useful even when no gauge group is present. The standard theory of the effective action begins with the "in out" amplitude written as a functional integral

$$e^{iW[J]} = N\int e^{i(S[\phi]+J_i\phi^i)}\mu[\phi]\,d\phi.$$

where $S[\phi]$ is the classical action, $\mu[\phi]$ is an appropriate integration measure, N is a normalization constant, and J_i are external sources coupled directly to the field components ϕ^i. (Summation-integration is understood over repeated indices). One then introduces the *mean* or *effective* field

$$\bar{\phi}^i = \langle \underset{\sim}{\phi}^i \rangle = \delta W/\delta J_i$$

and carries out the Legendre transform $\Gamma = W - J_i\bar{\phi}^i$, obtaining the effective field equations (with sources present)

$$\delta\Gamma/\delta\bar{\phi}^i = -J_i.$$

This procedure for defining the effective action is equivalent to an implicit scheme based on solving the functional integral equation

$$e^{i\Gamma[\phi]} = N\int e^{i\{S[\phi]+(\delta\Gamma[\overline{\phi}]/\delta\overline{\phi}^i)(\overline{\phi}^i-\phi^i)\}}\mu[\phi]d\phi,$$

in which the sources do not appear. This equation can be solved (for Γ) by iteration, with results that yield the usual Feynman rules. This very equation, however, points up the inelegance of standard procedures and their inability to deal with global questions. From the point of view of the geometry of the space of histories it is a ridiculous equation. The term involving $\overline{\phi}^i-\phi^i$ in the exponent of the integrand has no geometrical meaning whatever, particularly when the bundle of which the field is a cross section is not a vector bundle (e.g., in the case of the nonlinear σ model). The effective action defined in the above manner is plainly not invariant under changes of the basic field variables i.e., replacement of the ϕ^i by smooth nonlinear local functions of themselves.

Vilkovisky proposes to replace the above defining equation for Γ by the following:

$$e^{i\Gamma[\phi]} = N\int e^{i\{S[\overline{\phi}]+\Gamma_{;i}[\overline{\phi}]\sigma_{;}{}^{i}[\overline{\phi},\phi]\}}\mu[\phi]d\phi.$$

Here $\sigma[\overline{\phi},\phi]$ is the world function on the space of histories (!), based on an ultralocal metric for this space determined by the coefficient of the kinetic energy term in the classical action. Semicolons denote covariant functional differentiation with respect to $\overline{\phi}^i$. Since Γ and σ are scalars $\Gamma_{;i}$ and $\sigma_{;i}$ are just ordinary gradients, $\sigma_{;}{}^{i}$ being the contravariant form of the latter. Vilkovisky's choice is equivalent to introducing a preferred set of coordinates into the space of histories, namely Riemann normal coordinates with origin at the mean field $\overline{\phi}$. These coordinates are the components of the vector $\Phi^i=-\sigma_{;}{}^{i}[\overline{\phi},\phi]$ at $\overline{\phi}$. This vector points along the geodesic from ϕ to ϕ and has a length equal to the geodesic distance between $\overline{\phi}$ and ϕ.

A variant of Vilkovisky's idea had been introduced earlier for the nonlinear σ model by Honerkamp, but Vilkovisky has carried it forward to gauge theories, including the gravitational field. The key idea here is to recognize that a choice of gauge breaking term corresponds to a choice of coordinates in the orbits into which the gauge group decomposes the space of histories. By judiciously selecting a metric that respects the orbit decomposition Vilkovisky is able to construct an effective action that is independent of the choice of orbit coordinates and hence of the choice of gauge breaking terms.

In the gauge case Vilkovisky's metric is *non*local, but it contains a local component that is again determined by the coefficient of the kinetic energy term in the classical action. Since Vilkovisky's prescription is independent of the coordinates chosen in the space of histories and, in particular, of the orbit coordinates, it is very likely also free of the Gribov problem. A determination of this question, however, demands a study of the properties of the associated Riemann normal coordinate system, especially of the caustic problem for geodesics emanating from a given point. An investigation of this problem will soon be underway.
This problem has already been solved for the local component of Vilkovisky's metric in the case of quantum gravity.

Another question concerns the choice of measure $\mu[\phi]$ for the functional integration. One expects that μ should be the square root of the determinant of Vilkovisky's metric. This has indeed been verified in the non-gauge case, where the determinant has also been formally shown to equal $(\det G^{+}[\phi])^{-\frac{1}{2}}$ as required by unitarity. Here G^{+} is the advanced Green's function for the operator $S_{;ij}[\phi]$:

$$S_{;ik}G^{+kj}=-\delta_i^j \ .$$

In the case of gauge theories unitarity requires that this determinant be divided by another determinant involving the ghost Green's function. It is planned to check whether Vilkovisky's metric yields the required result.

The most exciting thing about Vilkovisky's prescription is that one can compute with it, obtaining results that are free of all gauge ambiguities. Vilkovisky and Barvinsky have already used it to compute the one-loop pole residues in the dimensional regularization of quantum gravity. Plans are underway to apply it to Yang-Mills theory in two-loop order. In this case it should not even be necessary to renormalize the ghost propagator, the renormalization rules thereby becoming still simpler than those used by Hart (1981) and by Abbott (1981). Finally, Vilkovisky's prescription should be of fundamental importance in theories that are not perturbatively renormalizable, such as quantum gravity and supergravity. It could be used to give completely unambiguous definitions of (and renormalization prescriptions for) operators such as the stress tensor. It could also be used to remove gauge ambiguities from summations of infinite subsets of Feynmann graphs.

3. CHANGING TOPOLOGY

The idea that space at the Planck scale might undergo topological fluctuations as a result of the quantum nature of the metric field was introduced by J.A.Wheeler in 1957. In DeWitt (1981a)it is stated that "Topology... is not a dynamical object, and to this date no one has come within twenty-five light years of making it so, despite a quarter century of handwaving about it". It is with this "scandal in physics" in mind that a study has been undertaken to come to grips with changing topology.

Such an undertaking faces formidable obstacles. To begin with, the spatial cross sections of a classical spacetime of Lorentzian signature cannot change topology without metric singularities occurring. One can, of course, attempt to get around this difficulty by rotating the time axis in the complex plane and "going Euclidean". However, most interesting metrics, above all those that contain gravitational wave packets, are not analytic and cannot be smoothly continued to a Euclidean region. Moreover, most metrics fail to remain real after the time axis has been rotated. Finally, one loses any sense of a "process" occurring and thereby any intuitive understanding of topological transitions. This is particularly true in the functional integral definition of transition amplitudes where one sees that the Euclidean approach is fraudulent in at least one basic respect: there is no homotopy relationship between topologies which allows one to determine the relative phases with which amplitudes for different topologies should be added together.

In the present study it was resolved to tackle the problem head-on within a Lorentzian framework. But then another formidable difficulty presents itself. The gravitational field, which is the quantum object that is expected to induce topological changes, has a highly complicated structure that is difficult to deal with. Moreover, the list of possible topologies for 3-space is daunting. Could one not begin with a spacetime of lower dimensions? Unfortunately the gravitational field ceases to be a dynamical object in lower dimensions. One can nevertheless perhaps still learn something from a model theory.

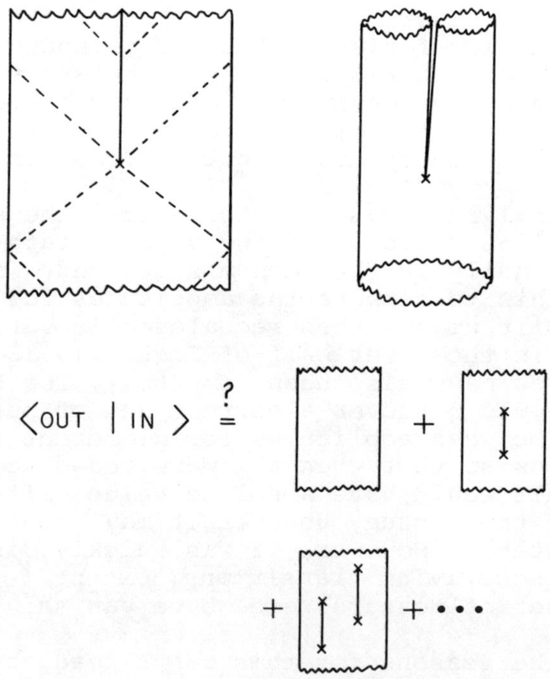

Fig. 1. Changing topology.

The model that has been selected for preliminary study is that of a massless scalar field in 1+1 dimensions. The initial topology that was chosen for spacetime is shown in Figure 1. It is called the "trousers" topology. A flat spacetime cylinder splits into two flat cylinders (i.e., an S^1 universe splits into two disjoint S^1 universes) by a simple reidentification of points to the future of the "crotch". The metric is chosen to be flat since flat spacetime is a solution of the gravitational field equations. The idea is that the model may give some insight into a WKB approximation for a true gravito-topological transition amplitude, with the fluctuating scalar field mimicking a gravitational field fluctuating around a solution of the classical field equations.

The "crotch" point is a singularity of the chosen spacetime but it is a mild one. The most natural boundary condition to impose on this singularity is to suppose that waves entering the "legs" from the "trunk", or vice versa, are propagated by a simple "shadow law" with the singularity inducing infinitely sharp discontinuities but not otherwise contributing to the waves. It is then a straightforward matter to introduce complete sets of mode functions in the trunk and in the legs, to define corresponding state vector spaces in the two regions, and to compute the Bogoliubov coefficients that relate these mode functions. All this has been done.

The original idea was then to compute the amplitude that the legs would be in the vacuum state if the trunk was originally in the vacuum state and, indeed, to go beyond this to compute the amplitudes for the S^1 universe to split up and then recoalesce in various way, as shown in the lower half of Figure 1. (Some of these amplitudes have also been computed). The hope was that one could discover a correct set of phase relationships between amplitudes for different topological transitions so that when all were added together the final result would have absolute value unity, as is required if the vacuum, however it may fluctuate, is to remain stable. However, it was quickly discovered that all nontrivial transitions, except for a precisely tuned set of measure zero, have vanishing amplitude.

Some of the reasons for this can already be understood by examining the simple trousers topology. Straightforward computation, using the Bogoliubov coefficients, shows that the probability for the legs to be found in the vacuum state if the trunk was in the vacuum state is zero. The splitting of the universe into two universes in fact induces the creation of an infinite number of particles containing an infinite amount of energy. The energy problem can already be understood at the classical level. The stress tensor for a scalar field, unlike the current vector, is *quadratic* in the derivatives of the field. The shadow boundary conditions induce sharp discontinuities in all waves (except those that have appropriate periodicities) and hence terms in the stress tensor involving the *square* of the delta function. These singular terms have their supports on the dotted lines shown in Figure 1.

It is of no help to remove the singular point by smoothing out the geometry in its vicinity. Any smoothing procedure that leaves the Lorentzian signature elsewhere untouched requires the introduction of a complex-valued metric in the crotch region, and this leads immediately to failure of the standard Bogoliubov coefficients and to loss of unitarity.

It looks as if nature itself resists topological changes. In the real world the gravitational field would react to uncontrolled topologically induced energy production by pinching off and damping the incipient transition. It is planned to test this conclusion by studying a more realistic model: the linearized gravitational field in a flat 3-torus. In this case there is a linearization constraint that must be satisfied: the total energy must be zero. Massless fields in stationary spacetime having compact spatial sections always possess zero-frequency modes in addition to Fock-space modes. Among the zero-frequency modes for the linearized gravitational field in a 3-torus is one that corresponds to a uniform dilation of the scale of the universe and that contributes negatively to the energy. The coefficient of this mode is not independent of the coefficients of the other modes but must adjust itself so that the total energy is zero. In the quantum theory this becomes a constraint on the allowed state vectors.

It is planned to calculate the Bogoliubov coefficients and transition amplitudes for a 3-torus that splits into two 3-tori. If the expectation value of the coefficient of the "out" dilation mode diverges in the "in" vacuum state then one will have direct evidence that the universe reacts to prevent topological change. There is no reason to suppose that these conclusions would need to be altered when spacetime is curved and such entities as wormholes try spontaneously to form. A singularity will always be present at the onset of topological change, and the shadow that it casts seems likely always to induce infinite particle production.

REFERENCES

Abbott, L.F.: 1981, "The Background Field Method beyond One Loop", *Nucl. Phys.*, B185, 189.

DeWitt, B.: 1981a, "A Gauge Invariant Effective Action", in C.J.Isham, R.Penrose and D.W.Sciama (eds), *Quantum Gravity 2, A Second Oxford Symposium*, Oxford University Press, Oxford.

DeWitt, B.: 1981b, "Approximate Effective Action for Quantum Gravity", *Phys.Rev.Lett.*, 47, 1647.

Hart, C.F.: 1981, Ph.D. Thesis, University of Texas.

Vilkovisky, G.A.: 1983, "The Unique Effective Action in Quantum Field Theory", *Phys. Lett.*, in press.

QUANTUM GRAVITY AND QUANTUM FIELD THEORY IN A CURVED SPACE
REPORT OF WORKSHOP D1

Abhay Ashtekar*

Physique Théorique, IHP,
11 Rue P. et M.Curie, 75231 Paris, France

and

Physics Department, Syracuse University,
Syracuse, NY 13210, USA

The workshop D1,'Quantum Gravity',was divided into two parts. The first and the longer part was devoted to a review of major developments that have occurred in recent years[(1)] while the second part was concerned more with directions for future. The review talks were given by Bernard Kay (Quantum Field Theory in curved space-times) Andy Strominger (Higher Derivative Quantum Gravity), Rafael Sorkin[(2)] (Global and Topological Aspects of Quantum Gravity), and, Lee Smolin (Conceptual problems in Quantum Gravity), and several directions for future work in the area were suggested by Jim Hartle, John Klauder, David Finkelstein and Roger Penrose.

1. REVIEWS

1.1. Quantum Field Theory in Curved Space-times

Kay gave a thorough review of the status of the field. Since there already exist several review articles (see, e.g. Avis, 1981; Birrel, 1981; Deutsch, 1981; Gibbons, 1983) and even a monograph (Birrel and Davis, 1982) on this subject, I will mention only the highlights of the talk.

B. Bertotti et al. (eds.), General Relativity and Gravitation, 453–470.

The talk emphasized the structural aspects of the theory. From this viewpoint, the problem underlying quantum field theory in curved space-times can be broken up into four parts: finding the algebra of field operators; identifying the set of physically interesting states; defining composite local observables involving products of fields at the same space-time point, such as the stress-energy operator; and, investigating whether one can single out preferred states which may be labelled as 'vacuum' or 'having n-particles'. Broadly, the current state of the art is such that the issues become more and more open-ended as one goes from the first to the fourth part. Thus, for fields satisfying linear equations - for concreteness, say a massive Klein-Gordon field $\phi(x)$ - one can introduce operator valued distributions, $\hat{\phi}(x)$ which generate the field algebra. Quantum states can be defined as positive linear functionals ω on this algebra which associate, to each product, $\hat{\phi}(x_1)\ \hat{\phi}(x_2)\ldots\hat{\phi}(x_n)$, with $x_1, x_2, \ldots, x_n$ all distinct, its 'expectation value' $\omega(\hat{\phi}(x_1)\hat{\phi}(x_2)\ldots\hat{\phi}(x_n))$. Given any one of these states, one can construct a Hilbert space (on which the operators have a natural action) such that $\omega(\hat{A})=\mathrm{Tr}\rho_\omega\hat{A}$ for all operators $\hat{A}$ in the algebra for some density matrix ρ_ω on the Hilbert space. However, there does *not* exist a single Hilbert space on which all states can be so represented. There is, however, a preferred class of states - the so called Hadamard states (Fulling et al., 1978) - for which the '2-point functions' $\omega(\hat{\phi}(x_1)\hat{\phi}(x_2)+\hat{\phi}(x_2)\hat{\phi}(x_1))$ have a specific form in the limit $x_1 \to x_2$, which mimicks the standard form of 2-point functions in flat space. Kay conjuctures that all Hadamard states are locally quasi-equivalent (Haag and Kastler, 1964) in the sense that, for any bounded region R of space-time, there is a Hilbert space representation such that $\omega(\hat{A}^R)=\mathrm{Tr}\rho_\omega\hat{A}^R$ for all Hadamard states ω and for all observables $\hat{A}^R$ in the algebra generated by $\hat{\phi}(x)$ with x in R. Next, let us consider the stress-energy operator whose classical expression involves products of fields evaluated at the same point. Since the quantum operators $\hat{\phi}(x)$ are distributions, one cannot take products $\hat{\phi}(x)\hat{\phi}(x)$. Hence to define the stress--energy operator, one must regularize its formal expression, which is divergent. This can be done by fixing a state and then regularizing the expectation value of the formal operator. However, the resulting expressions are not unique and the ambiguities are referred to as the 'finite renormalization ambiguities': two regularized expectation-values can differ by a term of the form:

$$a_1 g_{ab} + a_2 G_{ab} + a_3 {}^1H_{ab} + a_4 {}^2H_{ab} \qquad (1)$$

where a_i are arbitrary constants; g_{ab} the background metric; G_{ab}, its Einstein tensor; and ${}^1H_{ab}$ and ${}^2H_{ab}$ are, respectively, the variations of $\int\sqrt{-g}R^2$ and $\int\sqrt{-g}C_{abcd}C^{abcd}$ w.r.t. g_{ab}. Thus, the ambiguities are local in curvature. Their origin lies in the counter-terms $a_1\int\sqrt{-g}$, $a_2\int\sqrt{-g}R$, $a_3\int\sqrt{-g}R^2$ and $a_4\int\sqrt{-g}C_{abcd}C^{abcd}$ that are added to the action in the process of renormalization. Note that, even after regularization, $\hat{T}_{ab}$ may not belong to the field algebra. However, Kay conjectures that, on the Hilbert space associated with a bounded region R (whose existence was conjectured above), there exists an operator-valued distribution $\hat{T}_{ab}$ such that

$$\omega(\hat{T}^R_{ab}(x)) = \mathrm{Tr}\,\rho_\omega \hat{T}^R_{ab}(x)$$

for all Hadamard states ω. Finally, we come to the last part of the problem: definition of particle states. These can be defined unambiguously in stationary space-times (Ashtekar and Magnon, 1975; Kay, 1980). Similarly, one can unambiguously define 'in' and 'out' particle states in asymptotically flat or asymptotically stationary space-times. For zero rest mass fields, one can use null infinity to construct the 'in' and 'out' vacua (Hawking, 1975; Ashtekar, 1981b). Furthermore, in these contexts, a great deal is known about the scattering theory (Dimock and Kay, 1982, 1983). What is the situation in more general space-times? A number of definitions of particle states associated with Cauchy surfaces are known (see e.g. Parker, 1977; Dray and Renn, 1983). What one would like to have is a 'quasi-local' definition of vacuum and of particle states associated with open regions (say normal neighborhoods) of space-time and a detailed analysis of properties of these states. Another important open problem is that of defining the 'initial state' of the quantum field in cosmology.

So far the background metric was kept fixed. The next step is to analyse the effect of the quantum matter fields on the geometry. An obvious way to do this is via 'semi-classical' gravity where one imposes the equation $G_{ab} = 8\pi G\langle\hat{T}_{ab}\rangle$. In view of the finite renormalization ambiguities, one is led to consider, instead the equation

$$G_{ab} + \Lambda g_{ab} + \alpha\, {}^1H_{ab} + \beta\, {}^2H_{ab} = 8\pi G\langle\hat{T}_{ab}\rangle \qquad (2)$$

where Λ, α, β and G are to be thought of as running coupling constants. There is still some ambiguity about the expectation-value symbol; one can consider the expectation value in the in-in state, or, the matrix element in the in-out state. Let us, for concreteness, consider the in-in case. Then, a solution to the semi-classical theory would consist of a classical metric g_{ab} and a state ω on the field algebra $\hat{\varphi}$ constructed using this g_{ab} such that the field equation $(\Box - m^2)\hat{\varphi}=0$ as well as equation (2) above is satisfied. There are several reasons to believe that such a theory cannot be considered as 'fundamental' (Eppley and Hannah, 1977; Page and Geilker, 1981). However, one may wish to consider it as an approximation to a full quantum theory of gravity (involving 'higher derivatives') although even this viewpoint faces some difficulties (Horowitz and Wald, 1978; Horowitz, 1980; Hartle and Horowitz, 1981). Furthermore, the range of validity of this approximation is not completely clear. Nonetheless, it would be desirable to get a description of the black-hole evaporation process in this framework. By now, there exist several computations of the necessary $\langle \hat{T}_{ab} \rangle$ in this case (see, e.g. Bolashenko and Frolov, 1983; Frolov and Zel'nikov, 1983) and it would be instructive to find the semi-classical metric. The second obvious candidate for applications is cosmology. Here, the literature is even richer (see, e.g. Hartle, 1981; Parker, 1982; Hu, 1982, 1983), although the basic problem of the choice of the initial state is yet to be resolved satisfactorily.

Finally, considerable work has been done on interacting fields in curved backgrounds (see, e.g. Birrel, 1981; Birrel and Davis, 1982).

1.2. Higher Derivative Quantum Gravity

Since Strominger's talk was an expanded version of the abstract submitted to this workshop by Horowitz and Strominger (1983), my report will follow the abstract closely, (for details, see Strominger, 1983).

From the viewpoint of a conventional quantum field theorist, perhaps the most desirable property of a physical theory is that its action should lead to a renormalizable, unitary quantum theory. In the gravitational case, this viewpoint leads one to focus just on the Einstein-Hilbert action, setting aside geometrical and topological aspects of general relativity as secondary. Now, it is known since the sixties (DeWitt, 1967) that the resulting quantum field theory is unitary perturbatively. However, it is non-renormalizable. This leads one to look for modifications of the action. In recent years, a great deal of attention was given to modifications which lead to higher derivative gravity theories (see, e.g. the proceedings of the workshop A5). The higher derivative terms make the propagator behave as $1/p^4$, thereby increasing the convergence of Feynman diagrams. This makes the theory renormalizable perturbatively (Stelle, 1977). But now the perturbative unitarity is lost. However, there is some recent evidence that the unitary violations may disappear in a non-perturbative treatment and this has rejuvenated interest in these theories.

The most general category of higher derivative theories considered in this connection has the fourth order action:

$$S=-\int d^4x\sqrt{-g}\,(\Lambda+\frac{1}{\kappa^2}R+\beta R^2+\alpha C_{abcd}C^{abcd}). \qquad (3)$$

The loss of perturbative unitarity arises as follows. The classical theory allows small fluctuations with negative energy. To avoid such states in quantum theory, one is led to introduce an indefinite metric on the space of states. The negative norm states then lead to violation of unitarity in the perturbation theory. However, there exist non-perturbative treatments - 1/N-expansion, where N is the number of matter fields (Tomboulis, 1977, 1980) and 1/D-expansion, where D is the space-time dimension (Strominger, 1981; 1983) - which *can* restore unitarity and yield a renormalizable, unitary quantum theory with positive energy. However, the solution seems to be a forced one because it involves an essential use of expansions around flat space. The general feeling is that the theory may be relevant only for phenomenological calculations below Planck energy.

One can specialize by setting Λ and $1/\kappa^2$ equal to zero and obtain a scale invariant action:

$$S=-\int d^4x\sqrt{-g}\,(\beta R^2+\alpha C_{abcd}C^{abcd}). \qquad (4)$$

This may appear strange at first because, in the process of specialization, one has omitted even the Einstein-Hilbert term. How can one recover Einstein's theory in a suitable limit? The viewpoint is that the Einstein-Hilbert action is relevant only for low energy or long wavelength phenomena and is only an 'effective action' in this limit: quantum corrections to (4) break its scale-symmetry and induce the Einstein--Hilbert term which dominates the low-energy phenomena (for details, see, e.g. Adler, 1982). The analogy is to the weak interaction theory. There, the Fermi action dominates the low energy phenomena although the fundamental Salam-Weinberg action is very different from it. Note that, in this viewpoint, Einstein's theory is recovered as the long wave length limit *after* applying quantum corrections; it is not recovered within the classical theory or even in the classical limit of the quantum theory.

What is the situation with unitarity? Although the classical theory again allows small fluctuations with negative energy which lead, as before, to non-unitarity in the perturbative quantum treatment, it has been recently shown (Boulware et al., 1983) that, if $\alpha\cdot\beta>0$, all exact solutions of the theory have zero energy. This non-perturbative result opens up possibilities of new non-perturbative quantum treatments which do *not* involve an essential use of expansions around flat space. Overall, it is relatively easy to compute non-perturbative instanton effects in this theory because, unlike in the usual Einstein-Hilbert case, the action is manifestly positive and the equations admit non-trivial asymptotically Euclidean solutions. The first property makes computations easier - one does not have to make complex rotations of conformal factors (Hawking, 1979) while the second makes it easier to interpret the results physically. Finally, since the theory is renormalizable, one can actually compute the instanton determinants.

One can specialize even further and set $\beta=0$ in (4) and obtain a conformally invariant action:

$$S=-\int d^4x\sqrt{-g}C_{abcd}C^{abcd} \quad . \tag{5}$$

Again, analysis of Boulware et al. (1983) shows that, in the exact theory, there are no solutions with negative energy and one should be able to show unitarity without an essential use of a flat background. An additional attractive feature here is the full conformal invariance which, although broken by quantum fluctuations, will be restored at short distances since the theory is asymptotically free. Thus, the concept of length arises only as a low energy (long distance) symmetry breaking effect and ceases to be precise once we are in the regime below Planck length. Overall, the approach seems to be both practical and elegant.

1.3. Global and Topological Aspects of Quantum Gravity

The main part of Sorkin's talk was devoted to results on the consequences of choosing non-trivial spatial topologies. He began by recalling the distinction between the status of tensors and spinors: while tensors can be defined on a manifold without any extra structure - they constitute finite dimensional representations of GL(4) - spinors can be defined only when the manifold is equipped with a (conformal) metric. What are the ramifications of this distinction as far as physics is concerned? One may adopt the 'supersymmetry viewpoint' and simply ignore the distinction by treating bosons and fermions on the same footing. Alternatively, one may define GL(4)-spinors, classify them, and use these spinors in place of the SL(2,C)-ones (Friedman and Sorkin, 1980a; Šijački, 1983). Perhaps the most attractive possibility is to introduce half-integral spins without introducing *any* spinors; spin would then be a derived rather than a fundamental notion. There are a number of proposals along these lines in the framework of Yang-Mills theory (Jackiw and Rebbi, 1976; Hasenfratz and 't Hooft, 1976) in which spin arises from isospin. The basic idea here is that the classical configuration space has gauge degrees of freedom, whence the rotation group, SO(3), can admit a double valued representation: a rotation by 2π may map a configuration to a gauge-related one rather than to itself. A similar phenomenon can occur in general relativity in presence of non-trivial spatial topology.

To see this, consider the canonical quantization framework. Fix a 3-manifold Σ and let $\mathscr{C}$ denote the space of positive definite asymptotically flat 3-metrics q_{ab} on Σ. In the canonical quantization scheme, one considers wave functions, $\psi(q)$, on $\mathscr{C}$ and applies to them the quantum (analogs of classical) constants (see, e.g., Kuchar, 1981a and Ashtekar and Horowitz, 1982). There are two constraint equations, one vectorial and the other scalar. The role of the vector constraint is well understood (Higgs, 1958). Denote by $\mathscr{D}^\infty$ the group of diffeomorphisms of Σ which are asymptotically trivial (i.e., asymptotically identity), and by $\mathscr{D}^\infty_0$ its connected component of identity. Then the vector constraint requires that the wave functions be invariant under $\mathscr{D}^\infty_0$: $\psi(d\cdot q)=\psi(q)$ for all d in $\mathscr{D}^\infty_0$. Thus, the wave functions *can*, in particular, change under the action of diffeomorphisms which are not in the connected component of identity of $\mathscr{D}^\infty$, whence, if the topology of Σ is such that a rotation by 2π at infinity cannot be continuously deformed to identity,one can have states, $\psi(q)$, with half-integral spins (Friedman and Sorkin, 1980b). Thus, there is a beautiful intertwining between topology of Σ and spinorial states, constructed out of the 'purely bosonic' gravitational field. The group $G=\mathscr{D}^\infty/\mathscr{D}_0^\infty$ is an effective symmetry group (Sorkin, 1983). In the Yang-Mills theory, the analog of G is Abelian and its action leads to the θ-vacua. In the gravitational case, G is non-Abelian and depends on the topology of Σ; it is not universal. A detailed calculation of these groups is now available (Friedman and Witt, 1983). One can either consider the wave functions as being defined on $\mathscr{C}/\mathscr{D}_0^\infty$, or, on twisted bundles (with flat connection) over $\mathscr{C}/\mathscr{D}^\infty$ (Isham, 1981). In either case, irreducible representations of G dictate the 'number of quantum states'; the twisted bundles over $\mathscr{C}/\mathscr{D}^\infty$, for example, can be classified in terms of these representations. In the Yang-Mills theory, these global considerations lead to effects such as CP violation which have direct physical interest. In the gravitational case, such consequences are yet to be discovered. Also, all these analyses (Friedman and Sorkin, 1980b; Isham, 1981; Friedman and Witt, 1983) are incomplete in the sense that they essentially ignore the second (i.e. the scalar) quantum constraint. Indeed, for non-trivially constrained systems, not all representations are permissible and it may well be that the scalar constraint implies that quantum gravity does not admit the q-representation, whose existence is assumed here. However, since the arguments are essentially geometric - rather than functional-analytic - one would expect them to go through also in other representations.

Sorkin also briefly mentioned a number of other issues: gravitational instantons (see, e.g. Pope, 1981); the new differential structure on R^4 (Donaldson, 1983; Gompf, 1983); asymptotic quantization of radiative modes at null infinity (Ashtekar, 1981a) and the role of NUT or 'dual' 4-momentum in this framework (Ramaswami and Sen, 1981; Ashtekar and Sen, 1982); and, the result that, geometrodynamics does not admit any symmetry (Kuchař, 1981b). The last result poses a major obstacle in the definition of the Hilbert space of states in the usual (Kuchař, 1981a) canonical quantization program.

1.4. Conceptual Problems in Quantum Gravity

Smolin's talk questioned the basic premise on which all work in the field has been based, namely, the correctness of conventional quantum mechanics. It serves as an excellent bridge between the two parts of the workshop because, while he spoke of things that he has worked out in detail, the viewpoint is so novel that it opens up an entirely new direction for future work.

The overall viewpoint is that quantum mechanics is an incomplete theory and, therefore, attempts at 'quantization' of the gravitational field may be misguided right from the beginning. Smolin began by giving three distinct arguments - involving respectively, black hole evaporation, the Fulling-Rindler phenomenon associated with accelerated observers in Minkowski space, and, the fundamental difficulties associated with measurements of quantum states of gravitational radiation - to suggest that in the correct 'quantum theory of gravity', one would *not* be able to distinguish between the quantum fluctuations and the statistical fluctuations. (For details, see Smolin, 1983). The final suggestion itself is not new (Fenyes, 1952; Nelson, 1966, 1967, 1983). But, whereas previously there appeared to be no physical reason for identifying the two types of fluctuations, now the difficulties encountered in reconciling quantum mechanics with gravity seem to force one in that direction.

If one adopts the viewpoint that quantum fluctuations are simply statistical, one is led to ask for the physical origin of these fluctuations. Smolin feels that the most natural way to answer this question is via 'relational non-local hidden variable theories'. The non-local hidden variables are to be associated with pairs of particles in the universe. In any local observation, one cannot control or determine these non-local variables and this naturally leads to statistical fluctuations. To illustrate this point he presented a simple example. Consider a system of N particles and let its state be represented by a N×N matrix M_{ab}:

$$M_{ab}=L_a\delta_{ab}+N_{ab}$$

where L_a describe some local 'hidden' variables, and N_{ab}, non-local 'hidden' variables. The positions of the N particles are given by the eigenvalues, λ_a, of N_{ab} and these are observable. (All variables are known at this level and some become 'hidden' only in the conventional quantum description which would, e.g., deal only with the probability amplitudes, $\psi(\lambda)$. N_{ab} may be thought of containing information about variables associated with pairs of particles, in addition to their positions). Now, L_a and N_{ab} may evolve by deterministic equations. And yet, one may not be able to predict the evolution of λ_a, knowing only their initial values. Furthermore, under certain conditions, the evolution of the probability density, $\bar{\rho}(\lambda)$ *can be* shown to be governed by the Schrödinger equation, up to terms in $N^{-\frac{1}{2}}$.

Attractive as these ideas are, much work is clearly needed to test if they are indeed viable.

2. DIRECTIONS FOR FUTURE

Jim Hartle suggested that significant progress could be made by comparing different approaches to quantum gravity since problems that seem insoluble in one approach may well be manageable within another (Hartle and Kuchar, 1983). There are some recent instances in which this has happened: for example, the complex rotation of the conformal factor, needed in the path integral approach (Hawking, 1979), which appears to be rather mysterious within the approach, can be justified in the canonical framework (Hartle, 1980). Similarly, the problem of choice of time which plagues the canonical framework in the cosmological context seems to disappear in the path integral scheme (See Hawking's article in this volume). A number of other issues could be similarly simplified. One may, for example, ask the following questions: how does non-renormalizability of the covariant approach affect the canonical framework? How does the problem of constructing the Hilbert space in the canonical program show up in the covariant approach?

John Klauder pointed out that there exist simple models which mimick general relativity and whose quantization leads to surprising features. Consider, first, a model in particle dynamics with action

$$I=\int [p(t)\dot{q}(t)-p(t)q^2(t)]\,dt, \qquad (6)$$

subject to the condition $p(t)>0$. One is to think p as the 3-metric in relativity and q as the connection. (Here, and in what follows, one has the Palatini-like approach in mind in which the metric and the connection are treated as independent variables). For this model, the simple minded prescription,

$$\langle q'',t''|q',t'\rangle = N\int_{p>0} (\exp iI)\mathscr{D}p\mathscr{D}q \tag{7}$$

is in fact *incorrect*; the positivity requirement on p has subtle consequences. Field theoretic generalizations can be obtained by considering, for a scalar field g in Minkowski space, the action

$$I = \int g(x)\,[\dot{\Gamma}(x) - \Gamma^2(x)]\,d^4x \tag{8}$$

with $g(x)>0$, and, for a matrix-valued field, $g_{rs}(x)$,

$$I = \int g_{rs}(x)\left[\dot{\Gamma}_{rs}(x) - \Gamma_{ru}(x)\Gamma_{us}(x)\right]d^4x \tag{9}$$

where $g_{rs}(x)$ is required to be positive. (These positivity conditions mimick, of course, the condition that the 3-metric in relativity is positive definite). These models *are* soluble, although they are *not* renormalizable. Furthermore, g and Γ cannot be treated as canonical variables in quantum theory. (For details, see Klauder, 1970; Klauder and Aslaken, 1970). The close similarity of these models with the so-called strong coupling gravity (Isham, 1976) makes them all the more interesting. More generally, the analysis of non-renormalizable but exactly soluble models pursued by Klauder over the years is clearly extremely valuable for quantum gravity. (See Klauder, 1979, 1981 and references therein).

David Finkelstein suggested a completely new approach. Normally, one accepts the notions of set theory and topology and tries to replace only the fields on a given manifold by 'quantum fields'. That is, one keeps the set theory and topology intact and modifies only the 'higher' mathematical structures. Finkelstein suggested that one should rather begin with a modified, 'quantum set theory' then go on to 'quantum topology', and, finally, 'quantum physics'. He outlined a concrete proposal to construct these structures (Finkelstein and Rodriguez, 1983).

Roger Penrose ended the session by explaining that he expects a successful quantum theory of gravity to have a certain feature, and by pointing out that none of the ideas discussed above appear to lead in that direction. The desired feature is time-asymmetry. He feels that the second law of thermodynamics can be explained, or, at least understood, in terms of the 'Weyl curvature hypothesis' (Penrose, 1979) according to which the initial singularities have vanishing Weyl curvature whereas the final singularities are unconstrained, and that the correct quantum theory of gravity should 'explain' this time asymmetric constraint. The theory itself would be therefore time-asymmetric. Furthermore, he feels that this time asymmetry is intimately connected with the reduction of the wave packet. (For details, see Penrose, 1982). Thus, the successful theory would have not only to account for quantum effects associated with the gravitational interaction, but also dispel the mystery surrounding the arrow of time and the measurement theory in quantum mechanics. Clearly, we are still very far from these goals.

NOTES

* Alfred P.Sloan Research Fellow. Supported in part by the NSF contract PHY 80-08155 with Syracuse University.

(1) Originally, Chris Isham was to chair this workshop. He carried out the difficult task of going through the contributed abstracts and decided that it would be better to have four major reviews rather than a number of short contributions. Thus the workshop was essentially organized by him. Unfortunately, health problems prevented him from attending the conference and his physical absence in the workshop is the only reason why he is not the reporter of these proceedings.

(2) John Friedman and Rafael Sorkin prepared this talk jointly.

REFERENCES

Papers presented to the GR10 Conference and published in:

Bertotti, B., de Felice, F., and Pascolini, A. (eds): 1983, *10th International Conference on General Relativity and Gravitation. Contributed Papers.* Vol. 2., CNR, Roma.

Bolashenko, P.A., and Frolov, V.P.: 1983, "Quantum fluctuations in $S^2 \times S^2$ and the vacuum polarization near a Schwarzschild black hole", p. 1036.

Dimok, J., and Kay, B.S.: 1983, "Classical and quantum scattering on stars and black holes", p. 1062.

Dray, T., and Renn, J.: 1983, "Hamiltonian diagonalization and adiabatic particle definitions", p. 1064.

Finkelstein, D., and Rodriguez, E.: 1983, "Quantum set theory, quantum topology and Quantum Gravity", p. 1067.

Frolov, V.P., and Zel'nikov, A.I.: 1983, "Vacuum polarization of massive fields near a rotating black hole", p. 1070.

Horowitz, G.T., and Strominger, A.: 1983, "Higher derivative quantum gravity", p. 1083.

Šijački, D.J.: 1983, "Holonomic spinors in general relativity", p. 1164.

Sorkin, R.: 1983, "Remarks on diffeomorphisms as particles symmetries", p. 1164.

Other references

Adler, S.L.: 1982, "Einstein gravity as a symmetry - - breaking effect in quantum field theory", *Rev. Mod. Phys.*, 54, 729.

Ashtekar, A.: 1981a, "Quantization of the radiative modes of the gravitational field", in Isham et al., 1981, p. 416.

Ashtekar, A.: 1981b, "Asymptotic Quantization of the Gravitational Field", *Phys. Rev. Lett.*, 46, 573.

Ashtekar, A., and Horowitz, G.T.: 1982, "On the canonical approach to quantum gravity", *Phys. Rev.*, D26, 3342.

Ashtekar, A., and Magnon, A.: 1975, "Quantum fields in curved space time", *Proc. R.Soc. London*, A346,375.

Ashtekar, A., and Sen; A.: 1982, "NUT 4-momenta are forever", *J.Math. Phys.*, 23, 2168.

Avis, S.L.: 1981, "A review of some aspects of field theory on topologically non-trivial space-time", in Isham et al., 1981, p. 148.

Birrel, N.D.: 1981, "Interacting quantum field theory in curved space-time", in Isham et al., 1981, p. 164.

Birrel, N.D., and Davis, P.C.W.: 1982, *Quantum Fields in Curved Space-times*, Cambridge University Press, Cambridge.

Boulware, D., Horowitz, G.T., and Strominger, A.: 1983, "Zero-energy theorem for scale invariant gravity", *Phys. Rev. Lett.*, 50, 1726.

Deutsch, D.: 1981, "Particle and non-particle states in quantum field theory", in Isham et al., 1981, p. 131.

DeWitt, B.S.: 1967a, "Quantum theory of gravity. II. The manifestly covariant theory", *Phys. Rev.*, 162, 1195.

DeWitt, B.S., 1967b, "Quantum theory of gravity. III. Applications of the covariants theory", *Phys. Rev.*, 162, 1239.

Dimock, J., and Kay, B.S.: 1982, "Classical wave operators and asymptotic quantum field operators on curved space-times", *Ann. Inst. Henri Poincaré*, A37, 93.

Donaldson, S.K.: 1983, "An application of gauge theory to four dimensional topology", *J.Diff. Geo.*, 18, 269.

Eppley, K., and Hannah, E.: 1977, "The necessity of quantizing the gravitational field", *Found. Phys.*, 7, 51.

Fenyes, I.: 1952, "Eine wahrscheinlichkeitstheoretische Begründung und Interpretation der Quantenmechanik", *Z.Phys.*, 132, 81.

Friedman, J.L., and Sorkin, R.D.: 1980a, "The unitary multiplicity-free representation of $SL_4(R)$", *J.Math. Phys.*, 21, 1269.

Friedman, J.L., and Sorkin, R.: 1980b, "Spin ½ from

gravity", *Phys. Rev. Lett.*, 44, 1100.

Friedman, J.L., and Witt, D.: 1983, "Internal symmetry groups of quantum geons", *Phys. Lett.*, 120B, 324.

Fulling, S., Sweeney, M., and Wald, R.M.: 1978, "Singularity Structure of the Two-Point Function in Quantum Field Theory in Curved Spacetime", *Commun. Math. Phys.*, 63, 257.

Gibbons, G.: 1980, "Quantization about Classical Background Metrics", in J.Schmutzer (ed.), *Proceedings of the 9th International Conference on General Relativity and Gravitation*, VEB Deutscher verlag der Wissenschaften, Berlin (DDR), p. 165.

Gompf, R.: 1983, "Three exotic R^4's and other anomalies" *J.Diff. Geo.*, 18, 317.

Haag, R., and Kastler, D.: 1964, "An Algebraic Approach to Quantum Field Theory", *J.Math. Phys.*, 5, 848.

Hartle, J.B.: 1980, "Positive definite three geometries and conformal rotations", preprint UCSB TH-52, unpublished.

Hartle, J.B.: 1981, "Particle production and dynamics in the early universe", in Isham et al., 1981, p. 313.

Hartle, J.B., and Horowitz, G.T.: 1981, "Ground-state expectation value of the metric in the 1/N or semiclassical approximation to quantum gravity", *Phys. Rev.*, D24, 257.

Hartle, J.B., and Kuchar, K.: 1983, "The role of time in path integral formulations of parametrized theories", in S.M.Christensen (ed.), *Ahead of His Time: Bruce S. DeWitt*, Adam & Hilger, Bristol.

Hasenfratz, P., and 't Hooft, G.: 1976, "Fermion-Boson Puzzle in a Gauge Theory", *Phys. Rev. Lett.*, 36, 1119.

Hawking, S.W.: 1975, "Particle Creation by Black Holes", *Commun. Math. Phys.*, 43, 199.

Hawking, S.W.: 1979, "The path-integral approach to quantum gravity", in S.W. Hawking, and W. Israel (eds.), *General Relativity, an Einstein centenary survey*, Cambridge University Press, Cambridge, p. 746.

Higgs, P.W.: 1958, "Integration of Secondary Constraints in Quantized General Relativity", *Phys. Rev.Lett.*, 1, 373.

Horowitz, G.T.: 1980, "Semiclassical Relativity: the Weak Field Limit", *Phys. Rev.*, D21, 1445.

Horowitz, G.T.,and Wald, R.M.: 1978, "Dynamics of Einstein's Equation Modified by a Higher-Order derivative term", *Phys. Rev.*, D17, 414.

Hu, B.L.: 1982, "Quantum Field Theory and Relativistic Cosmology",in R.Ruffini (ed.), *Recent Developments of General Relativity*, vol. 2, North Holland, Amsterdam.

Hu, B.L.: 1983, "Symmetry Behavior at Finite Temperature in Dynamic Spacetimes", in G.Gibbons, S.W. Hawking, and S.Silkos (eds), *Proceeding of the Nuffield Workshop on the Very Early Universe*, Cambridge University Press, Cambridge.

Isham, C.J.: 1976, "Some Quantum Field Theory Aspects of the Superspace Quantization of General Relativity", *Proc. R.Soc. London*, A351, 209.

Isham, C.J.: 1981, "Topical θ-Sectors in Canonically Quantized Gravity", *Phys. Lett.*, B106, 188.

Isham, C.J., Penrose, R, and Sciama, D.W. (eds): 1981, *Quantum Gravity 2*, Oxford University Press, Oxford.

Jackiw, R.J., and Rebbi, C.: 1976, "Spin from Isospin in a Gauge Theory", *Phys. Rev. Lett.*, 36, 1116.

Kay, B.S.: 1980, "Linear Spin-Zero Quantum Fields in External Gravitational and Scalar Fields", *Commun. Math. Phys.*, 71, 29.

Klauder, J.R.: 1970, "Soluble Models of Quantum Gravitation", in M.Carmeli, S.I.Fickler, and L. Witten (eds), *Relativity*, Plenum, New York.

Klauder, J.R.: 1979, "New Measures for Nonrenormalizable Quantum Field Theory", *Ann. Phys.*,117,19.

Klauder, J.R.: 1981, "Scale Covariance, Nonrenormalizable Interactions, and High-Temerature Expansions", *Phys. Rev.*, D24, 2599.

Klauder, J.R., and Alasken, E.W.: 1970, "Elementary Model for Quantum Gravity", *Phys. Rev.*, D2, 272.

Kuchař, K.: 1981a, "Canonical Methods of Quantization", in Isham et al., 1981, p. 329.

Kuchař, K.: 1981b, "General Relativity: Dynamics Without Symmetry", *J.Math. Phys.*, 22, 2640.

Nelson, E.: 1966, "Derivation of the Schrödinger Equation from Newtonian Mechanics", *Phys. Rev.*, 150, 1079.

Nelson, E.: 1967, *Dynamical Theories of Brownian Motion*, Princeton University Press, Princeton.

Nelson, E.: 1983, "Quantum Fluctuations", to be published in W.Wyss (ed.),*The Proceeding of the 7th International Congress on Mathematical Physics*.

Page, D., and Geilker, C.D.: 1981, "Indirect Evidence for Quantum Gravity", *Phys. Rev. Lett.*, 47, 979.

Parker, L.: 1977, "The Production of Elementary Particle by Strong Gravitational Fields", in P. Esposito, and L.Witten (eds), *Asymptotic Structure of Space-Time*, Plenum, New York, p. 107.

Penrose, R.: 1979, "Singularities and Time-Asymmetry", in S.W.Hawking, and W.Israel (eds.), *General Relativity, an Einstein Centenary Survey*, Cambridge University Press, Cambridge, p. 581.

Penrose, R.: 1982, "Some Remarks on Gravity and Quantum Mechanics", in M.Duff, and C.J.Isham (eds), *Quantum Structure of Space-Time*, Cambridge University Press, Cambridge.

Pope, C.: 1981, "The Role of Instantons in Quantum Gravity", in Isham et al., 1981, p. 377.

Ramaswami, S., and Sen, A.: 1981, "Dual-Mass in General Relativity", *J.Math. Phys.*, 22, 2612.

Smolin, L.: 1983, "On the Nature of Quantum Fluctuations and their relation to gravitation and the principle of inertia", University of Mariland.

Stelle, K.S.: 1977, "Renormalization of higher-derivative quantum gravity", *Phys. Rev.*, D16, 953.

Strominger, A.: 1981, "Inverse-dimensional expansion in quantum gravity", *Phys. Rev.*, D24, 3082.

Strominger, A.: 1983, "Is there a quantum theory of gravity?", in S.M.Christensen (ed.), *Ahead of His Time: Bruce S. Dewitt volume*, Adam & Hilger, Bristol.

Tomboulis, E.: 1977, "1/N expansion and renormalization in quantum gravity", *Phys. Lett.*, 70B, 361.

Tomboulis, E.: 1980, "Renormalizability and asymptotic freedom in quantum gravity", *Phys. Lett.*, 97B,77.

SUPERGRAVITY AND THE KALUZA-KLEIN PROGRAM+
REPORT OF WORKSHOP D2

P. van Nieuwenhuizen*
Institute for Theoretical Physics
Utrecht University, The Netherlands

1. SUPERGRAVITY

Supergravity is undoubtedly one of the most active areas of research in gravity theory nowadays. It was invented in 1976 as a relativity theory in which the spin connection is the usual function of the tetrad (=vierbein fields(second-order formalism)) (Freedman et al.,1975), and shortly afterwards reformulated as a relativity theory with independent spin connection (Palatini formalism, first-order formalism) (Deser and Zumino, 1976). Supergravity can be viewed in two complementary ways:

(i) either it is gravity with a local Fermi-Bose symmetry, or

(ii) it is the gauge theory of Fermi-Bose symmetry. The latter has nothing to do with gravity, and there exist numerous Lagrangian field theory models with a rigid (i.e. constant) symmetry between fermions and bosons (Goldfand and Likhtam, 1971; Volkov and Akulov, 1972; Wess and Zumino, 1974, 1978). To promote this rigid symmetry to a local symmetry, one needs, of course, a new gauge field. This turns out to be a massless real vector-spinor field, called *gravitino*

$$\psi^a_\mu(x);\ \mu=1,2,3,4;\quad a=1,2,3,4. \tag{1}$$

In all models of supergravity, for each *fermion state* there is a corresponding *boson state*. In the

B. Bertotti et al. (eds.), General Relativity and Gravitation, 471–489.

simplest model, appropriately called simple supergravity, there are two bosonic states, namely the graviton with helicities ±2, and two fermionic states, namely the above gravitino with helicities ±3/2. The action for this model reads in second order formalism

$$\mathscr{L}=\mathscr{L}^{E}+\mathscr{L}^{RS}-\frac{1}{3}\ (S^2+P^2-A_m^2)+\mathscr{L}^4 \tag{2}$$

where $\mathscr{L}^E$ is the usual Hilbert-Einstein action of general relativity, $\mathscr{L}^4$ denotes 4-fermion couplings while $\mathscr{L}^{RS}$ is the Rarita-Schwinger action

$$\mathscr{L}^{RS}=-\frac{1}{2}\ \bar{\Psi}_\mu\gamma^{[\mu}\gamma^{\rho}\gamma^{\sigma]}D_\rho\Psi_\sigma \tag{3}$$

in which D_ρ is the gravitational covariant derivative. This is the unique action with positive energy for spin 3/2. The fields S (scalar), P (pseudoscalar) and A_m (axial-vector) are auxiliary fields. Their field equations read $S=P=A_m=0$. Due to them, there is for every bosonic *field* component a fermionic *field* component. Only with these auxiliary fields does there exist a representation of the local symmetry group as we shall discuss.

In first-order formalism one does not need $\mathscr{L}^4$. However, no auxiliary fields are known, and also the superspace approach to supergravity only exists in second-order formalism. The action is still invariant but the fields

$$e_\mu^m,\ \Psi_\mu^a \text{ and } \omega_\mu{}^{mn}$$

do not form a representation of a local symmetry group. If one uses second-order formalism, one needs extra 4-fermion couplings in the action and $\bar{\Psi}\Psi\varepsilon$ terms in $\delta\Psi_\mu$ (see below), while in first-order formalism one needs an extra law for $\delta\omega_\mu{}^{mn}$ which is somewhat complicated.

The local supersymmetry transformation rules, which leave $\mathscr{L}$ invariant up to a 4-divergence, read

$$\delta e_\mu^m=\frac{\kappa}{2}\ \bar{\varepsilon}\gamma^m\Psi_\mu$$

$$\delta\Psi_\mu=\frac{1}{\kappa}\ D_\mu\varepsilon+\bar{\Psi}\Psi\varepsilon \text{ and } S,P,A_m \text{ terms} \tag{4}$$

$$\delta S=\ldots,\ \delta P=\ldots,\ \delta A_m=\ldots$$

The 4-component spinor $\varepsilon^a(x)$ $(a=1,\ldots,4)$ is an anti-commuting real object, and $\varepsilon^a(x)$ and $\varepsilon^b(y)$ are independent when $x\neq y$ or $a\neq b$. Hence we have an infinite-dimensional Grassmann algebra

$$\varepsilon^a(x)\varepsilon^b(y)+\varepsilon^b(y)\varepsilon^a(x)=0. \tag{5}$$

On physical grounds one *needs* an infinite-dimensional algebra (otherwise the quantum theory will not be unitary - something which we cannot explain here).

Together with usual spacetime symmetries, the action has the following local symmetries:

(i) the diffeomorphisms δ_{gc}
(ii) the local Lorentz invariance δ_{lL}
(iii) the local supersymmetry in (4).

One can now evaluate the commutator of any two local symmetries on any of the fields e_μ^m (=vierbein fields), Ψ_μ^a, S,P,A_m, and one finds then that this local algebra closes (Ferrara and van Nieuwenhuizen, 1978; Stelle and West, 1978). If one would not have added S,P,A_m, closure would not hold (Freedman and van Nieuwenhuizen, 1976). One can guess that one needs these auxiliary fields by counting the number of bosonic and fermionic *field* components

$$\begin{aligned}
&16\ (e_\mu^m)-4\ (\text{diffeom})-6\ (\text{loc. Lorentz}) &&= 6\ \text{bosonic}\\
&16\ (\Psi_\mu)-4\ (\text{local supersymmetry}) &&= 12\ \text{fermionic}\\
&\ 6\ (S,P,A_m) &&= 6\ \text{bosonic.}
\end{aligned} \tag{6}$$

(Recall that if an action has a local gauge invariance, there is a field component which does not occur in the action, cf. electromagnetism with $A_\mu=\partial_\mu A$).

After simple supergravity was constructed, the so-called extended supergravities with N>1 gravitinos were constructed. The first of these, the N=2 model, contains *two* gravitino fields, a graviton and ... a photon. It realizes Einstein's dream of unifying gravity with electromagnetism (Ferrara and van Nieuwenhuizen, 1976). There are clearly again equal numbers of boson and fermion states. Also one can add a set of auxiliary fields such that the action remains invariant and the local algebra closes (de Wit and van Holten, 1979). In this model the first dramatic improvement in the quantum behaviour of supergravity was found. The ultraviolet divergences of the first--order (one-loop) radiative corrections to the S-matrix (not to the Green's function!) cancel (Grisaru et al., 1976). Also the two-loop corrections cancel (Grisaru, 1977). However, for the three-loop corrections a possible candidate divergence does exist (Deser et al., 1977) and it remains unclear whether in an actual computation (which is too hard to be done for the time being) this divergence is actually present.

In d=4 spacetime dimensions one can construct more and more complicated models with more and more gravitinos. From N=3 gravitinos onwards, there seems no longer to exist a set of auxiliary fields (Rivelles and Taylor, 1981) such that the local algebra closes (it only closes "on-shell", i.e., up to terms proportional to fermion (Freedman and van Nieuwenhuizen, 1976) or boson (Townsend and van Nieuwenhuizen, 1982) field equations). Nevertheless, invariant actions still exist. The one-and two-loop corrections still are finite, but the three-loop situation remains unclear. There exists a largest model with N=8 gravitinos (de Wit and Nicolai, 1982a,b). The reason that one cannot construct models with N>8 is that the irreducible representations in terms of states of the (expected) corresponding superalgebra would contain spin values exceeding two. Since it is known (Berends et al.,1979) that one cannot couple spin J>2 to gravity in a consistent fashion, we have the N=8 barrier.

The N=8 model is the model which should describe the world; the other models seem not to be large enough. However, this N=8 model in d=4 dimensions is quite complicated, and therefore it comes as a relief as well as a surprise that one can reformulate it as a very simple model in ... d=11 dimensions (Cremmer et al., 1978).

Why 11? Because

eight 4-component spinors = one 32-component spinor

and spinors in d=11 have 32 components.(In d=10 dimensions spinors have only 16 independent components. Spinors can be left- or right-handed in d=10). The kinetic action in d=11 consists of a sum of the Hilbert-Einstein action, the Rarita-Schwinger action (the same form as in (3), but γ^μ denote now eleven 32×32 Dirac matrices) and ... a Maxwell action for a 3-index antisymmetric real tensor

$$A_{\mu\nu\rho}(x) \tag{7}$$

called photon henceforth. In d=4 the simplest model (with one graviton and the gravitino) did not need a photon, but in d=11 the simplest model needs an extra photon. There are again equal numbers of boson and fermion *states*:

$$\begin{aligned}
&44 \text{ bosons for transverse traceless } e_\mu^m \\
&84 \text{ bosons for transverse } A_{\mu\nu\rho}\ (84=\tbinom{9}{3}) \\
&128 \text{ fermions for transverse traceless } \psi_\mu^a \\
&(m,\mu,\nu,\rho=1,\ldots,11;\ a=1,\ldots,32).
\end{aligned} \tag{8}$$

(By a traceless gravitino we mean that on-shell one has the irreducibility criterion $\gamma^\mu\psi_\mu^a=0$). However, no auxiliary fields seem to exist. Note that this model resembles the N=2 extended supergravity model in d=4, in that it contains a graviton and a "photon", but that it contains only one gravitino. (In d=4 one can omit the photon and one gravitino and one gets then the N=1 d=4 model).

This action in d=11 is actually the only action one can write down in d=11. One cannot even add a cosmological term (one can do so in d=4) without destroying the local supersymmetry of the action! (Nahm, 1978; Nicolai et al., 1981). Thus: there is a highest dimension in which supergravity models exist (d=11), and in that dimension there is only one model and its action is unique and contains a bosonic and fermionic graviton and "matter" field $A_{\mu\nu\rho}$. We write "matter" instead of matter because all 3 fields are really part of a "super" field in d=11, because they are all three related by local supersymmetry transformations(which resemble (4)). All fields belong to the "marble" (left-hand) side of the Einstein equations rather than its "low-grade wood" (right-hand) side. Nevertheless we will write $A_{\mu\nu\rho}$ on the right-hand side for notational convenience. The field equations read

Einstein: $$R_{MN}=-\frac{1}{6}\left(F_{M\ldots}F_N{}^{\ldots}-\frac{1}{12}\eta_{MN}F^2_{\ldots}\right) \tag{9}$$

Maxwell: $$D^M F_{MNPQ}= =\frac{i\sqrt{2}}{12\times 96}\,\varepsilon_{NPQR_1..R_4S_1..S_4}F^{R_1..R_4}F^{S_1..S_4} \tag{10}$$

gravitino: $\gamma^{[M}\gamma^{N}\gamma^{R]}D_N\psi_R + \frac{\sqrt{2}}{96}F_{ABCD}$

$$(\gamma^{MABCDN}\psi_N + 12\eta^{MA}\gamma^{BC}\psi^D) = 0. \tag{11}$$

The indices M, N, A, B... run from 1 to 11, F_{MNPQ} is the curl of the 3-index photon with strength 24 and

$$\varepsilon_{1\ldots 11} = \varepsilon^{1\ldots 11} = +1. \tag{12}$$

(Relativists are not afraid of using as coordinates $x^4=ict$, but particle physicists become very unhappy with $\varepsilon_{1234}=+1$, and prefer $\varepsilon_{0123}=+1$. We follow the conventions with $x^4=ict$, and $x^4=x_4$).

We come now to Kaluza's ideas. He tried to unify in 1921 gravity and electromagnetism as a gravity theory in d=5. The 5×5 metric contained in its 4×4 part the usual metric, but the fifth row and column contained the photon and a scalar. (The scalar field was often in those days put equal to zero, and this in turn caused inconsistencies, but we will keep all fields as they come and will not find inconsistencies. The problem lies elsewhere: we have not seen all these scalar fields in Nature). This unification is thus different from N=2 d=4 supergravity. Nowadays that we have experimentally seen vector-bosons ($W^{\pm}$,Z) as well as a large number of fermions, we expect that we must go higher then 5 dimensions, in order to accommodate all these extra fields. To which dimension? Which action? Supergravity gives a unique answer to both questions: d=11 and the unique supersymmetric action in it.

2. KALUZA-KLEIN SUPERGRAVITY

The Kaluza-Klein program splits into two parts (for two extensive reviews see: Duff and Pope, 1983; van Nieuwenhuizen, 1983):
(a) find a solution of the d=11 classical field equations which can be used as a vacuum for the final 4-dimensional theory. As we shall discuss, *as a result* space-time compactifies into a d=4 anti-de Sitter space-time times a d=7 space M_7;
(b) expand all fluctuations (all arbitrary!) of these fields about these background fields into spherical harmonics of M_7. Their coefficients are the final x-space fields, and one must then find among the mass-

less fields the fields of our d=4 world. Massive modes, having masses of the order of the Planck mass, should not be important for our low-energy world. (The observed masses in d=4 should be generated by spontaneous symmetry breaking, and shall not be discussed here; it is (presumably) typical particle-physics work rather than relativity work).
The rest of this section is an explanation of (a) and (b).

One requires that the background fields can be considered as a vacuum for an observer in d=4 dimension. This means that these fields have a maximal d=4 (anti-de Sitter or Poincaré) symmetry. Technically this means that there are Killing vectors $f_A{}^\mu$ under which all d=11 fields are invariant (up to gauge transformations where gauge means: local Lorentz for vierbein and gravitino, and Maxwell for the photon) and that the Lie brackets of $f_A{}^\mu\partial_\mu$ generate the anti--de Sitter algebra. The most general background solution is then

$$g_{MN}=\begin{pmatrix} g_{\mu\nu}(x)f(y) & 0 \\ 0 & g_{\alpha\beta}(y) \end{pmatrix} \tag{13}$$

$$F_{MNPQ}=\left\{e\varepsilon_{mnpq} \text{ or } F_{abcd}(y)\right\}, \quad \Psi_M=0 \tag{14}$$

where $g_{\mu\nu}(x)$ is the d=4 anti-de Sitter metric and where $\mu,\nu(m,n)$ and x^μ are d=4 curved (flat) indices and coordinates, while $\alpha,\beta(a,b)$ and y^α are the same objects but in a d=7 space.

The possibility to allow an arbitrary function f(y) in the background solution seems to have gone unnoticed so far. Physically it means that d=11 spacetime splits *locally* into a stacking of 4-dimensional anti-de Sitter universes, each with a different constant curvature (depending on y). Globally this space needsnot be a direct product of a 4 dimensional spacetime and a 7 dimensional space. (Consider as an example $(ds)^2=d\theta^2+f(\theta)d\varphi^2$. For $f(\theta)=\sin^2\theta$ we have S_2, but for $f(\theta)\neq\sin^2\theta$ presumably $S_1\times S_1{}^{(1)}$).These global aspects have not yet been studied (as far as I know) but this is clearly an important topic.

Summarizing: if one wants a background with a maximal d=4 symmetry (symmetry under translations and rotations in 4 directions, so d=4 homogeneous and isotropic), then almost all fields are fixed. Only the function f(y), the constant e and the functions $g_{\alpha\beta}(y)$, $F_{abcd}(y)$ are arbitrary. Note that if one takes f(y)=1, *one deduces compactification* into AdS(4)×M_7 (where AdS(4) is 4-dimensional anti-de Sitter space and M_7 any 7-dimensional manifold).

In addition to being an acceptable vacuum, the background must also satisfy the field equations. The reason is that we would like to expand in quantum fields (=fluctuations around background fields in a path-integral), and not have terms linear in quantum fields ("tadpoles"). (If one had such tadpole terms, one could replace the background field φ(back) by a new φ'(back)=φ(back)+<φ(quant)>, and this would mean that φ'(back) would be a solution of the field equations). A painful aspect of most work in present-day Kaluza-Klein theory, is that one always takes the classical field equations. It may very well be that if one took the effective action instead of the classical action (the sum of all one-particle irreducible N-point functions, a nonlocal object nobody can compute), the solutions of the field equations would be very different from the solutions of the classical field equations. Many people believe that quantum physics is a small ripple on a classical background, but results of exactly solvable models in 1+1 dimensions have shown that the quantum spectrum is completely different from the classical spectrum. Today it seems impossible to solve exactly 1+3 dimensional models, so let us hope that the classical approach we follow is not completely wasted time.

So the question arises, in how far is the remaining freedom in the background fields with maximal d=4 symmetry further fixed by requiring that these fields satisfy the classical field equations in (9), (10), (11)? The result is the following (for f(y)=1, the case f(y)≠1 is under study). The Einstein equations yield $R_{am}=0$, while R_{mn} and R_{ab} fix F_{abcd} and e as follows

$$R_{mn}(x)=\eta_{mn}\left\{\frac{2}{3}\,e^2+\frac{1}{72}\,(F_{abcd}(y))^2\right\} \tag{15}$$

$$R_{ab}(y)=\eta_{ab}\left\{-\frac{1}{3}\,e^2+\frac{1}{72}\,(F_{abcd}(y))^2\right\} -\frac{1}{6}\,F_{acde}(y)F_b{}^{cde}(y). \tag{16}$$

Furthermore, the d=11 Maxwell equations are satisfied

if and only if $F_{abcd}(y)$ satisfies

$$D^aF_{abcd}=\frac{-e\sqrt{2}}{24}\varepsilon_{bcdefgh}F^{efgh}. \qquad (17)$$

One class of solutions consists in taking $F_{abcd}=0$. This class is called the Freund-Rubin (FR) class since they first proposed this solution of d=11 supergravity (Freund and Rubin, 1980). Note that in the FR class the d=4 metric is maximally symmetric and fixed (up to an overall scale) but that *any* solution of $R_{ab}=\Lambda\eta_{ab}$ yields a solution of d=11 supergravity. Typical solutions are: the round seven-sphere of Duff and Pope (1983) and numerous coset spaces (D'Auria et al.,1984; Castellani and Romans, 1983; Castellani et al., 1983a) including the physically interesting coset spaces $SU_3\times SU_2\times U_1/SU_2\times U_1\times U$ (Witten, 1981; Castellani et al., 1984). Also solutions of $R_{ab}=0$ yield solutions (for example, the space $K_3\times T_3$ where T_3 is the three-torus) (Duff et al., 1983); in this case one has evidently e=0 and $F_{abcd}=0$.

A second class of solutions is due to Englert (1982). He found a function $F_{abcd}(y)$ which satisfied both the Einstein and the Maxwell equations. The Maxwell equations reduce, in differential form, to

$$\delta dA=k{*}dA \text{ ; } k \text{ a constant.} \qquad (18)$$

Since $\delta\sim{*}d{*}$, one has ${*}dA\sim A$ or $dA\sim{*}A$. In other words, F_{abcd} is dual (with the d=7 ε symbol) of A_{abc}. If $A_{abc}(y)$ is transverse($D^aA_{abc}=0$ or $\delta A=0$) this means that A_{abc} must be an eigenfunction of the Hodge-de Rham operator $\Delta=\delta d+d\delta$ (the d'Alembertian in d=7). Given a compact space M_7, one can find the spectrum of Δ on three-forms. However, this is not enough, one still must satisfy the Einstein equations. In particular$(F_{abcd}(y))^2$ must be a constant (y-independent), see (15).

The Englert class of solutions uses as ingredients Killing spinors. A Killing spinor on M_7 is a spinor satisfying (Walker and Penrose, 1970; Duff, 1983; Duff and Pope, 1983)

$$D_\alpha\eta=ic\gamma_\alpha\eta \quad ; \; c \;\text{ a constant.} \qquad (19)$$

By iterating this equation one finds two integrability conditions (Englert et al., 1983) from

$$[D_\alpha, D_\beta]\,\eta \equiv 2c^2\gamma_{[\alpha}\gamma_{\beta]}\,\eta = \frac{1}{4}\,R_{\alpha\beta ab}\gamma^a\gamma^b\eta: \tag{20}$$

(i) M_7 must be an Einstein space.
(ii) $C_{abcd}\gamma^{[c}\gamma^{d]}=0$, where C_{abcd} is the d=7 Weyl tensor and $\gamma^{[c}\gamma^{d]}$ the SO(7) generators. (Since in d=7 a spinor has 8 components, the γ^c are 8×8 matrices). It was recently shown (Warner and van Nieuwenhuizen, 1984) that further integrability conditions exist, which must be satisfied; these fix in general the sign of c. (The absolute value of c follows from (20)). If one has such a Killing spinor, (which is the case if M_7 is such that *all* integrability conditions are satisfied), then

$$F_{bcde} = a\bar{\eta}\gamma_{[b}\gamma_c\gamma_d\gamma_{e]}\eta; \quad \text{a a constant} \tag{21}$$

is a solution of the Maxwell and Einstein equations, provided that

$$a(8c-\sqrt{2}e)=0, \quad 24e^2=-\left(\frac{1}{3}\,e^2+\frac{5}{3}\,a^2\right) \tag{22}$$

$$R_{mn}=\eta_{mn}\left(\frac{2}{3}\,e^2+\frac{7}{3}\,a^2\right), \quad R_{ab}=-\eta_{ab}\left(\frac{1}{3}\,e^2+\frac{5}{3}\,a^2\right). \tag{23}$$

Given the manifold M_7, c follows from (19), e from (22), a from (22) as well and (23) fixes the "radius" of the d=4 anti-de Sitter spacetime. Moreover, since an η exists M_7 is an Einstein manifold, and (23) fixes its over all scale. Hence indeed if M_7 has a spinor, there is an Englert solution. Clearly, M_7 is compact if M_4 is anti-de Sitter spacetime ($R<0$ on the sphere in our conventions).

One should not deduce from these results that supergravity *predicts* that our spacetime is 4 dimensional. One could have taken e purely imaginary. In this way, the signs in R_{ab} and R_{mn} would be interchanged, and it would be M_4 which is compact and M_7 which is open. In this case maximal d=7 symmetry would imply that $F_{abcd}(y)=0$ and $F_{mnrs}(x)$ is completely arbitrary, but F_{mnrs} is obviously proportional to a d=4 ε-symbol times a function g(x,y) and the Bianchi identities fix g(x,y)=constant. Hence in this case the Freund-Rubin and Englert ansätze coincide. Preliminary work seems to indicate that this

$$11=7+4 \quad \text{instead of} \quad 11=4+7 \tag{24}$$

is also possible. Interesting puzzles found in this model are: the massless scalar excitations in d=7 propagate with a *non*conformal field equation (in d=4 their field equation is conformal) and a new square root of antisymmetric tensors is found based on *a new duality concept in odd (4k-1) dimensions* (Pilch et al., 1983a,b). I mentioned these d=7 results, because they show how different physics in odd dimensions is from physics in even dimensions. Supersymmetric theories may give further clues why we live in d=4.

Let us now briefly mention a class of background solutions of the d=11 field equations which do not have a maximal d=4 symmetry. As an example, consider the metric of a d=4 black hole, and at each point of this four-dimensional spacetime one has a seven-sphere whose radius R depends on its distance from the origin of the black hole (Bais et al., 1984)

$$(ds)^2 = B(r)dt^2 + A(r)dr^2 + r^2 d\Omega_2 + R(r)^2 d\Omega_7 . \tag{25}$$

If one also allows a similar freedom in the photon curls

$$F_{mnrs} = ie(r)\varepsilon_{mnrs}, \quad F_{abcd} = a(r)\bar{\eta}\gamma_{[a}\gamma_b\gamma_c\gamma_{d]}\eta \tag{26}$$

then one can plug these ansätze into the coupled Maxwell-Einstein equations and find a solution. Interestingly enough, if $a(r\to\infty)$=constant, one has far away an Englert solution but if $a(r=0)$ would tend to zero, one would have a Freund-Rubin solution at r=0. Thus one would have an interpolating solution, with $a(r)$ playing the same role as the Higg's field in the monopole solution of 't Hooft-Polyakov. However, there are many questions raised by this kind of *local compactification*: what is the global topology (cf. our discussion of the function $f(y)$ in the metric g_{mn}). Clearly, for relativists there is much interesting work to do here.

There are, of course, very many solutions of the d=11 field equations (recall that any solution of the Einstein equations in compact d=7 space gives a solution). The question arises which solution one should take. In particular, could some solution "tunnel" to another solution, and if so, which solution is the lowest-lying (Witten, 1982)? Another aspect is the symmetry group of M_7. If M_7 is a coset space G/H, the background fields will have a group of space symmetries G, and the final d=4 field theory (see below) will have a local Yang-Mills group G. So, some people argue that $G=SU_3\times SU_2\times U_1$ should be taken. As Witten noticed (Witten, 1981; Castellani et al., 1984), in that case H is fixed namely H is the maximal subgroup $SU_2\times U_1\times U_1$ of G. (Recall: the dimension of M_7 will be the dimension of G-dimension of H, where dimension=number of generators. Hence dimG/H=(8+3+1)-(3+1+1)=7!). If one would take $G=SU_5$, even the maximal H would not be large enough to yield only 7 dimensions. Indeed, $H(\max)=SU_4\times U_1$ and now dimG/H=24-(15+1)=8. Since no d=12 supergravity exists (is that clear?) the grand unified group does not exist according to Kaluza-Klein supergravity. However, the mass spectrum of the $SU_3\times SU_2\times U_1$ spaces (there are many ways to embed H into G) does not seem to contain left-handed neutrinos, and also quarks have not been found (D'Auria and Fré, 1983). This is a very serious problem although, maybe when the new accelerators have been built, right-handed "neutrinos" with heavy masses will be found. In that case one should still explain why left-particles are so much lighter. Other people prefer the round S_7 as background (Duff and Pope, 1983) since it is the most symmetrical choice. They may be right, but in that case one does not even find straight-away a local $SU_3\times SU_2\times U_1$ Yang-Mills group, but only a local SO(8) (recall that S_7=SO(8)/SO(7)). There exists a scenario for getting nevertheless bound states with an SU_5 (or other) symmetry group, but it is at this stage rather a "scenario".

We must now discuss the fluctuations, because they will yield the final d=4 fields. Let us assume that M_7 is a coset manifold G/H. If one has a coset manifold M_7, one can expand every field into a complete set of spherical harmonics (Salam and Strathdee, 1982). For example, denoting the fluctuations of, for example $g_{\mu\nu}(x)$ by $h_{\mu\nu}(x,y)$, this *arbitrary* field $h_{\mu\nu}(x,y)$ can be expanded as follows (Castellani et al., 1983b,c)

$$h_{\mu\nu}(x,y)=\sum H_{\mu\nu}^{N_1}(x)\,Y^{N_1}(y) \tag{27}$$

where N_1 labels all scalar spherical harmonics. On S_7, for example, they correspond to the Young diagrams

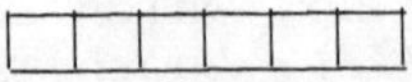

The mode Y^{N_1}=constant corresponds to the graviton.

There are also vector, tensor, etc., spherical harmonics needed. For example

$$h_{\mu\alpha}(x,y)=\sum B_{\mu}^{N_7}(x)Y_{\alpha}^{N_7}(y)+B_{\mu}^{N_1}(x)D_{\alpha}Y^{N_1}(y). \qquad (28)$$

If and only if $Y_{\alpha}^{N_7}(y)$ is a Killing vector of M_7, $B_{\mu}^{N_7}(x)$ is a massless Yang-Mills field. One can fix the internal diffeomorphisms by the de Donder or Lorentz gauge $D^{\alpha}h_{\alpha\mu}=0$, and in this gauge all $B_{\mu}^{N_1}=0$. Having thus expanded all d=11 field fluctuations (namely, $h_{\mu\nu}$, $h_{\mu\alpha}$, $h_{\alpha\beta}$, Ψ_{μ}, Ψ_{α}, $A_{\mu\nu\rho}$, $A_{\mu\nu\alpha}$, $A_{\mu\alpha\beta}$, $A_{\alpha\beta\gamma}$) into G/H spherical harmonics, one then plugs these expansions into the field equations. One collects all terms with a given spherical harmonic, and thus obtains a set of d=4 field equations. For example, one finds

$$\left(D^{\mu}(D_{\mu}B_{\nu}^{N_7}-D_{\nu}B_{\mu}^{N_7})+(\Delta_{y}+6)B_{\nu}^{N_7}\right)Y_{\alpha}^{N_7}(y)=0. \qquad (29)$$

Since Δ_y is diagonal on Y^{N_7} one thus finds the spectrum. Methods have been and will be further developed to obtain the spectrum of *all* invariant operators on M_7 (of which the Hodge-de Rham operator is only one) (D'Auria and Fré, 1983). One often finds coupled field equations at the linearized level, which one must diagonalize. The matrices involved in this diagonalization process are even N_i dependent when the modes are massive.

We have repeatedly been using the words "massless modes", but it is not clear at all what massless means. In an anti-de Sitter spacetime in d=4, a massless gravitino needs in the action an extra term which looks like a mass term. For example in the action of simple supergravity in (2) one must add (Townsend, 1977)

$$\mathcal{L}(\text{cosm})=12g^2+g\bar{\Psi}_{\mu}\gamma^{[\mu}\gamma^{\nu]}\Psi_{\nu}. \qquad (30)$$

However, in the N=8 d=4 model the spin 1/2 fields do not need such an extra mass term, although the scalars need such a term (de Wit and Nicolai, 1982a,b). The d=4 scalar equation is conformally invariant but the d=7 scalar equation is not, as we saw. In the N=4 d=4 model (Freedman and Schwarz, 1978), on the other hand, the spin 1/2 fields get an extra mass term (but this model is really a spontaneously broken model). So far, the strategy to determine what the massless modes in a model are, has been to see which modes are related to the graviton by local supersymmetry transformations. However, it is clear that one should be able to answer this question without the help of supergravity. Presumably, the properties of Green's functions, which are quite different in different dimensions, should be used to determine what particles are massless.

The recent Kaluza-Klein work is fun, and one learns (and simplifies) a lot of existing differential geometry. It is a rigorous and clear field of research (but since these concepts are known to be complementary in successful physical theories, this may not be such a good thing). There loom Two Terrible Threats:

(i) the chirality problem. So far no left-handed neutrinos have been found;

(ii) the cosmological constant problem. Let us explain this.

In the classical solutions of the Freund-Rubin or Englert class, the only free parameter is e; it determines the "radius" R of M_7. One would like this radius to be small, since then the excited modes will have very high masses. Also, the Yang-Mills coupling constant is proportional to $g=\kappa/R$, and since g is experimentally of order unity, R must be of the order of the Planck length. To achieve this one must make the d=7 cosmological constant $\Lambda(7) = -\frac{1}{3} e^2 - \frac{5}{3} a^2$ in (23)

very large in absolute value. But then $\Lambda(4) = \frac{2}{3}e^2 + \frac{7}{3} a^2$ will also be enormous. Here one sees how painful the result is that one cannot add to the d=11 action a cosmological constant without destroying supersymmetry: one cannot use such a freedom to fine-tune $\Lambda(4)$ to zero. Obviously, this is a very serious problem. It may at a very deep level be connected to quantum rather than classical gravity. At this moment of writing we do not know whether the free function f(y) in $g_{\mu\nu}(x,y)$, discussed earlier, can help.

Einstein, when he heard of Kaluza's work, was fascinated by it (and kept Kaluza's paper which he had to referee for a very long time on his desk). Einstein might also have been fascinated by the discovery of supergravity. After all, if particles are singularities of the gravitational field, as he believed, there should be a fermionic gravitational field since there exist fermions. Combining these two fascinating fields yields a beautiful new area of research. Let us hope that Nature is aware of it.

APPENDIX

Letter of A.Einstein to Dr. Kaluza, found by C. Hoenselaers and printed for the first time in de Sabbata and Schmutzer, 1983.

14.X.21

Most revered Dr. Kaluza,

I am having second thoughts about having restrained you from publishing your idea on a unification of gravitation and electricity two years ago. Your approach seems in any case to have more to it than the one by H. Weyl. If you wish I shall present your paper to the academy after all, provided you send it to me. That, however, could happen only at the end of November, as I shall be absent from Berlin until then.

With my distinguished respects

A. Einstein
Haberlandstr. 5
Berlin

As he wrote in a postcard dated 9.XII.1921. Einstein presented the mentioned paper to the academy on 8.XII.1921.

NOTES

+ This work was supported in part by NSF Grant #PHY 81-09110 A-0.

* On leave from Institute for Theoretical Physics, SUNY at Stony Brook Long Island, NY 11794 USA.

(1) I thank C.N.Pope for this example.

REFERENCES

Bais, F.A., van Baal, P., and van Nieuwenhuizen, P.: 1984, "Local Compactification and Black Holes", *Nucl. Phys.*, B, to be published.

Berends, F.A., de Wit, B., van Holten, J.W., and van Nieuwenhuizen, P.: 1979, "On Field Theory for Massive and Massless spin-$\frac{5}{2}$ particles", *Nucl.Phys.*, B154, 261.

Biran, B., Englert, F.,de Wit, B., and Nicolai, H.: 1983, "Gauge N=8 Supergravity and its Breaking from Spontaneous Compactification", *Phys.Lett.*, 124B, 45.

Castellani, L., and Romans, L.J.: 1983, "N=3 and N=1 Supersymmetry in a new Class of Solutions for d=11 Supergravity", Calt-68-1053 preprint.

Castellani, L., Romans, L.J., and Warner, N.P.: 1983a, "Symmetries of Coset Spaces and Kaluza-Klein Supergravity", Calt-68-1055 preprint.

Castellani, L., D'Auria, R., Fré, P., Pilch, K., and van Nieuwenhuizen, P.: 1983b, "Fermion Masses for the Englert Solution on the Round 7 Sphere", Stony Brook preprint, 83/60.

Castellani, L., D'Auria, R., Fré, P., Pilch, K., and van Nieuwenhuizen, P.: 1983c, "The Bosonic Mass Formula for the Freund-Rubin Solutions of d=11 Supergravity on General Coset Manifolds", Stony Brook preprint 83/60.

Castellani, L., D'Auria, R., and Fré, P.: 1984, "$SU_3 \times SU_2 \times U_1$ from d=11 Supergravity", *Nucl. Phys.*, B, to be published.

Cremmer, E., Julia, B., and Scherk, J.: 1978, "Supergravity Theory in 11 Dimensions", *Phys. Lett.*, 76B, 409.

D'Auria, R., and Fré, P.: 1983, "On the Fermion Mass Spectrum of Kaluza-Klein Supergravity", ITP-SB-83-57 preprint.

D'Auria, R., Fré, P., and van Nieuwenhuizen, P.: 1984, "N=2 Matter Coupled Supergravity from Compactification on Coset G/H Possessing an Additional Killing Vector", *Phys. Lett.*, to be published.

De Sabbata, V., and Schmutzer, E. (eds): 1983, *Unified Field Theory of more than 4 Dimensions, Including Exact Solutions*, World, Scient. Publ. Co., Singapore.

Deser, S., Kay, J., and Stelle, K.: 1977, "Hamiltonian Formulation of Supergravity", *Phys. Rev.*, D16,2448.

Deser, S., and Zumino, B.: 1976, "Consistent Supergravity", *Phys. Lett.*, 62B, 335.

de Wit, B., and Nicolai, H.: 1982a, "N=8 Supergravity with Local SO(8)×SU(8) invariance", *Phys. Lett.*, 108B, 285.

de Wit, B., and Nicolai, H.: 1982b, "N=8 Supergravity", *Nucl. Phys.*, B208, 323.

de Wit, B., and van Holten, J.W.: 1979, "Multiplets of Linearized SO(2) Supergravity", *Nucl. Phys.*, B155, 530.

Duff, M.J.: 1983, "Supergravity, the Seven-Sphere, and Spontaneous Symmetry Breaking", *Nucl. Phys.*, B219, 389.

Duff, M.J., and Pope, C.N.: 1983, "Kaluza-Klein Supergravity and the Seven-Sphere", in S.Ferrara, J.G. Taylor, and P.van Nieuwenhuizen (eds), *Supersymmetry and Supergravity '82*, World Scient. Publ. Co., Singapore.

Duff, M.J., and Nilsson, B.E., and Pope, C.N.: 1983, "Compactification of d=11 Supergravity on K3×T3", *Phys. Lett.*, 129, 39.

Englert, F.: 1982, "Spontaneous Compactification of Eleven-Dimensional Supergravity", *Phys. Lett.*, 119B, 339.

Englert, F., Rooman, M., and Spindel, P.: 1983, "Supersymmetry Breaking by Torsion and the Ricci-Flat Squashed Seven-Spheres", *Phys. Lett.*, 127B, 47.

Ferrara, S., and van Nieuwenhuizen, P.: 1976, "Consistent of Supergravity with Complex Spin-$\frac{3}{2}$ Gauge Field", *Phys. Rev. Lett.*, 37, 1669.

Ferrara, S., and van Nieuwenhuizen, P.: 1978, "The Auxiliary Fields of Supergravity", *Phys. Lett.*, 74B, 333.

Freedman, D.Z., and Schwarz, J.H.: 1978, "N=4 Supergravity Theory with Local SU(2)×SU(2) Invariance", *Nucl. Phys.*, B137, 333.

Freedman, D.Z., and van Nieuwenhuizen, P.: 1976, "Properties of Supergravity Theory", *Phys.Rev.*, D14, 912.

Freedman, D.Z., van Nieuwenhuizen, P., and Ferrara, S.: 1975, "Combined Supersymmetric and Gauge-Invariant Field Theories", *Phys. Rev.*, D12, 2286.

Freund, P.G.O., and Rubin, M.A.: 1980, "Dynamics of Dimensional Reduction", *Phys. Lett.*, 97B, 233.

Goldfand, Y.A., and Linkhtam, E.P.: 1971, "Extension of the Algebre of Poincaré Group Generations and Violation of P Invariance", *JETP Lett.*, 13, 323.

Grisaru, M.T.: 1977, "Two-Loop Renormalizability of Supergravity", *Phys. Lett.*, B66, 75.

Grisaru, M.T., van Nieuwenhuizen, P., and Vermaseren, J.A.M.: 1976, "One-Loop Renormalizability of Pure Supergravity and of Maxwell-Einstein Theory in Extended Supergravity", *Phys. Rev. Lett.*, 37, 1662.

Nahm, W.: 1978, "Supersymmetries and their Representations", *Nucl. Phys.*, B135, 149.

Nicolai, H., Townsend, P.K., and van Nieuwenhuizen, P.: 1981, "Comments on 11-Dimensional Supergravity", *Nuovo Cimento Lett.*, 30, 315.

Pilch, K., van Nieuwenhuizen, P., and Townsend, P.K.: 1983a, "Compactification of d=11 Supergravity on S_4 (or 11=7+4)", ITP-SB-83-51 preprint.

Pilch, K., van Nieuwenhuizen, P., and Towsend, P.K.: 1983b, "Self Duality in Odd Dimensions", *Phys. Lett.*, B, to be published.

Rivelles, V.D., and Taylor, J.G.: 1981, "Off-Shell Extended Supergravity and Central Charge", *Phys. Lett.*, 104B, 131.

Salam, A., and Strathdee, J.: 1982, "On Kaluza-Klein Theories", *Ann.Phys.*, 141, 316.

Stelle, K.S., and West, P.C.: 1978, "Maximal Auxiliary Fields for Supergravity", *Phys. Lett.*, 74B, 330.

Townsend, P.K.: 1977, "Small-Scale Structure of Spacetime as the Origin of the Gravitational Constant", *Phys. Rev.*, D15, 2795.

Townsend, P.K., and van Nieuwenhuizen, P.: 1982, "Which Field Representations are Allowed in Interacting Locally Supersymmetric Field Theories?", *Phys. Lett.*, 109B, 179.

van Nieuwenhuizen, P.: 1983, "An Introduction to Simple Supergravity and the Kaluza-Klein Program", in B.F. DeWitt and R.Stora (eds) *Proceedings of the 1983 Les Houches Summer School*, North-Holland, Amsterdam.

Volkov, D.V., and Akulov, V.P.: 1972, "Possible Universal Neutrino Interaction", *JETP Lett.*, 16, 438.

Walker, M., and Penrose, R.: 1970, "On Quadratic First Integrals of the Geodesic Equations for Type {22} Spacetimes", *Commun. Math. Phys.*, 18, 265.

Warner, N., and van Nieuwenhuizen, P.: 1984, "Integrability Condition of Killing Spinors", *Commun. Math. Phys.*, to be published.

Wess, J., and Zumino, B.: 1974, "Supergauge Transformations in four Dimensions", *Nucl. Phys.*, B70,39.

Wess, J., and Zumino, B.: 1978, "Superfield Lagrangian for Supergravity", *Phys. Lett.*, 74B, 51.

Witten, E.: 1981, "Search for a Realistic Kaluza-Klein Theory", *Nucl. Phys.*, B186, 412.

Witten, E.: 1982, "Instability of the Kaluza-Klein Vacuum", *Nucl. Phys.*, B195, 481.

THE QUANTUM MECHANICS OF THE UNIVERSE

Steven W. Hawking

Department of Applied Mathematics and
Theoretical Physics, University of Cambridge,
Silver Street, Cambridge CB3 9EW, England

Classical general relativity predicts that the universe had a singular origin. I show that the singularity can be removed by quantum mechanics, just as in the case of the classical model of the atom. I propose that the quantum state of the universe is defined by a path integral over compact positive definite metrics. I show that in a simple model this boundary condition leads to a wave function which can be regarded as a superposition of wave functions peaked around classical oscillating solutions with a long inflationary period.

Because gravity is long range and always attractive, it dominates over all other interactions and causes the universe to be unstable: it has to contract or expand. In the classical theory of general relativity, this instability leads to singularities of infinite curvature, like the Big Bang, at which the concepts of space and time and the laws of physics break down (Hawking and Ellis, 1973). There is an obvious similarity between this collapse of the universe and the collapse of the atom which was predicted by classical physics at the beginning of the century. The problem of the atom was solved by quantum mechanics, indeed, it was one of the main motivations for its invention. In this talk I shall try to show that quantum mechanics may also solve the problem of space-time singularities. I shall put forward a proposal for the quantum state of the universe which does not have any

B. Bertotti et al. (eds.), General Relativity and Gravitation, 491–498.

singularities or arbitrary initial conditions but which seems, at least in simple models, to lead to a universe like the one we observe.

I shall adopt the Euclidean approach to quantum gravity (Hawking, 1979a,b). I shall assume that the probability for a Euclidean, i.e. positive definite, metric $g_{\mu\nu}$ and a matter field configuration ϕ is proportional to

$$\exp(-\tilde{I}[g_{\mu\nu},\phi]) \tag{1}$$

where $\tilde{I}$ is the Euclidean action

$$\tilde{I}[g_{\mu\nu},\phi] = \frac{m_p^2}{16\pi}\left\{-\int_{\partial M} 2Kh^{\frac{1}{2}}d^3x - \int_M (R-2\Lambda+16\pi m_p^{-2}L(g_{\mu\nu},\phi))g^{\frac{1}{2}}d^4x\right\} \tag{2}$$

where h_{ij} is the 3-metric on the boundary ∂M and K is the trace of the second fundamental form of the boundary. The surface term is necessary because the curvature scalar R contains second derivatives of the metric which have to be removed by an integration by parts. This produces a surface term which is cancelled by the surface term in equation (2) (York, 1972; Gibbons and Hawking, 1977).

The physics of the universe is governed by probabilities of the form (1) for all metrics $g_{\mu\nu}$ and matter field configurations ϕ that belong to some class C. The specification of the class C determines the quantum state of the universe. There are two, and only two, natural choices for the class C. They are:

(a) Compact metrics $g_{\mu\nu}$ and matter fields ϕ that are regular on them.
(b) Metrics that are asymptotic to a non-compact space of maximal symmetry such as flat Euclidean space or Euclidean anti-de Sitter and matter fields that are asymptotically zero.

Metrics of type (b) define the vacuum state that is used in particle scattering calculations where one sends particles in from infinity and measures what comes back to infinity. However, in cosmology one is interested in observations, not at infinity, but in a finite region in the interior of the spacetime. It does not really matter whether or not this finite region is connected to an infinite asymptotic region. If one used metrics of type (b), there would be two kinds of metrics that contributed to the probability of an observable:

(i) Connected asymptotically Euclidean or anti-de Sitter metrics.

(ii) Disconnected metrics which consisted of a compact part that was connected to the region of observation and a separate asymptotically Euclidean or anti-de Sitter part.

One cannot exclude disconnected metrics from C because they can be approximated arbitrarily closely by connected metrics in which the disconnected parts were joined by thin tubes with negligible action. It can be shown that the contribution to the probability from metrics of kind (ii) dominates over that from kind (i). This is not the case for particle scattering calculations because disconnected compact parts do not contribute to the scattering of particles from infinity. Thus even if one adopted boundary conditions of type (b) for the quantum state of the universe, the result, as far as observations in a finite region were concerned, would be almost the same as if one had adopted boundary conditions of type (a). It therefore seems more natural to postulate that the quantum state of the universe is defined by metrics of type (a). This incorporates the idea that the universe is completely self-contained and has no external asymptotic region. One could say "The boundary conditions of the universe are that it has no boundary". It should be emphasised that this is a *proposal* for the boundary conditions of the universe. The real test is whether this proposal leads to predictions that are in agreement with observations. I shall show that this is the case, at least in a simple toy model. One is lead to a picture in which there are no singularities and in which the universe oscillates with a long period of exponential expansion. To be more precise, the quantum state of the universe that arises from these boundary conditions can be interpreted as a superposition of states that are peaked around different oscillating universes in the classical limit.

In interpreting the quantum mechanics of the universe one is usually interested, not in the probability of an entire 4-metric $g_{\mu\nu}$, but in the probability of observing some more restricted set of quantities. Such probabilities can be deduced from the basic probability (1) by integrating over the unobserved quantities. A particularly important case for cosmology is the probability $P[h_{ij},\phi]$ of finding a closed compact

3-submanifold S which divides the 4-manifold M into two parts, M_+ and M_-, and on which the induced 3-metric is h_{ij} and the matter field configuration is ϕ:

$$P[h_{ij},\phi]=\int_C d[g_{\mu\nu}]d[\phi]\exp(-\tilde{I}[g_{\mu\nu},\phi]) \tag{3}$$

where the integral is over all 4-metrics $g_{\mu\nu}$ and matter field configurations ϕ that belong to the class C and which induce the given 3-metric and matter field configuration on S. One can factorize the *probability* P into the product of two *amplitudes*:

$$P\left[h_{ij},\phi\right]=\Psi_+\left[h_{ij},\phi\right]\Psi_-\left[h_{ij},\phi\right] \tag{4}$$

where

$$\Psi_\pm=\int_{C_\pm} d\left[g_{\mu\nu}\right]d[\phi]\exp(-\tilde{I}[g_{\mu\nu},\phi])\ . \tag{5}$$

The integrals are over the classes $C_\pm$ of metrics and matter fields on the two halves $M_\pm$ of the 4-manifold M. In the case of boundary conditions of type (a), Ψ_+ and Ψ_- are real and equal. I shall drop the subscripts $\pm$ and shall call them "The Wave Function of the Universe" (Hartle and Hawking, 1983).

The wave function $\Psi[h_{ij},\phi]$ does not depend on time explicitly but only implicitly through its dependence on the 3-metric h_{ij} which determines the position of S in M. This means that it is unchanged if one moves the surface S a distance N along the unit normals to S in the path integral (5) i.e.

$$\int d\left[g_{\mu\nu}\right]d[\phi]\frac{\delta\tilde{I}}{\delta N}\exp(-\tilde{I})=0. \tag{6}$$

This implies that Ψ must obey a functional differential equation, the Wheeler-DeWitt equation, which is the analogue of the Schroedinger equation (Hartle and Hawking, 1983; Hawking, 1983a,b):

$$\left[-G_{ijkl}\frac{\delta^2}{\delta h_{ij}\delta h_{kl}}-h^{\frac{1}{2}}\left(R(h)+2\Lambda+ 8\pi m_p^{-2}T_{nn}\left(\frac{\delta}{\delta\phi},\phi\right)\right)\right]\Psi\left[h_{ij},\phi\right]=0 \tag{7}$$

where T_{nn} is the projection of the Euclidean energy momentum tensor normal to the surface S and the time derivatives of the matter fields ϕ are replaced by functional derivatives $\frac{\delta}{\delta\phi}$. The quantity G_{ijkl} is

$$G_{ijkl}=\tfrac{1}{2}h^{\frac{1}{2}}(h_{ik}h_{jl}+h_{il}h_{jk}-h_{ij}h_{kl}). \tag{8}$$

It can be regarded as a metric on superspace, the space of all 3-metrics. It has signature (- + + + + +). The Wheeler-DeWitt equation can therefore be regarded as a hyperbolic wave equation on superspace with $\sqrt{h}$ as the time coordinate. The boundary conditions at small values of $\sqrt{h}$ can be deduced from the choice of compact metrics for the class C that defines the quantum state of the universe. One uses the saddle point approximation for the wave function Ψ:

$$\Psi\left[h_{ij},\phi\right]\cong N\textstyle\sum_i A_i\exp(-B_i) \tag{9}$$

where N is a normalization factor, B_i are the actions of compact solutions of the field equations that are bounded by S and which agree with the given 3-metric and matter field configuration on S. The prefactors A_i are given by the determinants of the operators governing small fluctuations around the classical solutions. I shall ignore them in what follows and concentrate on the more important quantities B_i.

In order to find the wave function of the universe one simply has to solve the Wheeler-DeWitt equation (7) with the boundary conditions (9). This is rather hard to do because the Wheeler-DeWitt equation is an infinite dimensional functional differential equation. However, one can get some idea of the solution by considering the equation restricted to a finite dimensional submanifold of superspace, called minisuperspace. In other words, one freezes out all but a finite number of the degrees of freedom of the gravitational and matter fields. The Wheeler-DeWitt equation then becomes a finite dimensional wave equation and can be solved with the boundary conditions (9) using standard techniques.

I shall discuss here a simple minisuperspace model which consists of a homogeneous isotropic universe containing a massive scalar field ϕ. Further details and other models can be found in Hartle and Hawking, 1983; Hawking, 1983a,b; Moss and Wright, 1983. The Wheeler-DeWitt equation is

$$\frac{1}{2}\left[\frac{1}{a^p}\frac{\partial}{\partial a}a^p\frac{\partial}{\partial a} - a^2 - \frac{1}{a^2}\frac{\partial^2}{\partial \phi^2} + a^4 m^2 \phi^2\right]\Psi[a,\phi] = 0 \qquad (10)$$

where a is the radius of the universe, ϕ is the scalar field and m is the mass.The index p represents some, but not all, the uncertainty in the factor ordering in the Wheeler-DeWitt equation. Its value will not be important in what I have to say.

The causal structure of the Wheeler-DeWitt equation is seen more easily by introducing new coordinates

$$x = a\sinh\phi \qquad y = a\cosh\phi \qquad (11)$$

In these coordinates, the second derivatives take the form $\partial^2/\partial y^2 - \partial^2/\partial x^2$ thus, the characteristics are lines at 45° in the (x,y) plane. The region a>0 is mapped into the interior of the future light cone of the origin. The surfaces of constant ϕ are straight lines through the origin and the surfaces of constant a are spacelike hyperbolae within the light cone.

If one knew the solution on the light cone of the origin, one could solve the wave equation by standard techniques. The light cone is the surface a=0, $\phi=\pm\infty$. It is therefore rather hard to apply the boundary conditions (9) there but one can use (9) to estimate the form of the wave function near the light cone for $|\phi|>>1$, $a<(m|\phi|)^{-1}$:

$$\Psi(a,\phi) \cong N\exp\left[\frac{1}{3m^2\phi^2}\left(1-(1-m^2\phi^2a^2)^{3/2}\right)\right] . \qquad (12)$$

The potential $-a^2+a^4m^2\phi^2$ in the Wheeler-DeWitt equation (10) changes sign on the surface $a=(m|\phi|)^{-1}$ which is spacelike for $|\phi|>1$. For a slightly greater than $(m|\phi|)^{-1}$ the wave function will oscillate:

$$\Psi(a,\phi)\cong N\exp\left(\frac{1}{3m^2\phi^2}\right)\cos\left[(m^2\phi^2a^2-1)^{3/2}/(3m^2\phi^2)-\frac{\pi}{4}\right].\quad(13)$$

One can solve the Wheeler-DeWitt equation with the boundary conditions (12) and (13) near the light cone. Because the wave function is rapidly oscillating one can use the WKB method. One writes the wave function in the form

$$\Psi(a,\phi)=C(a,\phi)\cos S(a,\phi)\quad(14)$$

where $S(a,\phi)$ is a rapidly varying phase and $C(a,\phi)$ is a slowly varying amplitude. The trajectories of ∇S in the (a,ϕ) plane correspond to classical Lorentzian solutions $(a(t),\phi(t))$ of the field equations. The semi-classical approximation for large $|\phi|$ corresponds to a classical solution which starts with $\phi(t)=\phi_1, d\phi/dt=0$, at a minimum radius $a(t)=(m|\phi_1|)^{-1}$ at $t=0$ and then expands in a de Sitter-like manner. The mass term tends to make $|\phi(t)|$ decrease with a timescale $3H/m^2$. However, the expansion rate $H=m|\phi_1|$. Thus if $|\phi_1|>1$ the universe will expand by a factor of order $\exp(3\phi_1^2)$ before $|\phi|$ decreases significantly. For large $|\phi_1|$ this gives a long inflationary period.

Eventually $|\phi(t)|$ decreases to zero and starts oscillating in time. The universe goes over to a matter dominated phase with $a(t)\propto t^{2/3}$. The universe will expand to a maximum radius of order $1/m\phi_1\exp(6\phi_1^2)$ and then recollapse. A classical solution that starts with a general value of ϕ_1 will collapse to a singularity but there are special values of ϕ_1 for which the universe oscillates indefinitely. The wave function Ψ can be interpreted as a superposition of wave functions that are peaked around these oscillating non-singular solutions. Such a wave function would require the "Many Worlds" interpretation of quantum mechanics.

To summarize, quantum mechanics seems to solve the problem of the collapse of the universe just as it solved the problem of the collapse of the atom. If the quantum state of the universe really is defined by a path integral over compact positive definite metrics, then we will have solved the problem of the boundary conditions of the universe. All we will then need will be a complete set of local laws governing the physical fields, and we should be able, at least in principle, to calculate the probability of anything in the universe. In practice, of course, the calculations will be too complicated in any but the very simplest of cases.

REFERENCES

Gibbons, G.W., and Hawking, S.W.: 1977, "Action integrals and partition functions in quatum gravity", *Phys. Rev.*, D15, 2752.

Hartle, J.B., and Hawking, S.W.: 1983, "Wave Function of the Universe", *Phys. Rev.*, D, to be published.

Hawking, S.W.: 1979a, "The Path-integral approach to quantum gravity", in S.W.Hawking, and W.Israel (eds), *General Relativity: An Einstein Centenary Survey*, Cambridge University Press, Cambridge, p. 746.

Hawking, S.W.: 1979b, "Euclidean Quantum Gravity", in S.Deser, and M.Levy (eds), *Recent Developments in Gravitation*, Plenum Press, New York, p. 145.

Hawking, S.W.: 1983a, "Quantum Cosmology", in *Les Houches lectures*, to be published.

Hawking, S.W.: 1983b, "The Quantum State of the Universe", University of Cambridge preprint.

Hawking, S.W., and Ellis, G.F.R.: 1973, *The Large Scale Structure of Space-Time*, Cambridge University Press, Cambridge.

Moss, I.G., and Wright, W.A.: 1983, "The Wave Function of the Inflationary Universe", Newcastle University preprint.

York, J.M.: 1972, "Role of Conformal Three-Geometry in the Dynamics of Gravitation", *Phys. Rev. Lett.*, 28, 1082.

Highlights of the Conference

HIGHLIGHTS OF THE CONFERENCE

John Archibald Wheeler

Center for Theoretical Physics
University of Texas
Austin, Texas 78712, USA

1. RELATIVITY AND GRAVITATION: THE GREAT TRADITION

As we come to the end of this fascinating conference on gravitation, we are reminded how close this historic city lies, geographically as well as spiritually, to the founders of our subject. On the town wall itself, even if perhaps in not quite the right place, we read an inscription which is certainly right in spirit, "Da questa torre Galileo molta via de cieli svelò". Only a few miles from here at Selasca on Lake Maggiore on July 20, 1866 Bernhard Riemann died at age 39, devoting his last days to a unified theory of gravity and electromagnetism and, as an independent enterprise, giving us that system of numbers for characterizing multiply-connected topologies which is named after Betti, but identified with the symbol, R_n, that honours Riemann. He inspired Einstein, that founder of our subject, who tells us in his own words, "Only the genius of Riemann, solitary and uncomprehended, had already won its way by the middle of the last century to a new conception of space, in which space was deprived of its rigidity, and in which its power to take part in physical events was recognized as possible." And it is not far from Padova to Bern where Einstein gave us, in 1905, special relativity, an event celebrated in the same place fifty years later by the first of these conferences, initiated - along with our organization - by the statesmanship of André Mercier.

The summarizer of that first in this series of conferences was Wolfgang Pauli. His words deserve as

B. Bertotti et al. (eds.), General Relativity and Gravitation, 501–509.

much attention now as they did then, "Ich glaube, das Wichtigste, das wir sonst gehört haben, war das Referat von Lichnerowicz über das *Cauchysche Anfangwertproblem* in den nichtlinearen Feldgleichungen der allgemainen Relativitätstheorie. Ich lege sehr grossen Wert auf das Studium dieser Probleme, weil ich bestimmt meine, dass es auch bei der Feldquantisierung, auf die ich zu sprechen komme, eine wesentliche Rolle spielen wird". Happily for us, our wise organizers made the initial value problem topic number one in the series of five workshops entitled A1, A2, A3, A4, A5, on the initial value problem, on exact solutions and singularities, on equations of motion, gravitational radiation and the asymptotic structure of space-time, on numerical relativity, and on classical and quantum alternative approaches to gravitational theories.

Highlights of the "A" and other workshops have been so well summarized by our distinguished rapporteurs, M.A.H. MacCallum, I. Novikov, P. Bender and P. van Nieuwenhuizen, and the plenary addresses were so splendid, that I can limit myself to a modest undertaking in my own "summary of a summary". I have gone around day after day and asked one participant and another, what report do *you* think is the most interesting, or controversial, or important? And can you put your point in a single simple sentence? But for the conference as a whole I got my summary sentence from Einstein himself, "... at any given moment out of all conceivable constructions (theories) a single one has always proved itself absolutely superior to all the rest..."

The central idea of general relativity we see in simpler terms with each passing year. Today we say, "Space tells mass how to move, and mass tells space how to curve". Tomorrow we shall surely grasp it still more simply. But now after almost seventy years, general relativity proves itself ever richer in inspiration to both experiment and theory, and both for great groups and for isolated individuals.

2. MAKING THE MOST OF METRICS

Isolated individuals? MacCallum was too modest to describe in his summary of the A workshops what he and his associates are doing for the isolated worker on the mathematics of relativity; but more than one collegue has spoken of its significance. Years ago the great astronomer Harlow Shapley loaned out telescopes in large numbers to isolated amateur astronomers all over the world. In return they undertook to provide measurements on selected variable stars. Those measurements have since proved of decisive value to the progress of astronomy. What about the *metrics* produced by scores of deeply interested investigators all over the world? We now see taking shape a new machine, which we might call the "MacCallum centrifuge" (Figure 1). Feed metrics into it! It will separate them into categories, "silver", "gold", "platinum", ... Then each can be most clearly categorized and put to the best use. Thanks to it, everyone will be able to contribute more effectively than ever to the common enterprise that we call "general relativity".

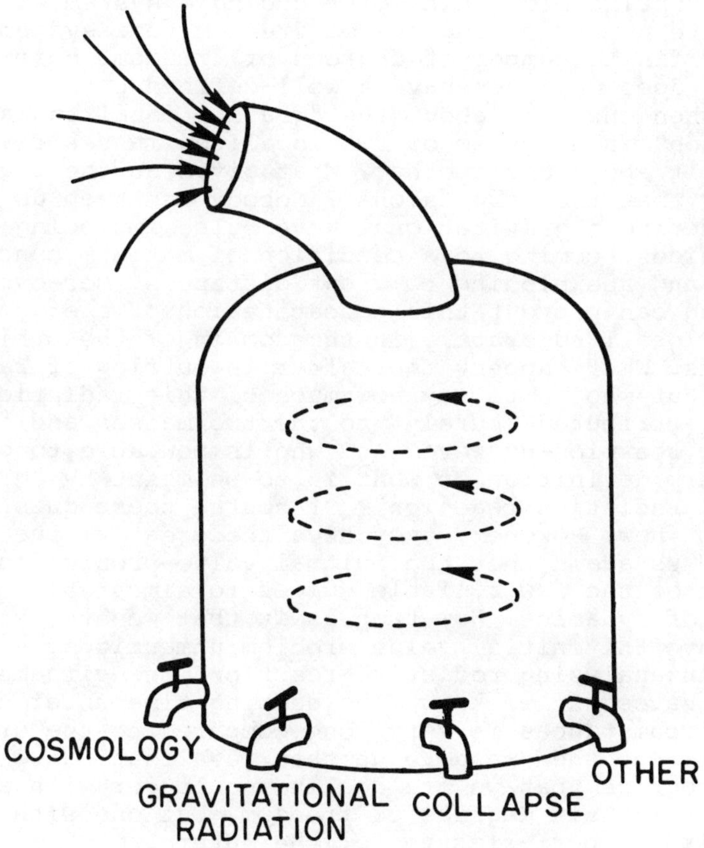

Fig. 1 Symbolic representation of metrics of the most varied provenance being reduced to standard form, classified, and made available for applications in gravitation physics ("MacCallum centrifuge").

3. GRAVITATIONAL RADIATION: PROBLEMS AND PROSPECTS

Gravitational radiation was explored in depth in the A workshops and also in the instructive plenary addresses of Ronald Drever and Thibaud Damour.

Marvellous advances are being made in gravitational wave detectors. A resonance detector? Or a laser to measure a pulse? Which will win? We are seeing a horse-race in which there will be many winners, including all of us!

The theory of gravitational radiation continues to open new avenues for advance and new issues to be unravelled. Among these issues is this: How can one ever, even in principle, think of reducing a system with an infinite number of degrees of freedom to a system with only a finite number of degrees of freedom? More specifically, does one even have a well-defined problem in hand when one asks about the "gravitational radiation reaction" in a system of two localized masses revolving in orbit about one another? What gives us the right to say that they are "alone"? Nobody can keep us from considering a gravitational wave pulse imploding from many sides, coming to a condition of maximum concentration, and reexploding to great distances. Moreover, nothing can prevent this concatenation of energy from being localized exactly in the domain of the orbital motions. What happens thereafter is outflow of radiation. But who is to say how much of this radiation is to be attributed "purely" to the two masses and "purely" to the reexploding wave? And who is now able to give a sharp definition of what is to be meant by "gravitational radiation reaction"? Of course these questions are not new. However, they have the great virtue to remind us again that the initial value problem furnishes one of the few reliable guides to almost all questions of dynamics. How then is it that we ever manage to leave the initial value problem unmentioned and unused in analyzing radiative reaction? One circumstance alone saves us, we know. The damping time under everyday circumstances is very long compared to the orbital period. But when we give up that idealization we have to recognize that we are dealing no longer with a system with six degrees of freedom, but one with an infinite number - lesson for the future.

4. "WEATHER" AS A CENTRAL TOPIC OF ASTROPHYSICS

Igor Novikov gave us an illuminating survey of the workshops B1, B2, B3, dealing with the early universe, the physics and astrophysics of black holes, cosmological parameters and cosmological models. That domain of considerations conducted us from symmetries to perturbations; and from perturbations to hydrodynamics and radiation flow and from there to what I can only call "weather". Weather at the cosmological scale, we are coming more and more to appreciate, is enormously richer than weather on Earth.

One particularly fascinating example of weather out of the many we have heard about is the machinery for taking a star apart and making it into "food" for other stars. Neither Bicknell and Gingold nor Carter and Luminet use the word "helium bomb" for their process, but that is what it often is. The tidal deformation produced by a passing black hole enormously compresses and heats a normal star and thereby detonates it. It is largely or entirely torn apart. We learn from this work a new lesson about the black hole. It not only eats. It also cooks and cuts up its own food!

Another example of weather is provided by the joint action of gravitation and radiation inside a cloud of molecules with a mass of $10^5 M_\odot$ to $10^6 M_\odot$, such as we see in our galaxy. Fragmentation takes place, calculations indicate and observations verify. How sensitive (question of Jean Loup Puget of Meudon and other colleagues) is the mass of the typical fragment to the conditions of density and temperature in and around the cloud? Are the conditions in the outer reaches of this and other galaxies such as to promote the formation of objects with masses of the order of $0.1 M_\odot$ or less? Does a large part of the mass today reside in dark objects of this kind? Is there something to be learned out of weather at this level about the "mystery of the missing mass"?

5. BETTER CHECKS THEN EVER ON THE SOLAR SYSTEM

Peter Bender's summary climaxed a week of revelations about unprecedented recent advances in precision determination of the parameters of the solar system. I may only recall, as one example out of many, that, thanks to 1136 Viking determinations of distance, we now know the distance to Mars within ± 7m. And as Peter Bender tells us, "there is not much you can do to the solar system without producing an unforgivable change in the Mars data". It is a special comfort that at least two great laboratories serve as independent consciences of the solar system. Each has its own repertoire of data. Each has its own way of handling the data. Each has its own way of stating the conclusions from the analysis. And we have essential agreement! Not to have two independent laboratories would be to have only half a laboratory!

There is another payoff from this improved knowledge of the solar system, but a very important one. From the Viking data, plus present day knowledge about nucleosynthesis, Vittorio Canuto and Ronald Hellings conclude that the constant of gravitation, expressed relative to the constants of atomic physics, cannot have changed by more than 20 percent in the last 20×10^9 years.

6. WAVE FUNCTION "FOR THE UNIVERSE"?

Quantum gravity, quantum field theory in curved space and supergravity formed the subject of workshops D1, D2, summarized by Peter von Nieuwenhuizen, and were also dealt with in part in the plenary lectures. Surely the calculation by Stephen Hawking and James Hartle of the wave function in its dependence on the 3-geometry, $\Psi({}^{(3)}G)$, marks a really important step forward in understanding the content and consequences of quantum gravity. So, too, is the instructive calculation of Alexander Vilenkin describing the spontaneous creation of a model universe by quantum tunneling out of "nothing at all".

These new advances give occasion to reflect on old questions. If it is impossible, because of the finite speed of light, to get any signal whatsoever from the antipodal regions of a closed model universe (similar here and now to our own) how can the question even have a meaning whether the universe is closed or open? And if the wave function of a system describes, as quantum theory tells us, not the system, but our knowledge about the system, then how do we have any way to attach a meaning to the term, "the wave function of the universe"? This question emphasizes the issues of principle we face in bringing quantum theory and relativity into union. Happily we know that the sharpening of a question is often the first step in the answering of the question!

7. ALL FIELDS AS "MANIFESTATIONS" OF THE GEOMETRY OF CURVED EMPTY SPACE

If there is one lesson from our conference which is easier to encapsulate than any other, it is the message that

$$4 + 7 = 11.$$

Einstein would surely be happy, not to mention Kaluza and Klein, to see his old dream come to life in a new form: the forces and fields of nature no longer regarded as foreign intruders in space, but manifestations of the curvature of space. Curved empty space and nothing more! A space with seven more dimensions than the familiar space! These seven dimensions curved up into a 7-sphere of fantastically small dimension! Excitation of the "electromagnetic field" and the "chromodynamic field" interpreted as "vibrations" of this empty-space geometry! The success of our distinguished colleagues pursuing this line of research, the impressive way that the 7-sphere "construction proves itself absolutely superior to all the others", and the unity that this development brings to physics marks it as one of the most exciting developments of our time. Never have the worlds of the very large and the very small been brought into closer connection. Surely rich developments lie ahead.

8. WHAT UNDERLIES GEOMETRY?

We find ourselves led at the end to a still deeper issue. Is not the augmentation in dimensionality from four to eleven the first stage in going from four to infinity? And is not the march from four to infinity the first step in going from four dimensions to no dimensionality at all? Surely no theory of geometry that presupposes geometry will ever explain geometry - and dimensionality. They must go back to a deeper foundation. As Plato tells us in his *Timaeus*, "That which is to take all forms must have no form".

What *is* the elementary building substance or building block or building unit which lies behind and beneath all we see and know? The ultimate entity, we know, is not uncrunchable particles. It is not fields, as fields, we are now also learning, not least from "4 + 7 = 11". It cannot be geometry as we know geometry - we have just argued -because that would presuppose dimensionality. Only one place is evident to turn for an ultimate building unit.

Quantum theory presents us with the strangest object in all of physics, an entity which has no localization in space and time, the "elementary quantum phenomenon".

The elementary quantum phenomenon is typified by the photon or electron that enters an apparatus through a slit and unpredictably triggers one counter or another. We have learned from the Einstein-Podolsky-Rosen experiment and Bohr's considerations that we have no right to speak about what the photon or electron is doing in all its travel from point of entry to point of action. In today's words, "no elementary quantum phenomenon is a phenomenon until it is a registered ('observed', 'indelibly recorded') phenomenon, 'brought to a close' by 'an irreversible act of amplification'". Three features of Bohr's phenomenon signal it as a building unit prior to, and even more basic than particles and fields, time or geometry. First, unlocalized and unlocalizable in space and time, it has a status above space and time. Second, it has a yes, no character, than which nothing is simpler. Third, all the information we acquire we get, directly or indirectly, from elementary quantum phenomena.

9. TOWARDS ERA III, THE INFORMATION-THEORETIC FORMULATION OF PHYSICS

Physics, like history, has its eras. In Era I, thanks to Galileo and Kepler, we learned to describe motion with a time coordinate as well as space coordinates. Era II, the era of law, began with Newton and led through Maxwell and Einstein to chromodynamics and "4 + 7 = 11". Era I gave us curves and was founded on geometry, Era II gave us differential equations, first ordinary, then partial. Are we not seeing today the gradual opening of an Era III, in which the mathematical tool is information theory and the ultimate physical entity is the elementary quantum phenomenon?

As I reflect on what we have heard and learned in this inspiring conference, and try to catch some small glimpse of the road ahead one question concerns me more that any other. What conceivable alternative whatsoever do we have in hand to serve as the ultimate building unit for all of physics except the elementary quantum phenomenon? And if there is no alternative, then what is the combinatoric principle that puts together billions upon billions of these information-theoretic elements to generate "everything out there", a principle "so compelling and so obvious" that we don't see it at all?

Are we going to have to infringe again on the world of philosophers to make headway with our problems? In older times, we had to take over from them the problem of motion. In our own century, Einstein established dominion over the subject of space and time. Tomorrow will it not be existence itself that comes under the purview of physics?

10. THE EVER GROWING SCOPE OF PHYSICS WITH THE GREATEST REVOLUTION YET TO COME

Physics has a wonderful feature. The greater its advances, the more clearly does it demand of us still greater advances. No longer are we content to accept the central principle of our subject, quantum theory, as "given". We seek to derive it from still deeper, still more compelling considerations. No longer are we satisfied to accept the "laws" and the initial conditions as two forever distinct elements. We demand to see them as two specializations, extreme specializations to be sure, but specializations, of a more general entity. And if philosophy is too important to be left entirely to philosophers, the exciting things that we have learned at this meeting - so many, so very many, of which I could not mention here - have taught us more: Astrophysics is too important to be left entirely to the astrophysicists. Nuclear physics is too important to be left entirely to the nuclear physicists. Elementary particle physics is too important to be left entirely to the elementary particle physicists. Mathematics is too important to be left entirely to the mathematicians. We know, too, not least from this meeting itself, that we have gained immensely from the partnership of many outstanding experimentalists in the work of gravitation physics. Most of all we know that the greatest revolution in physics - call it Era III or call it what we will - is yet to come.

NOTE

* Preparation for publication assisted by University of Texas Center for Theoretical Physics and by NSF Grant PHY 8205717.

INDEX